Search and Discovery

A Tribute to Albert Szent-Györgyi

Contributors

Seymour S. Cohen
G. Csako
Arpad I. Csapo
Setsuro Ebashi
Tamas Erdos
J. Gergely
Charles B. Huggins
H. E. Huxley
Irvin Isenberg
Herman M. Kalckar
Benjamin Kaminer
Michael Kasha
Hans Krebs
K. Laki
Fritz Lipmann
Laszlo Lorand
Joseph Loscalzo
Linus Pauling
John Platt
Alberte Pullman
Bernard Pullman
A. H. Reddi
George H. Reed
Albert Szent-Györgyi
Andrew G. Szent-Györgyi
Annemarie Weber
George Weber
E. F. Wilson
S. T. Yancey

Search and Discovery

A Tribute to Albert Szent-Györgyi

EDITED BY

BENJAMIN KAMINER

Department of Physiology
School of Medicine
Boston University Medical Center
Boston, Massachusetts

ACADEMIC PRESS New York San Francisco London 1977

A Subsidiary of Harcourt Brace Jovanovich, Publishers

ACADEMIC PRESS, INC.
111 Fifth Avenue, New York, New York 10003

United Kingdom Edition published by
ACADEMIC PRESS, INC. (LONDON) LTD.
24/28 Oval Road, London NW1

Library of Congress Cataloging in Publication Data

Main entry under title:

Search and discovery.

Includes bibliographies and index.
1. Biological chemistry–Addresses, essays, lectures.
2. Szent-Györgyi, Albert, Date I. Szent-Györgyi, Albert, Date II. Kaminer, Benjamin.
[DNLM: 1. Biochemistry–Congresses. QU4 S439 1975]
QP514.2.S4 599′.01′92 76-52721
ISBN 0–12–395150–X

PRINTED IN THE UNITED STATES OF AMERICA

To see what everyone has seen
and think what no one has thought

Albert Szent-Györgyi

Contents

SUBMOLECULAR BIOLOGY

CELL GROWTH AND CANCER

EPILOGUE

List of Contributors

Numbers in parentheses indicate the pages on which the authors' contributions begin.

*SEYMOUR S. COHEN** (313), Department of Microbiology, University of Colorado School of Medicine, Denver, Colorado

G. CSAKO (303), Institute of Pathophysiology, Medical School, University of Debrecen, Hungary

ARPAD I. CSAPO (117), Department of Obstetrics and Gynecology, Washington University School of Medicine, St. Louis, Missouri

SETSURO EBASHI (77), Department of Pharmacology, Faculty of Medicine, University of Tokyo, Bunkyo-ku, Tokyo, Japan

TAMAS ERDOS (151), Centre National de la Recherche Scientifique, Laboratoire d'Enzymologie, Gig-sur-Yvette, France

J. GERGELY (91), Department of Muscle Research, Boston Biomedical Research Institute, Department of Neurology, Massachusetts General Hospital, and Department of Biological Chemistry, Harvard Medical School, Boston, Massachusetts

CHARLES B. HUGGINS (273), Ben May Laboratory for Cancer Research, University of Chicago, Chicago, Illinois

H. E. HUXLEY (63), MRC Laboratory of Molecular Biology, Cambridge, England

IRVIN ISENBERG (195), Department of Biochemistry and Biophysics, Oregon State University, Corvallis, Oregon

*Present address: Department of Pharmacology, S.U.N.Y. Stony Brook, Stony Brook, New York.

HERMAN M. KALCKAR (33), Department of Biological Chemistry, Harvard Medical School and the John Collins Warren Laboratories of the Huntington Memorial Hospital of Harvard University at the Massachusetts General Hospital, Boston, Massachusetts

BENJAMIN KAMINER (161), Department of Physiology, School of Medicine, Boston University Medical Center, Boston, Massachusetts

MICHAEL KASHA (219, 339), Institute of Molecular Biophysics and Department of Chemistry, Florida State University, Tallahassee, Florida

HANS KREBS (3), Metabolic Research Laboratory, Nuffield Department of Clinical Medicine, Radcliffe Infirmary, Oxford University, Oxford, England

K. LAKI (303), National Institute of Arthritis, Metabolism and Digestive Diseases, National Institutes of Health, Public Health Service, U. S. Department of Health, Education, and Welfare, Bethesda, Maryland

FRITZ LIPMANN (17), The Rockefeller University, New York, New York

LASZLO LORAND (177), Department of Biochemistry and Molecular Biology, Northwestern University, Evanston, Illinois

JOSEPH LOSCALZO (99), Department of Biochemistry and Biophysics, School of Medicine, University of Pennsylvania, Philadelphia, Pennsylvania

LINUS PAULING (43), Linus Pauling Institute of Science and Medicine, Menlo Park, California

JOHN PLATT (55), Mental Health Research Institute, University of Michigan, Ann Arbor, Michigan

ALBERTE PULLMAN (231), Institut de Biologie Physico-Chimique, Fondation Edmond de Rothschild, Paris, France

BERNARD PULLMAN (251), Institut de Biologie Physico-Chimique, Laboratoire de Biochimie Théorique Associé au C.N.R.S., Paris, France

A. H. REDDI (273), Ben May Laboratory for Cancer Research, University of Chicago, Chicago, Illinois

GEORGE H. REED (99), Department of Biochemistry and Biophysics, School of Medicine, University of Pennsylvania, Philadelphia, Pennsylvania

ALBERT SZENT-GYÖRGYI (329), Institute for Muscle Research, Marine Biological Laboratory, Woods Hole, Massachusetts

ANDREW G. SZENT-GYÖRGYI (107), Department of Biology, Brandeis University, Waltham, Massachusetts

ANNEMARIE WEBER (99), Department of Biochemistry and Biophysics, School of Medicine, University of Pennsylvania, Philadelphia, Pennsylvania

GEORGE WEBER (279), Laboratory for Experimental Oncology, Indiana University School of Medicine, Indianapolis, Indiana

E. F. WILSON (303), National Institute of Arthritis, Metabolism and Digestive Diseases, National Institutes of Health, Public Health Service, U. S. Department of Health, Education, and Welfare, Bethesda, Maryland

S. T. YANCEY (303), Laboratory of Chemical Pharmacology, National Cancer Institute, National Institutes of Health, Bethesda, Maryland

Preface

This volume is dedicated to Albert Szent-Györgyi and stems from a Symposium, "Search and Discovery," held in his honor at Boston University School of Medicine. As the participants recalled their associations with "Albi" or "Prof," as Szent-Györgyi is affectionately known to many, an incredible warmth and air of excitement permeated the auditorium—indeed a reflection of the radiance and inspiration of this remarkable man.

The papers presented scanned the fields of his major contributions and interests: metabolism, vitamin C, molecular mechanisms of muscle contraction, submolecular biology and cell growth, and cancer; and the social interrelations of science were also not neglected. These milestones form the basis of this volume.

Szent-Györgyi was born in Budapest on September 16, 1893. In an autobiographical sketch entitled "Lost in the 20th Century" (*Annual Review of Biochemistry* **32**, 1963), Szent-Györgyi, on referring to Gowland Hopkins' influence on his scientific development, writes "Research is not a systematic occupation but an intuitive artistic vocation." Those of us who had the privilege of working in Prof's laboratory witnessed a blend of the artistic with the scientific. His profound insights and adventurous spirit led to exciting new ideas and generated a lust for discovery.

When he established his Institute for Muscle Research at the Marine Biological Laboratory in 1947 he enjoyed fishing in the Woods Hole waters. Asked why he always used such a big hook, he replied "I think it more exciting not to catch a big fish, than not to catch a small one." In the laboratory he follows the same philosophy by using a "big hook," and, indeed, he has made some big catches.

Prof's influence and impact extend beyond the confines of the laboratory and are far-reaching. Throughout his life he has been intensely concerned with the serious problems of mankind. He opposed Hitler and Stalin, and recently was outspoken against the involvement of the U.S. Government in the Vietnam War.

In a recent little book "The Crazy Ape" Szent-Györgyi sees hope in the youth of the world and calls upon them to organize and exercise their democratic power to create a new world. There is a closeness and rapport between this octogenarian and youth. Cicero wrote "For just as I approve of the young man in whom there is a touch of age, so I approve of the old man in whom there is some flavor of youth. He who strives to mingle youthfulness and age may grow old in body, but old in spirit he will never be." In Prof we have a man who remains youthful both in body and in spirit. When he turned 80 he ventured, for the first time, onto water skis, and now he continues to enjoy water-skiing every summer in addition to his regular swims around Penzance Point in Woods Hole. It was a joy to watch his physical and intellectual vibrance at this Symposium as is beautifully portrayed by Michael Kasha in the Epilogue of this volume.

His past successes (and failures) are never a hindrance to him. With confidence he pronounces "I never look back, I only look forward." In his latest little book "Electronic Biology in Cancer" (1976, Dekker, New York) Szent-Györgyi writes "In my next book I hope to be able to show how this knowledge can be used to arrest or prevent cancer." As Robert Louis Stevenson said "To travel hopefully, is better than to arrive safely."

Benjamin Kaminer

Acknowledgments

First I wish to thank my wife Freda, my daughter Lauren, and my son Brian for the encouragement they gave me to organize the Symposium. Irvin Isenberg and Andrew Szent-Györgyi offered helpful advice and made useful suggestions during the planning stages. Thanks are also due to Jane McLaughlin who has been a devoted research associate to Dr. Albert Szent-Györgyi for over 25 years. I also thank Estelle ("Kepie") Engel and Walter Bonner for their interest and support. The encouragement of John Sandson and Richard Egdahl is also appreciated. Sally McNulty and Andrea Morris attended with efficiency and grace to all the numerous details, and their help is deeply appreciated. Woodland Hastings, John Edsall, and John Sandson added warmth to the proceedings with their remarks during the luncheons. A number of students helped with numerous organizational details, and special thanks are due to Rebecca Biegon (and her husband Glen), Kurt Dasse, David Sack, Susanne Churchill, Andrew Wexler, and Thomas Morris. Thanks are also due to Anthony Gorman, and Don Giller. Jerome Glickman, Terry Field, and their staff competently attended to the closed-circuit television and videotaping of the Symposium. The arrangements for the postsymposium gathering at the Marine Biological Laboratory were facilitated through Homer Smith, who has always kindly given "Prof" special attention at the Marine Biological Laboratory. Denis Robinson acted as host, and his after-dinner remarks and the reminiscences of Teru Hayashi and George Wald were enjoyed by all. The event ended with a delightful chamber concert by Robert Allen, Agi Jambor, Jelle Atema, and Michael Gallagher.

Brian Kaminer, Delbert Philpott, Bradford Herzog, and Kip are responsible for the photographs published. The assistance of the staff of Academic Press is gratefully acknowledged.

The Symposium was sponsored by the National Science Foundation, Muscular Dystrophy Association, Inc., Ciba-Geigy Corporation, Hoffman-LaRoche, Inc., Merck, Sharp & Dohme Research Laboratories, G.D. Searle & Co., and the Upjohn Company.

Hans A. Krebs

Charles Huggins

Michael Kasha

Benjamin Kaminer

Albert Szent-Györgyi

Koloman Laki

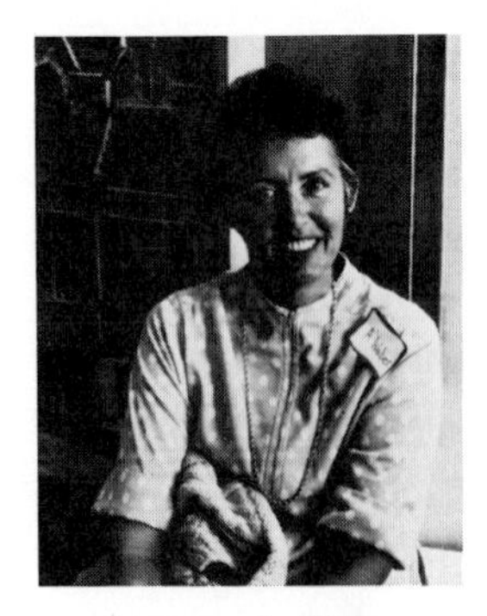
Annemarie Weber

Arpad Csapo

Laszlo Lorand and Linus Pauling

Irvin Isenberg

Fritz Lipmann

John Gergely

Andrew Szent-Györgyi, Albert Szent-Györgyi, Irvin Isenberg, and Eva Szent-Györgyi

Hugh E. Huxley

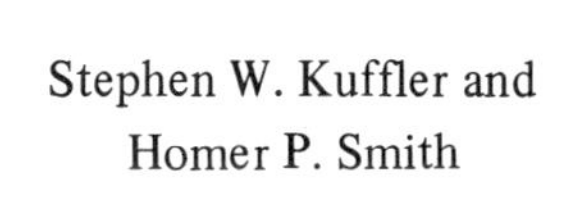

Stephen W. Kuffler and Homer P. Smith

Bernard Pullman

John T. Edsall

Seymour S. Cohen

George Weber

Richard Bersohn

Carl F. Cori

Dewitt Stetten, Jr.

George Wald (holding early sample of vitamin C given to him by Szent-Györgyi)

John Platt

Setsuro Ebashi, Alberte Pullman, and Linus Pauling

Herman Kalckar

Mathew Meselson, Mrs. Linus Pauling, Linus Pauling, and Irvin Isenberg

Woodland Hastings and Tamas Erdos

Some Recollections of Albert Szent-Györgyi

JOHN T. EDSALL*

My first glimpse of Albert Szent-Györgyi came a little over forty years ago, when he gave a talk, here in the Biological Laboratories, on pure vitamin C and its properties. I remember particularly his description of his arguments with Arthur Harden, the eminent editor of the *Biochemical Journal,* regarding the naming of the compound. Its chemical structure then was still obscure. In view of his ignorance, and of the fact that vitamin C was somehow related to the carbohydrates, Albert proposed to christen it temporarily "ignose." Harden sternly rejected the proposal, and Albert offered "godnose," which was also summarily rejected. Harden then proposed "hexuronic acid" and it appeared in the published paper under that name, although in fact this proved incorrect when the structure was worked out.

This little episode reveals some of the great differences in the outlook of various eminent scientists. Of Arthur Harden, one of the great pioneers in the study of cell-free fermentation, F. G. Hopkins and C. J. Martin wrote, after his death, in a short biographical article: "He was a superb critic but unduly suspicious of adventures into the realm of imagination." In short, Arthur Harden and Albert Szent-Györgyi lived in different worlds.

*Biological Laboratories, Harvard University, Cambridge, Massachusetts.

Albert came back to Harvard—and this I think was just about forty years ago—to give three lectures, which centered on the biochemistry of respiratory processes and the role of the dicarboxylic acids. This work opened some entirely new vistas in our understanding of biological oxidations; it was a major step toward the unraveling of the tricarboxylic acid cycle by Krebs some two years later.

I had no suspicion at that time that Albert's next major point of attack would be the biochemistry of muscle, and especially of muscle proteins, a field in which I once had worked myself. Indeed, here in the United States, we knew little or nothing of that story during the war years when communication with Hungary was cut off. That story of creative scientific achievement, in the midst of war and the constant perils from the Nazi control of the country, came to us gradually after the war ended and communication was restored. The scientific achievement transformed our ideas of the role of muscle proteins in the contractile process; what von Muralt and I and others had called myosin was revealed as a complex of two proteins, myosin and actin, both essential for the structure of the muscle fiber and for contraction. That work of Szent-Györgyi and Straub and their collaborators, and the slightly earlier demonstration by Engelhardt and Ljubimova in Moscow that myosin was an enzyme with ATPase activity, really represented the beginning of the modern biochemistry of muscle. All the great developments in muscle biochemistry since have grown out of those discoveries.

I do not need to elaborate here on the great scientific stimulus that Albert has given to so many of us in the years since he settled in Woods Hole, with his ventures into submolecular biology and other unconventional enterprises. There is plenty of testimony to that in this volume. I would like to pay tribute, however, to his stand for humanity, in opposition to the madness of the nuclear arms race, and other misuses of science and technology that threaten the future of the earth and of mankind. He has been, and is, a firm and courageous upholder of the values that we need to cherish, and for that I thank him most deeply.

METABOLISM

1

Errors, False Trails, and Failures in Research

HANS KREBS

I must begin by referring to one of the reasons for my special affectionate attachment to Albert Szent-Györgyi. At a very critical time of my career, at the beginning of the Hitler era in the spring of 1933, when I was in difficulty, he proved to be a true friend.

I had met Albert Szent-Györgyi first in 1929 at the International Physiological Congress held at Boston. Early in 1933 I wrote to him from Freiburg im Breisgau at Szeged in Hungary where he was Professor of Biochemistry, begging for a sample of ascorbic acid for research purposes. This was at that time a very scarce chemical, made only in research laboratories. It took some weeks before I received a reply because Szent-Györgyi was traveling, visiting research centers in England and Holland where he had worked earlier. During this interval the political scene in Germany had gravely blackened. His reply in German is reproduced in Fig. 1. The translation is as follows:

The Hague, 12 April 1933

Dear Colleague:

I am glad to know that you are interested in ascorbic acid. Unfortunately, your letter reached me while traveling and it will be three weeks before I shall be back home again. I will then send you ascorbic acid immediately. Should you no longer require it, please let me know as our supplies are at present limited. If I do not hear from you I will send the substance promptly after my return.

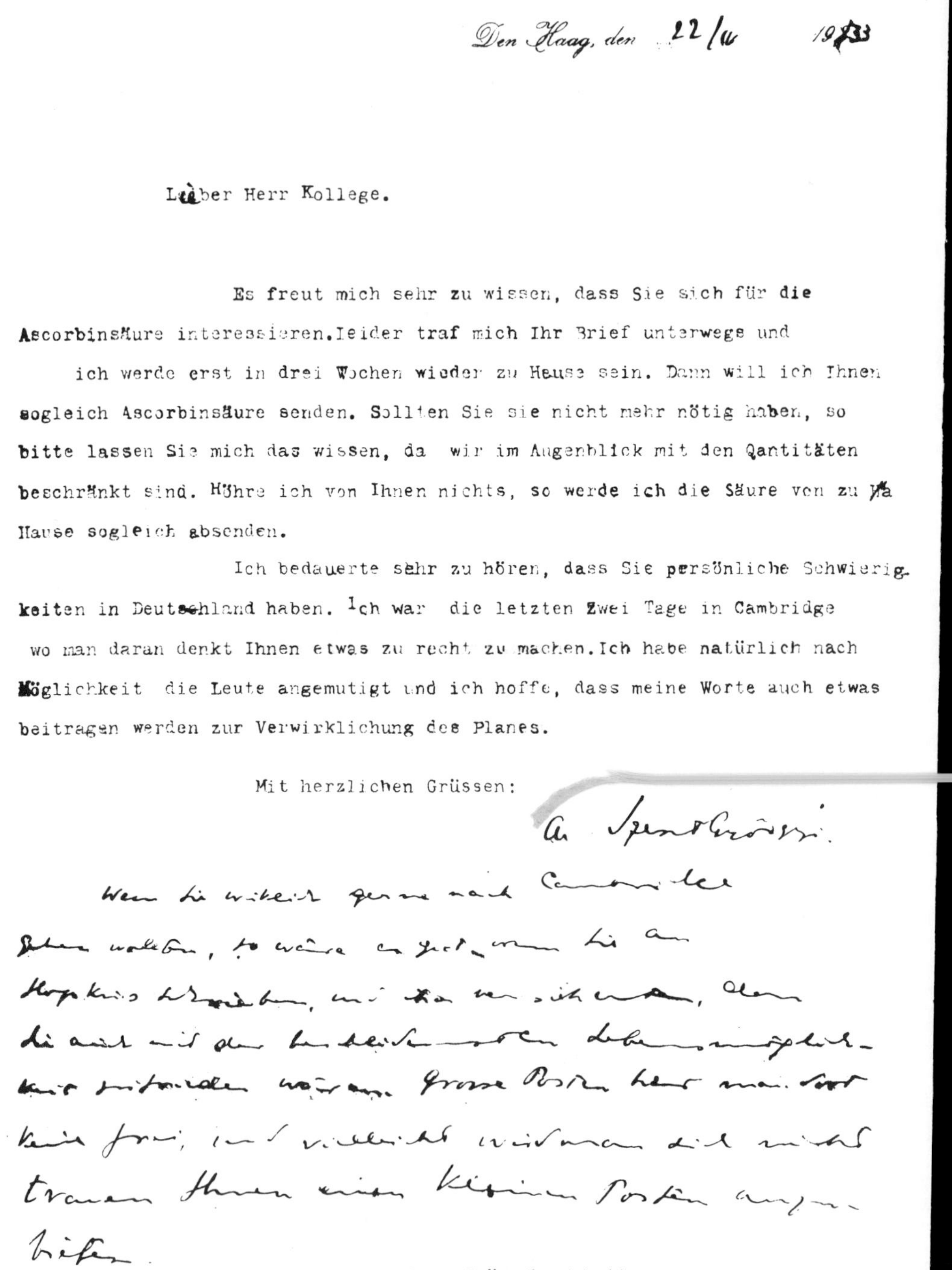

Den Haag, den 22/4 1933

Lieber Herr Kollege.

Es freut mich sehr zu wissen, dass Sie sich für die Ascorbinsäure interessieren. Leider traf mich Ihr Brief unterwegs und ich werde erst in drei Wochen wieder zu Hause sein. Dann will ich Ihnen sogleich Ascorbinsäure senden. Sollten Sie sie nicht mehr nötig haben, so bitte lassen Sie mich das wissen, da wir im Augenblick mit den Qantitäten beschränkt sind. Höhre ich von Ihnen nichts, so werde ich die Säure von zu Hause sogleich absenden.

Ich bedauerte sehr zu hören, dass Sie persönliche Schwierigkeiten in Deutschland haben. Ich war die letzten Zwei Tage in Cambridge wo man daran denkt Ihnen etwas zu recht zu machen. Ich habe natürlich nach Möglichkeit die Leute angemutigt und ich hoffe, dass meine Worte auch etwas beitragen werden zur Verwirklichung des Planes.

Mit herzlichen Grüssen:

A. Szent-Györgyi.

Wenn Sie wirklich gerne nach Cambridge gehen wollten, so wäre es gut, wenn Sie an Hopkins schreiben, und ihn versichern, dass Sie auch mit den bescheidensten Lebensmöglichkeiten zufrieden wären. Grosse Posten hat man dort keine frei; und vielleicht wird man sich nicht trauen Ihnen einen kleinen Posten anzubieten.

Fig. 1. Copy of Szent-Györgyi's original letter.

I am very sorry to hear that you have personal difficulties in Germany. During the last few days I was in Cambridge where people have in mind helping you somehow. Of course, I have encouraged them as much as possible and I hope that my words will have contributed a little toward the realization of the plans.

With kind regards,
A. Szent-Györgyi

P.S. If you really would like to come to Cambridge it would be best if you wrote to Hopkins and told him that you would be content with very modest opportunities. Senior posts are not available and perhaps he might be diffident to offer you a junior position.

Therefore, ask Hopkins (if you would like to come to Cambridge) to give you an opportunity there. If it does not embarrass you, do refer to my encouragement.

On the very same day I took up the suggestion and wrote to Hopkins, and 10 weeks later I was installed at Cambridge. So it was Albert Szent-Györgyi's kindness and considerateness, over 42 years ago, which decisively influenced my life at the most awkward, almost catastrophic, stage and I lived happily in England ever after, needless to say, with everlasting memories of deep-felt gratitude to Albi. When a few years ago I had an occasion of showing Albi his 1933 letter, he was amazed that I had preserved "this scrap of paper." To me of course his was not a "scrap of paper" but a very important treasured document, testifying to some of Albi's endearing qualities.

Another reason for my special affection for Albert Szent-Györgyi, which I am sure I share with many, is my close affinity to his outlook on life and my admiration for his courage and eloquence in expressing his philosophy. In the Bulletin of the Atomic Scientists (April 1975), he published what he called "A Little Catechism" setting out in 21 succinct paragraphs his passionate concern for real peace and the welfare of fellow man, his moral courage which gives him strength to speak out in support of unpopular truths, and his profound wisdom in discerning the essentials of life. It is reproduced here with the permission of the Editor and Publishers of the Bulletin of the Atomic Scientists.

Bitten Sie also Hopkins, wenn Sie gerne nach Cambridge möchten, Ihnen eine Möglichkeit dazu zu geben. Wenn Ihnen dies ohne Weiteres nicht unangenehm ist, beziehen Sie sich auf meine Ermunterung.

Fig. 1. *continued*

A LITTLE CATECHISM

ALBERT SZENT-GYÖRGYI

Inflation is mostly but a form of governmental grand larceny, the government's spending the citizen's savings.

There can be no healthy economy when one-third of the revenue is spent on instruments and organizations of destruction, when armaments are the biggest and best business, when millions of young men are kept in uniform and when workmen continue to produce means of devastation.

Armies are not instruments of peace, but of war. Every army is a threat to peace; the greater the army, the greater the threat.

Armies mean power and power corrupts. Armies corrupt governments, making them wish for more power and a bigger army.

If you want your country to be strongest, ask first: strongest in what? Do you want to have the strongest capability to kill, destroy, terrorize, overkill? Or, the strongest country in economy, fairness, goodwill, helpfulness, knowledge, health and happiness?

Peace is mostly but disguised domination, be it Pax Romana, Pax Germanica or Pax Americana.

It is untrue that there have always been wars because man is bloodthirsty. Wars have always occurred because there always have been individuals or small groups of people willing to sacrifice others' lives for their own profit or ambition.

Anything that can happen will happen and anything that can go wrong will go wrong (Murphy's Law).

We either have to stop proliferating or else move to a better and bigger globe.

There can be no balance with death control (modern medicine) without birth control.

A world transformed by science can be run only by the spirit which created science, the search for truth and putting two and two together with a cool head, without fear, greed and lust for domination.

The basic rule of coexistence is: don't do to others what you don't want to be done to yourself.

If you want to be called a democracy, don't support corrupt dictators or military juntas.

Killing a fellowman is murder regardless of uniform, language, creed, color or slogans.

There will be peace when we will look upon instruments of murder and destruction with revulsion, instead of national pride.

The present crisis will get increasingly worse until dog eats dog, instead of man helps man.

Mankind with atomic forces in his hand and greed, fear, and lust for domination in his heart is destined to eliminate itself.

Rich is he who has more money than desire, and poor is he who has more desire than money. Today's unskilled laborer has more money than the princes of a few hundred years ago. What makes us poor is the desire for more. The key to happiness is not to get more, but to enjoy what we have and to fill the empty frame of our lives instead of enlarging it.

Life is a late by-product of the forces which created the universe. Life can be wiped out without causing a major disturbance to the universe. Human life can be made lasting by construction, not by destruction. It can be made enjoyable by health, happiness, beauty and knowledge.

Wealth comes from excellence, and not excellence from wealth (Plato). The present crisis is moral and intellectual; economics are secondary.

The present world crisis is not the sum of single national recessions. It is a global phenomenon which cannot be corrected by any local action. It can be corrected only by a global revolution: a revolution by man against outdated ideas and government attitudes, a revolution which will liberate man from the reign of present terror and allow him to use his wonderful abilities for his advantage–improving his life instead of destroying it. Such a revolution would make armies superfluous, liberating man from a terrific burden. It could inaugurate a second, golden age of man.

The development of science and technology are incompatible with the present outdated political ideas and human relations. It is this disharmony which led to two world wars and–if not corrected will lead to a third world war, to a collapse of civilization and, possibly, to the disappearance of man before the end of this century.

I believe it is instructive and healthy to look back at one's research efforts after intervals of, say, 10, 20 or 30 years, and in the light of subsequent developments to take stock of one's published–and unpublished–material. Doing this I find that some pieces of my own research have stood the test of time inasmuch as they have proved relevant to the progress of the subject. They turned out to be essential links in the development of the subject. Other pieces proved unimportant to the progress, and still others were downright erroneous–erroneous not in terms of the accuracy of experimental measurements but in respect to conceptual matters, i.e., design and interpretation of experiments. Quite a few of my research efforts never reached the stage of publication and in a sense were therefore failures. Let me add at once that such "failures", up to a point, are of course a regular accompaniment of research; they are unavoidable if one explores new territories.

ERRORS AND MISCONCEPTIONS IN THE WORK LEADING TO THE CONCEPT OF THE TRICARBOXYLIC ACID CYCLE

My first papers on the tricarboxylic acid cycle, published in *Enzymologia* (Krebs and Johnson, 1937) and *The Lancet* (Krebs, 1937), referred to the intermediary stages of the oxidation of *carbohydrate*, because I believed at the time, as it turned out incorrectly, that I was studying the oxidation of carbohydrate. I used as experimental material pigeon breast muscle, which had been introduced into biochemistry by Albert Szent-Györgyi. It is a material which shows an exceptionally high rate of oxygen consumption after it is minced and suspended in a saline medium. The material thus presented the phenomenon I wanted to study–the intermediary stages of oxidations–in a relatively stable and easily accessible form. It was a mistake, however, to assume that it was carbohydrate that underwent oxidation.

The basis of the belief that carbohydrate was the chief substrate of oxidation in muscle was the conviction that, in general, carbohydrate is *the* fuel of muscular contraction. Another reason for believing that muscle oxidizes mainly carbohydrate was Meyerhof's (1930) concept of the Pasteur effect in muscle: that the primary source of energy is lactic acid formation and that oxidations are mainly concerned with supplying the energy for the resynthesis of carbohydrate from lactate, a view generally accepted although subsequently proved incorrect. At rest and during moderate exercise the oxidation of fat is the main source of energy in muscle. Only during severe exercise do glycolysis and lactate oxidation contribute in a major way to the supply of energy.

In my own experiments I was already puzzled that addition of lactate did not increase the oxygen consumption of minced muscle and that the lactate present (which arose during the process of mincing and was very substantial) did not disappear. In fact all my critical experiments were carried out with unphysiologically high concentrations of *pyruvate* as substrate, and of course we now know that pyruvate is an effective precursor of acetyl coenzyme A. Acetyl-CoA was discovered some 8 years after the formulation of the tricarboxylic cycle concept.

Another error in the area of the tricarboxylic cycle field of which I was guilty concerns the occurrence of the cycle in microorganisms. Like many other investigators I was sceptical about the occurrence of the cycle in *Escherichia coli* and in yeast because these two organisms failed to metabolize added citrate, in contrast to some animal tissues where the ready oxidizability of citrate had been an essential piece of evidence. Until the middle 1950's it was not appreciated that microorganisms, and for that matter some cells of higher organisms, may have permeability barriers for small molecules which prevent the entry of citrate into the cell. The concept of special transport mechanisms–the permeases–was introduced by Monod in the 1950's (Cohen and Monod, 1957). It was eventually Swim and Krampitz (1954) who, with the help of isotopic carbon, demonstrated conclusively that the cycle can occur at rapid rates within the *E. coli* cell.

Another interesting misconception in the area of the tricarboxylic acid cycle concerns the interpretation by Szent-Györgyi of his work between 1934 and 1936 (Gözsy and Szent-Györgyi, 1934; Szent-Györgyi, 1935, 1936), a conceptual error which was responsible, at least in part, for the fact that people today talk of the Krebs cycle and not of the Szent-Györgyi cycle. It was Szent-Györgyi who discovered the catalytic role of C_4-dicarboxylic acids in the oxidations of pigeon breast muscle. Szent-Györgyi was at that time interested in hydrogen transport from organic molecules to molecular oxygen and his interpretation of the catalytic role of fumarate was based on the correct observation that oxaloacetate is very readily reduced to malate in pigeon breast muscle, even aerobically. He suggested that the malate/oxaloacetate system might act as one of the hydrogen carriers between the fuels of respiration and molecular oxygen.

At that time, it must be remembered, information on electron transport was very rudimentary because the roles of the pyridine nucleotides, flavoproteins and cytochromes had not yet been clearly established.

Szent-Györgyi expressed the view that the functions of malate and oxaloacetate were not to serve as a fuel but as catalysts. In fact he was not at all concerned with the intermediary metabolism but preoccupied with hydrogen transport.

But of course, despite this conceptual error, his papers were important contributions to biochemical thought and it is interesting to recall the formal justification for Szent-Györgyi's Nobel Prize as stated by the Nobel Committee (1938): "for his discoveries in connection with the biological combustion processes, with special reference to Vitamin C and the catalysis of fumaric acid". We now know that neither vitamin C nor fumarate is directly concerned with biological combustion processes, that the catalytic role of C_4-dicarboxylic acids is that of the tricarboxylic acid cycle.

FALSE INTERPRETATION OF ISOTOPE EXPERIMENTS RELATED TO THE TRICARBOXYLIC ACID CYCLE

In 1941 Wood *et al.* and Evans and Slotin published observations on the fate of CO_2 which had been incorporated into the carboxyl group of oxaloacetate by the pyruvate carboxylase reaction. The findings appeared to be inconsistent with my original formulation of the tricarboxylic acid cycle. The experiments consisted of adding isotopic carbon in the form of bicarbonate to pigeon liver, isolating α-oxoglutarate as 2,4-dinitrophenylhydrazone and locating the isotopic carbon in the α-oxoglutarate molecule. The distribution proved contrary to expectations in that the isotope was detectable only in the carboxyl group adjacent to the α-oxo group, the "α-carboxyl." It was expected that the two carboxyl groups should contain equal amounts of the isotope because the probability of the isotope appearing in either α- or in the γ-carboxyl would be equal if citrate, a symmetrical molecule, is the precursor of α-oxoglutarate (see Fig. 2). The cyclic scheme was, therefore, modified by the assumption that the condensation of oxaloacetate with a pyruvate derivative yields primarily *cis*-aconitate which was taken to be directly converted to isocitrate while the formation of citrate is due to a slow side-reaction. There was some experimental evidence in favor of this modification. However, in 1948 Ogston showed that there was a fallacy in the interpretation of the isotopic experiment and that, in fact, citrate does not necessarily behave as a symmetrical molecule when attached to an enzyme. An asymmetrical enzyme which attacks a symmetrical compound may, under certain well-definable conditions, distinguish between the identical group of the substrate, namely, when the substrate is rigidly attached

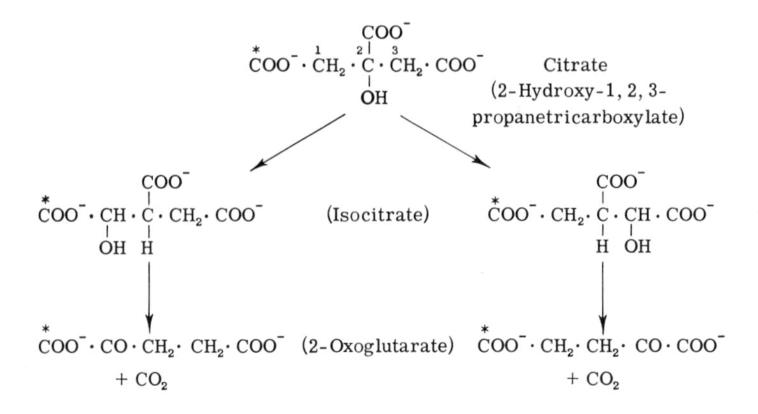

Fig. 2. Fate of labeled citrate as postulated on the assumption that citrate behaves as a symmetrical molecule. The labeled carbon atoms are indicated by an asterisk. The carbon atoms of citrate are numbered on the basis of describing citrate as 2-hydroxy-1,2,3-propanetricarboxylate. Citrate labeled with isotopic carbon in either the 1 or the 3-carboxyl group forms two kinds of isocitrate molecules which differ in respect to the relative position of the hydroxyl group to the carbon-labeled carboxyl. One leads to 1-carboxyl-labeled oxoglutarate, the other 5-carboxyl-labeled oxoglutarate.

to the enzyme, and rigidly attached means at least three combination points. Thus for 7 years, wrong conclusions were drawn from isotope data because of the failure to appreciate the ability of enzymes to make symmetrical molecules, such as citrate, behave in an asymmetrical way. Ogston's important discovery, incidentally, also explained how optically active substances generally arise from inactive ones, e.g., L-lactate from pyruvate, L-amino acids from α-oxo acids, L-malate from fumarate or oxaloacetate.*

FALSE TRAILS CONCERNING GLYCOLYSIS

A false trail in the field of anaerobic carbohydrate metabolism of vertebrates was the idea that there may be two pathways of glucose degradation in some

*This account of the asymmetrical behavior of citrate corresponds to the historical development. Today the subject would be formulated directly in terms of the concept of "chirality" of citrate. Citrate belongs to a category of "symmetrical" molecules where the two symmetrical pieces possess a nonsuperimposable mirror–image relationship to each other, like a left and right hand. An asymmetrical enzyme can therefore differentiate between the two pieces. It is not necessary to postulate a rigid attachment of citrate to the enzyme (Ogston, 1958; Cahn, 1964; Cahn *et al.*, 1966; Cornforth, 1974).

tissues, e.g., in embryonic tissue, brain, and tumor cells, one being the same as in muscle, the other a pathway not involving phosphorylated intermediates. This hypothesis was still vigorously pursued in 1937 (Needham and Lehmann, 1937) and considered as reasonable by distinguished biochemists. The evidence for the nonphosphorylating route was essentially of a negative kind in that a number of observations made on muscle tissue could not be confirmed for the other materials. There was no positive suggestion regarding the nature of the alternative pathway. At that time the close similarity of the pathways of glycolysis in materials as different as yeast cells and muscle was already firmly established. It is therefore difficult retrospectively to appreciate that the hypothesis was considered to be attractive.

A FALSE TRAIL IN THE STUDY OF PROTEIN SYNTHESIS

In 1950 it occurred to me that problems of protein synthesis could be studied in tissue slices by examining systematically the factors which modify the rate of protein synthesis. The experimental work was carried out at my suggestion by a Ph.D. student, Lowell E. Hokin (1951a,b). In planning the work, we thought that the chance of detecting a net increase of a protein would be good if digestive glands were used as experimental protein synthesizing material. These glands are equipped to synthesize enzymes at rapid rates and many of the enzymes can be quantitatively determined. Experiments on the formation of pepsin by gastric mucosa proved unsuccessful but the formation of amylase by pigeon pancreas slices proved a suitable system. Pancreas slices produce amylase on aerobic incubation and the rate of synthesis can be increased approximately 3-fold when the mixture of amino acids is added to the incubation medium.

But our approach, although it gave publishable results (Hokin, 1951a,b), and although it was reasonable in the light of contemporary knowledge, did not get us very far because it was wrongly conceived in light of later developments. The wrong concept was the assumption that the action of ordinary enzymes might entirely explain protein synthesis. The limitation of this assumption had not yet been clearly formulated in 1950. Indeed the view that ordinary enzymes are involved in peptide synthesis was supported by the demonstration that bonds of this type are formed under the influence of typical enzymes, e.g., in the synthesis of glutathione (Bloch and Anker, 1947; Bloch, 1949), of hippuric acid (Chantrenne, 1951) and of glutamine (Speck, 1947; Elliott, 1948). And indeed ordinary enzymes, i.e., specific proteins, do in fact participate in protein synthesis, for example in activating amino acids and converting them into aminoacyl adenylates (Hoagland, 1955). However, it was not properly understood that even if we can attribute the synthesis of an enzyme to the catalytic power of another enzyme we cannot continue the line of argument indefinitely. There must come a stage when an enzyme is not synthesized by a traditional type of enzyme but by something fundamentally different. It is now elementary knowledge that the great majority of peptide chains, and all enzymes, are synthesized by mechanisms in which DNA, messenger RNA, transfer RNA and ribosomes are key catalysts, in cooperation with ordinary enzymes which synthesize aminoacyl-tRNA. Exceptions of protein synthesis not involving DNA and RNA are certain low molecular antibiotics studied especially by Lipmann (1971).

ERRORS IN CLINICAL MEDICINE

A field where errors abounded–and I dare say still abound–is clinical medicine. This does not imply that physicians are less wise and less competent than other people. Errors are forced upon the physician because a sick patient demands treatment, irrespective of whether the science of medicine can supply information on the nature of the disease, and irrespective of whether accurate diagnosis is possible. The physician cannot tell the patient "we do not know" and "we do not understand what is wrong with you." He must act and prescribe a therapy. So he must have a working hypothesis which guides his action. He is thus forced to work in an area which is no longer simple science (where one may simply confess ignorance) but in a situation where a tentative and even wrong hypothesis is better than none at all. But not all errors and misconceptions in medicine stem from this particular situation, for example, the older ideas on the nature of coeliac disease, a condition first described in 1888 by Gee. At first it was believed that the disease was caused by a chronic infection (Herter, 1908), but in the early part of this century it became evident that the severity of the disease depends on the nature of the diet, and in 1924 Haas discovered that a diet containing large quantities of bananas can virtually cure the disease while this diet is eaten. Thus the banana diet was the chief prescription until about 1950. In that year Dutch pediatricians (Dicke, 1951; Dicke *et al.*, 1953; Weijers and van de Kamer, 1953) discovered that the curative action of banana is not at all due to any constituent of the banana but to the absence from the banana diet of wheat or rye flour. Naturally, when bananas form the bulk of the diet other staple constituents are omitted. We now know that coeliac disease is due to the inability of predisposed individuals to digest the gluten of flour, and that incompletely digested gluten is toxic to the small intestine.

An area of many misconceptions was a belief in the curative effects of many plant preparations and of certain spring waters which contain salts or gases, or are radioactive. Many of the hundreds of plant preparations which were still recommended as cures at the beginning of this century have by now been abandoned in rational therapy. Visiting so-called watering places or spas and drinking the special waters is still popular on the continent of Europe although radioactivity has disappeared from the advertisements since the early 1950's. No doubt a restful period at these comfortable places, combined with moderate exercise, dietary control and perhaps the laxative effects of some spa waters, can be beneficial, but the interpretation of cause and effect is somewhat analogous to the banana story in coeliac disease: beneficial effects were attributed to the wrong causes.

These are but a few examples of misconceptions in the history of medicine. It would be easy to enlarge the list. The situation was clearly expressed in the 1920's by Dean Edsall of Harvard in a Commencement Address when he told the medical graduates: "You have learned a great deal in the past 4 years. Unfortunately half of what you have been taught is wrong, and we do not know which half." Presumably this referred especially to the concepts of the nature of diseases and the effectiveness of various therapeutic measures. Today, 50 years later, one hopes that much less than half of what is taught in medicine is wrong but we must still remain skeptical; there is likely to be substantial residue of erroneous concepts for the very reasons I have given.

SCIENTIFIC REVOLUTIONS

My concern in this essay is with relatively minor errors and false trails—those which Kuhn (1962) calls "normal science." The history of science records numerous big and small errors of which the members of the scientific community were collective victims. The correction of big errors has sometimes been referred to as a "revolution." An example is the Copernican revolution which replaced Ptolemy's geocentric cosmology by a heliocentric cosmology. Popper (1975) recently referred to "revolutions" arising from the work of Darwin, Faraday, Clerk Maxwell, J. J. Thompson, from the discovery of x-rays, radioactivity and of isotopes which all shattered erroneous views held collectively by scientists. A more recent error of this kind in which many leading physicists—Otto Hahn (1962, 1965), Lise Meitner, Enrico Fermi, Mme. Joliot-Curie, Emilio Segrè—were involved between 1934 and 1938 was the belief that on bombardment of uranium with neutrons, isotopes of uranium with a higher atomic weight or new elements ("transuranium elements") would be formed. This assumption was based on the generally accepted view that slow neutrons would be taken up by the atomic nucleus so that its mass would increase by one unit and that the new nucleus might be unstable and lose beta particles with the formation of a new element. Many experiments were in agreement with this concept, but in 1939 Hahn and his collaborators established (see Hahn, 1962, 1965) that this interpretation was wrong and that the findings must be explained by the fission of uranium, a process which up to then had been regarded as impossible by leading physicists, e.g., by Rutherford (see Hahn, 1962, 1965).

GENERAL REMARKS

The experiences which I have discussed illustrate general features of science. First, neither the progress of the individual scientist nor the progress of corporate scientific knowledge is a straight upward line but a path with wrong turnings leading to dead ends. Often published papers may convey the impression that research consists of an unbroken series of discoveries; in fact very rarely is there unbroken progress in pioneering research. As a rule there are many downs as well as ups but published papers are not meant to be laboratory reports and therefore contain positive results only.

Negative experiments, disappointing though they may at first appear, are often valuable by invalidating a plausible working hypothesis. Negative experiments have been likened to the scaffolding in the construction of a building. A scaffold and its paraphernalia are essential but they disappear after the completion of the building. Likewise when presenting a paper, the scientist omits the detours and negative experiments, although they helped him to establish the final results.

Second, several of the errors and false trails originated from a false conceptual background, i.e., from false assumptions of a general kind which were not questioned in the planning and interpretation of the experiments. To recall examples: the assumptions that carbohydrate is *the* fuel of muscle respiration; that symmetrical molecules behave symmetrically in the presence of enzymes because they do so in ordinary chemical reactions; that all processes in living cells are catalyzed by ordinary enzymes. Some false conclusions, like the curative factor in the banana, arose from lack of appreciation that the "experiment" involved the change of more than one variable—addition of bananas being one and omission of other items being another. Thus a lesson to be learned is the importance of checking continuously the validity of the basic assumptions, the correctness of which may be too readily taken for granted.

The experimental scientist usually insists that experimenting rather than theorizing is his primary function. But he can go too far. Success in research comes from a balanced mixture of experimentation and of imaginative critical theorizing. We must constantly review whether our experiments ask the right kind of questions and whether the conclusions we draw from the results are not prejudiced by erroneous premises.

REFERENCES

Bloch, K. (1949). *J. Biol. Chem.* **179**, 1245–1254.
Bloch, K., and Anker, H. S. (1947). *J. Biol. Chem.* **169**, 765–766.
Cahn, R. S. (1964). *J. Chem. Educ.* **14**, 116–125.

Cahn, R. S., Ingold, C. K., and Prelog, V. (1966). *Angew. Chem.* **78**, 413–447.
Chantrenne, H. (1951). *J. Biol. Chem.* **189**, 227–233.
Cohen, G. N., and Monod, J. (1957). *Bacteriol. Rev.* **21**, 169–194.
Cornforth, J. W. (1974). *Tetrahedron* **30**, 1515–1524.
Dicke, W. K. (1951). *Ned. Tijdschr. Geneeskd.* **95**, 124–129.
Dicke, W. K., Weijers, H. A., and van de Kamer, J. H. (1953). *Acta Paediatr. (Stockholm)* **42**, 223–231.
Elliott, W. H. (1948). *Nature (London)* **161**, 128–129.
Evans, E. A., Jr., and Slotin, L. (1941). *J. Biol. Chem.* **141**, 439–450.
Gee, S. (1888). *St. Bartholomew's Hosp. Rep.* **24**, 17–20.
Gözsy, B., and Szent-Györgyi, A. (1934). *Hoppe-Seyler's Z. Physiol. Chem.* **224**, 1–10.
Haas, S. V. (1924). *Am. J. Dis. Child.* **28**, 421–437.
Hahn, O. (1962). *Naturwiss. Rundsch.* **15**, 43–47.
Hahn, O. (1965). *Naturwiss. Rundsch.* **18**, 86–91.
Herter, C. A. (1908). "Infantilism from Chronic Intestinal Infection." Macmillan, New York.
Hoagland, M. B. (1955). *Biochim. Biophys. Acta* **16**, 288–289.
Hokin, L. E. (1951a). *Biochem. J.* **48**, 320–326.
Hokin, L. E. (1951b). *Biochem. J.* **50**, 216–220.
Krebs, H. A. (1937). *Lancet* **2**, 736–738.
Krebs, H. A., and Johnson, W. A. (1937). *Enzymologia* **4**, 148–156.
Kuhn, T. S. (1962). 'Structure of Scientific Revolutions," *Found. Unity Sci.*, Vol. 2, No. 2.
Lipmann, F. (1971). *Science* **173**, 875–884.
Meyerhof, O. (1930). "Die chemischen Vorgänge im Muskel." Springer-Verlag, Berlin and New York.
Needham, J., and Lehmann, H. (1937). *Biochem. J.* **31**, 1210–1254.
Nobel Committee. (1938). Les Prix Nobel en 1937. Stockholm.
Ogston, A. G. (1948). *Nature (London)* **162**, 963.
Ogston, A. G. (1958). *Nature (London)* **181**, 1462.
Popper, K. R. (1975). *In* "Problems of Scientific Revolution" (R. Harré, ed.), pp. 72–101. Oxford Univ. Press (Clarendon), London and New York.
Speck, J. F. (1947). *J. Biol. Chem.* **168**, 403–404.
Swim, H. E., and Krampitz, L. O. (1954). *J. Bacteriol.* **67**, 419–425 and 426–434.
Szent-Györgyi, A. (1935). *Hoppe-Seyler's Z. Physiol. Chem.* **236**, 1–20.
Szent-Györgyi, A. (1936). *Hoppe-Seyler's Z. Physiol. Chem.* **244**, 105–116.
Weijers, H. A., and van de Kamer, J. H. (1953). *Acta Paediatr. (Stockholm)* **42**, 91–112.
Wood, H. G., Werkman, C. H., Hemingway, A., and Nier, A. O. (1941). *J. Biol. Chem.* **139**, 483–484.

2

On Some Unusual Nucleotide Polyphosphates*

FRITZ LIPMANN

The compounds I will discuss are nucleotides and are unrelated except for what seems to me to be an unusual binding of the phosphates. I will begin by discussing cyclic AMP,† and will then turn to the guanosine polyphosphates, initially picturesquely called magic spot compounds by Cashel and Gallant (1969), magic spot indicating that some uncertainty existed about their structure and manner of synthesis which Sy and Lipmann (1973) began to study some time ago. At the end, I will briefly mention a relatively recently discovered new activity of GTP in protein synthesis (Adams and Cory, 1975; Muthukrishnan *et al.,* 1975; Abraham *et al.,* 1975) when it serves to cover up the 5′-terminal of eukaryote mRNA by forming a G5′ppp5′NpN link.

*These studies have been supported by grant GM-13972 from the National Institutes of Health.

†The abbreviations used are ppGpp, guanosine 5′-diphosphate-3′-diphosphate; pppGpp, guanosine 5′-triphosphate-3′-diphosphate; an asterisk over a p, e.g., $\overset{*}{p}$, denotes a ^{32}P-label in that position; cAMP, cyclic adenosine 3′,5′-monophosphate.

CYCLIC AMP

I became interested in cAMP very early because, contrary to some other opinions, I thought that this strange ring phosphate had to have an energy-rich end in the 3′-attachment (Fig. 1). It was known that the phosphorolytic split by phosphorylase of the straight 5′–3′-phosphate bridge in polynucleotides yielded N5′P~P, i.e., an energy-rich phosphate, at sufficiently high inorganic phosphate concentrations (Grunberg-Manago *et al.,* 1955). I was quite happy, therefore, to see an abstract (Greengard *et al.,* 1969) on the reversibility of the adenylate cyclase reaction (cf. Fig. 1), which was followed by two additional papers (Hayaishi *et al.,* 1971; Takai *et al.,* 1971). They used a cyclase isolated from *Brevibacterium liquefaciens,* a uniquely suitable enzyme that was soluble and could be prepared in quite large quantities. Hayaishi *et al.* (1971) point out that since the splitting reaction yielded two separate molecules, the reverse reaction was very dependent on the concentration of the reactants and required a considerable amount of enzyme and long incubation times, an important facet of the reaction deserving attention considering the reversibility equilibria (Table 1).

In the hope of getting more information on adenylate cyclase, we decided to try to isolate this enzyme from *Escherichia coli* (Tao and Lipmann, 1969) since Sutherland's laboratory (Makman and Sutherland, 1965) had shown that *E. coli* could form rather large amounts of cAMP. When *E. coli* was homogenized lightly, much of the enzyme went down with the particulate fraction (Tao and Lipmann, 1969; Ide, 1969; Tao and Huberman, 1970), but washing the residue yielded a reasonable amount of soluble cyclase.

The characteristics of cyclase isolated from different organisms vary extensively. For example, our enzyme was strongly inhibited by pyrophosphate (Table II) and, in contrast to the stimulation by fluoride that had been found

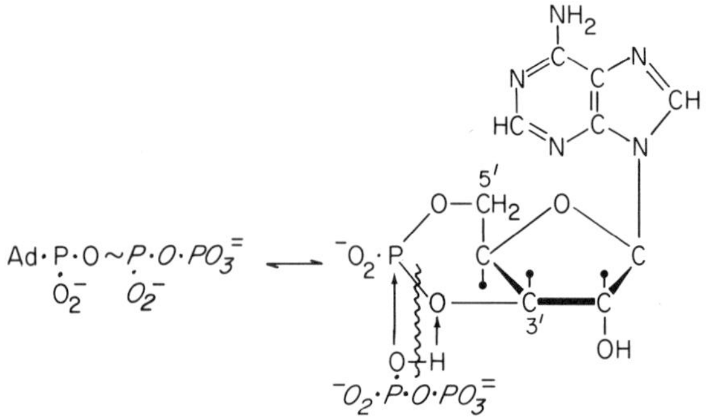

Fig. 1. Pyrophosphorolysis of cAMP. Reverse reaction of adenylate cyclase (Lipmann, 1971).

TABLE I
Estimated Change of Free Energy of cAMP Hydrolysis[a]

Reaction	ΔG^0_{obs} (kcal $mole^{-1}$)
Cyclic AMP + $PP_i \rightleftharpoons$ ATP	−1.6
ATP + $H_2O \rightleftharpoons$ 5′-AMP + PP_i	−10.3
Sum: cyclic AMP + $H_2O \rightleftharpoons$ 5′-AMP	−11.9

[a]From Hayaishi *et al.* (1971).

with animal tissue cyclases, it was inhibited by fluoride (Tao and Lipmann, 1969). The combination of both completely wiped out the reaction. We were rather disappointed when we could not demonstrate reversibility. This was attributed to the strong binding of pyrophosphate, the product of the split of ATP, which might prevent reversal, as was extensively discussed in an earlier publication (Lipmann *et al.,* 1971). However, in their discussion of various published negative attempts to show reversal, including ours, Hayaishi *et al.* (1971) refer to the concentration dependence requiring rather high concentrations of the reactants in the reverse reaction due to the joining of two molecules: cAMP + PP_i → ATP. I guess it is equally important that reversal demands considerable amounts of enzyme and long incubation times to approach equilibrium. Looking back at our experiments, I think we probably used sufficiently high concentrations of reactants but comparatively very low concentrations of enzyme and short incubation periods. This gave a good forward reaction but was probably not suitable for showing reversal, so that I am inclined now to believe that Hayaishi *et al.* were right and that it was the use of unfavorable conditions that prevented us from seeing a reverse reaction; the strong inhibition by PP_i, however, may make this enzyme particularly unsuitable. The energy-rich nature

TABLE II
Effect of Fluoride and Inorganic Pyrophosphate[a]

	[^{32}P]cAMP formed (pmole)
No addition	410
+ NaF	84
+ $NaPP_i$	28
+ NaF, $NaPP_i$	0

[a]The reaction mixture contained 2.5 μmoles of Tris-HCl, pH 8.5, 1 μmole of $MgSO_4$, 15 nmoles of [α-^{32}P]ATP, 28 μg of protein, with 0.25 μmole of NaF and 0.1 μmole of $NaPP_i$ added or eliminated, in a final volume of 50 μl. For further details, see Tao and Lipmann (1969).

of the 3′-terminal of the cyclic-bonded phosphate is confirmed by the calculation by Hayaishi *et al.* from their data of the free energy of hydrolysis of cAMP. As shown in Table I, the value obtained was nearly 12 kcal. The data indicate that the 3′-terminal of the phosphate bond in cAMP is one of the most highly energy-charged bonds metabolically available.

Release by Binding of cAMP of an Inhibitory Protein Complexed with Protein Phosphokinase

I will now turn to experiments illustrating the manner by which cAMP in animal tissues acts as secondary carrier of activity for a large array of hormones (Robison *et al.*, 1971). Quite generally, the hormones appear first to activate a membrane-bound cyclase, which produces the cAMP that then activates protein phosphokinases which carry phosphate from ATP to a specific protein acceptor. Tao and Salas in our laboratory (Tao *et al.*, 1970) found the effect of cAMP on phosphokinase in reticulocytes to be due to the binding to a regulatory protein; thereby, the cAMP caused the release of a protein tightly bound to and blocking the activity of kinase. In parallel with these experiments, similar results were reported by Gill and Garren (1970) in experiments on the adrenal cortex, and by

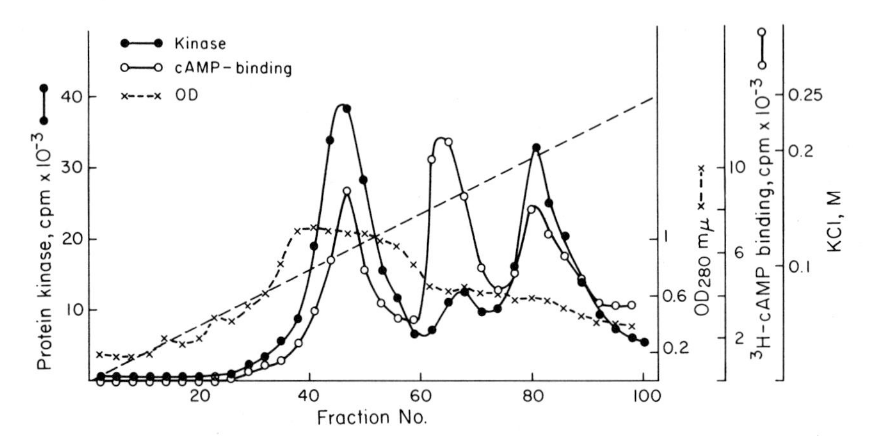

Fig. 2. DEAE-cellulose chromatography of protein kinase from adrenal cortex (Lipmann, 1971). About 800 μg of a 0–50% ammonium sulfate fraction obtained from an adrenal cortex 100,000 *g* supernatant was applied to a 2.3 × 21 cm DEAE-cellulose column equilibrated with 20 m*M* Tris-HCl, pH 7.5, 1m*M* dithiothreitol. The column was washed with 160 ml of the same buffer, and the enzymes were then eluted with a linear 0–0.25 *M* KCl gradient in the same buffer (450–450 ml). Fractions of 10 ml were collected at a flow rate of 0.6 ml/min. Protein kinase and [^{3}H]cAMP binding activities were determined as described (Fig. 3 in Lipmann, 1971). The specific activity of [γ-^{32}P]ATP was 21.1 cpm/pmole, and in the kinase test, calf thymus histone was used as phosphate acceptor.

Reimann *et al.* (1971) with muscle protein kinase. I will illustrate a few, partly unpublished experiments we did with Maria Salas, quoted in a review by Lipmann (1971), using adrenal cortex homogenates, which confirm and somewhat expand on the ones reported by Gill and Garren (1970).

In Fig. 2, the DEAE-cellulose chromatography of phosphokinase activity is compared with the binding of cAMP. This chromatogram shows three peaks: the first, phosphokinase I, shows coincidence of the binding protein and kinase when tested in the presence of cAMP; the second is one that shows very little phosphokinase but high binding activity, the latter being typical for adrenal cortex extracts; the third, kinase II, is similar to I. For further experiments, we used kinase I which is a purer fraction. In Fig. 3 it is shown that sucrose gradient density centrifugation in the presence of cAMP separates the binding protein and phosphokinase. Although the separation is not complete, one can isolate free protein kinase from part of the second peak. The inhibitory protein was isolated from the second peak in Fig. 2. Then, to illustrate the cAMP action, it is shown that our separated kinase preparation alone produced relatively low activity

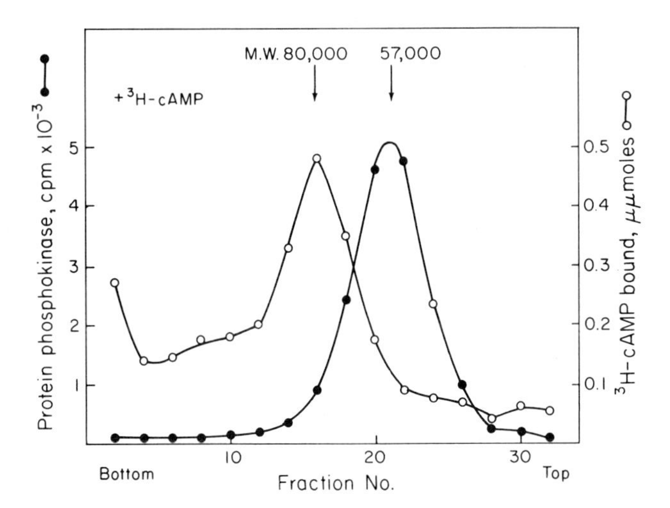

Fig. 3. Sucrose gradient sedimentation with cAMP present of protein kinase I activity and cAMP binding activity (Lipmann, 1971). Adrenal cortex protein kinase I (0.12 ml, 7.6 mg/ml), 8 m*M* theophylline, and 10^{-6} *M* [^{3}H]cAMP were incubated at 0°C for approximately 10 min, and then layered over a sucrose gradient containing 1.2×10^{-6} *M* [^{3}H]cAMP in addition to the buffer described in Fig. 9 of Lipmann (1971). The centrifugation was carried out in a tube parallel to those of the same Fig. 9. Two-drop fractions were collected, and the binding of [^{3}H]cAMP was determined by taking an aliquot of 0.05 ml of each fraction and, after addition of 33 m*M* Tris-HCl, pH 7.5, 10 m*M* $MgCl_2$, and 67 m*M* KCl, filtering through Millipore. The filter was washed as described in Fig. 3 of Lipmann (1971), and protein kinase activity was determined as described in the same figure. Closed circles, protein kinase activity; open circles, [^{3}H]cAMP bound.

(Table III, column 2, line 1) but was stimulated by the addition of cAMP (column 3). Addition of increasing concentrations of the separated cAMP binding protein increasingly reduced the activity of our kinase (column 1), but this inhibition was reversed by the addition of cAMP.

These results show how cAMP, when produced by hormone-induced cyclase activity, induces phosphate transfer by releasing protein kinase inhibitor through the binding of cAMP to the inhibitor protein. This activation of a protein phosphokinase appears to be the main route by which cAMP acts in animal tissues, as was first shown by Krebs's group (Walsh *et al.,* 1968) for the cAMP effect on muscle phosphorylase; there they discovered that a relatively unspecific cAMP-dependent protein phosphokinase phosphorylates a specific kinase which, in a cascade mechanism, activates phosphorylase by a second phosphate transfer that is independent of cAMP.

Altogether, this compound, which I call here an unusual "poly"-phosphate where phosphate is doubly bound to the same molecule (see Fig. 1), is a very strange chemical creature. I would like to end this part of my discussion by referring to cAMP in Fig. 4, taken from Sutherland's book (Robison *et al.,* 1971). Its spatial configuration indicates that the six-membered phosphate ring is joined with the furanose ring of the ribose; this creates a crammed situation in the phosphate ring, making the energy-rich bond. It might be responsible for the tendency of this molecule to bind strongly and specifically to proteins. Since the strain in the ring phosphate seems to be independent of the nucleotide base attached to the ribose, the present prominently discussed activities of cyclic guanosine monophosphate should likewise be connected with the strained ring phosphate bond.

TABLE III
Effect of Binding Protein on Protein Kinase Activity

Inhibitory protein[a] (μg)	Protein kinase activity[b]		Stimulation by cAMP (%)
	–cAMP	+cAMP	
0	250	476	90
4	144	485	236
8	130	457	251
16	103	457	343

[a]The inhibitory cAMP binding protein was isolated from the second peak in Fig. 2 and inhibitory-free protein kinase from the descending part of peak II in Fig. 3. Protein kinase activity was measured as described in the legend to Fig. 2.

[b]Picamoles of ^{32}P incorporated into histone.

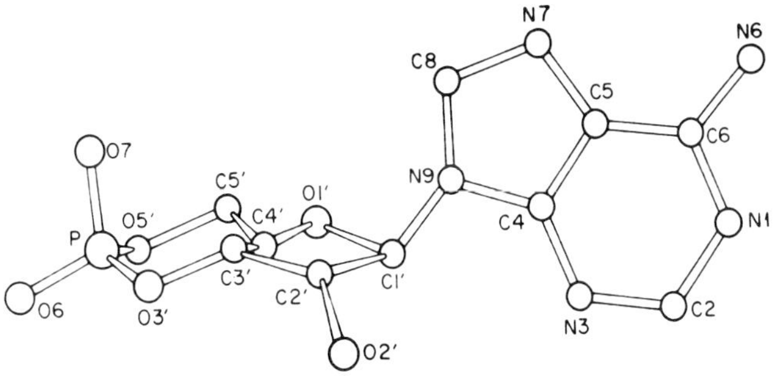

Fig. 4. Conformation of the ribose phosphate ring system in nucleoside 3′,5′-cyclic phosphate (Robison *et al.*, 1971). Comparing this with Fig. 1, the interlocking of the two rings through the common C3′ and C4 leads to appreciable strain in the 3′,5′-phosphodiester.

GUANOSINE TETRA- AND PENTAPHOSPHATES

I will now turn to a group of guanosine polyphosphates. These compounds were initially called magic spot compounds I and II, and are now referred to as ppGpp and pppGpp. Cashel and Kalbacher (1970) partially identified them as guanosine 5′-di- or triphosphates containing an additional diphosphate in 2′- or 3′-position. They are specifically formed in "stringent" strains of *E. coli*; these strains are deficient in amino acids and, in response to deprivation of one of the amino acids, they cause an inhibition of RNA synthesis in addition to the expected cessation of protein synthesis, which is in contrast to their counterparts, the "relaxed" strains. In view of earlier work on the role of GTP in protein synthesis, I became interested in exploring the exact mechanism by which these compounds are synthesized.

On the Mechanism of Pyrophosphate Transfer

An approach to the analysis of their mode of synthesis was opened through the discovery by Haseltine *et al.* (1972) that from ribosomes of stringent, but not of relaxed, strains of *E. coli* a protein fraction could be washed off by 0.5–1.0 *M* NH_4Cl, which, on re-addition to the ribosome, caused added ATP to transfer phosphates to GDP or GTP. With this *in vitro* system, it became relatively easy to approach the mechanism of synthesis of magic spot compounds. Kjeldgaard and his colleagues soon found (Pedersen *et al.*, 1973) that it was also necessary to add a mRNA as well as a matching but *uncharged* tRNA. Although it had been assumed that the pyrophosphate hung on to a 2′- or 3′-hydroxyl of the ribose in guanylic acid was transferred as a unit, we thought that the proof for this was not compelling and we set out first to test thoroughly the manner in which the two phosphates are added since it is known that pyrophosphate may be added either in a two-step reaction from the γ-phosphates of two ATP's (Chaykin *et al.*, 1958; Henning *et al.*, 1959), or by a pyrophosphoryl transfer (Khorana *et al.*, 1958). The latter mechanism was shown by Khorana *et al.* (1958) to apply to synthesis of 5′-monophosphate-1′-pyrophosphoryl ribose. There, characteristically, from ATP marked in the β-position, this β-phosphate will appear on the accepting ribose hydroxyl, indicating the transfer of pyrophosphate as a unit from one ATP. Therefore, we tested transfer with γ- and β-marked ATP to see in what position the phosphate was attached. In all these experiments for the analysis we used thin layer chromatography (Sy and Lipmann, 1973), which very nicely separates the different compounds.

Comparing β– and γ-marked ATP as a donor and using the *in vitro* reaction of Haseltine *et al.* (1972), it could be shown (Fig. 5) that the β-marker is attached directly to the carbon while the γ-marker is not. This can be demonstrated by splitting the guanosine nucleotide-attached pyrophosphate with a zinc-activated yeast inorganic pyrophosphatase (Schlesinger and Coon, 1960) which, in contrast to magnesium pyrophosphatase, is able to attack nucleotide-bound polyphosphate. This procedure showed that the marked phosphate remains with the molecule only in the case of the β-marked ATP, whereas in the case of the γ-marking, the marked phosphate is split off as P_i (see Sy and Lipmann, 1973). This excludes a two-step reaction and proves a one-step pyrophosphoryl transfer from ATP.

Our second aim was to decide in what position on the ribose the pyrophosphoryl group was inserted; for this purpose, we used the conveniently marked pG$\overset{*}{p}$ obtained from ppG$\overset{*}{p}$p by split with zinc pyrophosphatase as described. The pG$\overset{*}{p}$ was then tested with Kaplan's specific rye grass 3′-nucleotide phosphatase (Shuster and Kaplan, 1953). Since 3′- or 2′-linked phosphate is rather sensitive to acid hydrolysis, we used pG$\overset{*}{p}$ obtained by acid hydrolysis of ppG$\overset{*}{p}$p as a check, being aware, however, of the fact that during such a procedure there would be an isomerization of the 2′- and 3′-positions. Figure 6 shows the time course of hydrolysis with the rye grass 3′-nucleotidase, and it can be seen that using for hydrolysis the product obtained with the enzymatic method (upper curve) thus retaining the original position, the nucleotidase completely hydrolyzes the radioactively marked phosphate, proving it to be attached in 3′-position. On the other hand, the pG$\overset{*}{p}$ isolated from the acid hydrolysate showed that just half of the radioactively marked phosphate was hydrolyzed by the 3′-specific enzyme; thus, the acid hydrolysis had indeed isomerized and now only half of it remained in 3′-position.

For identification of the function of the stringent factor, it became quite important when Sy *et al.* (1973) observed that there was a very slow transfer of pyrophosphate from ATP to GTP or GDP with stringent factor in the absence of

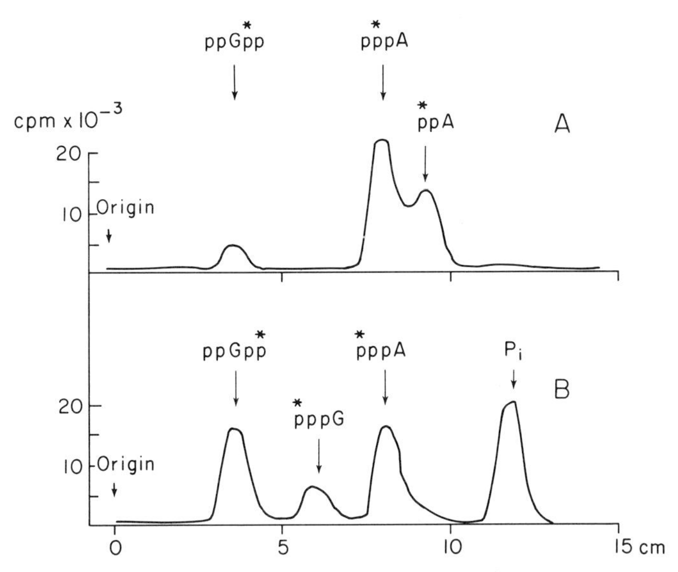

Fig. 5. Synthesis of ppGpp labeled with ^{32}P in various positions (Sy and Lipmann, 1973). The label was introduced by use of the following reactions: A, 1 m*M* GTP, 1m*M* [β-^{32}P]ATP (60 Ci/mol), and B, 1 m*M* GTP, and 1 m*M* [γ-^{32}P]ATP (130 Ci/mol). Reaction mixtures contained in 50 μl: 20 m*M* Tris-acetate (pH 7.8), 2 m*M* dithiothreitol, 10 m*M* $Mg(OAc)_2$, 145 μg of washed ribosomes, and 39 μg of 0.5 *M* NH_4Cl ribosomal wash. Incubations were performed at 37°C for 1 hr and stopped by chilling and addition of 1 μl of 88% formic acid. The precipitate containing the ribosomes was removed by centrifugation, and 1 μl of the supernatant was chromatographed on polyethylenimine cellulose thin layer sheets using 1.5 *M* KH_2PO_4 (pH 3.4) as developing solvent. The chromatograms were then scanned in a Varian radioscanner. Marker nucleotides, indicated by arrows, were visualized under ultraviolet light. The transfer reactions realized in the two sets of experiments are formulated in the following equations: (A) ppG + $p\overset{*}{p}pA \rightarrow ppG\overset{*}{p}p$ + pA; (B) ppG + $\overset{*}{p}ppA \rightarrow ppGp\overset{*}{p}$ + pA.

the ribosomal system; as shown in Fig. 7, the non-ribosomal activity of the factor could be dramatically increased by the addition of 20% methanol. For identification of the stringent factor as the enzyme catalyzing the transfer, we were interested to find out whether the enzymatically active protein acting in conjunction with the ribosomal system was identical to the one that causes pyrophosphoryl transfer in the absence of ribosomes. To test this, both activities were compared by centrifugation in a sucrose density gradient. It appears from Fig. 8 that the molecular weight of the enzyme is coincident for both conditions. The higher activity recorded on the left in the ribosomal system appears in the upper curve, and the one without ribosome in the lower curve is recorded on the right. We conclude that the same enzyme is acting in ribosomal and nonribosomal systems and has a molecular weight of approximately 75,000. The

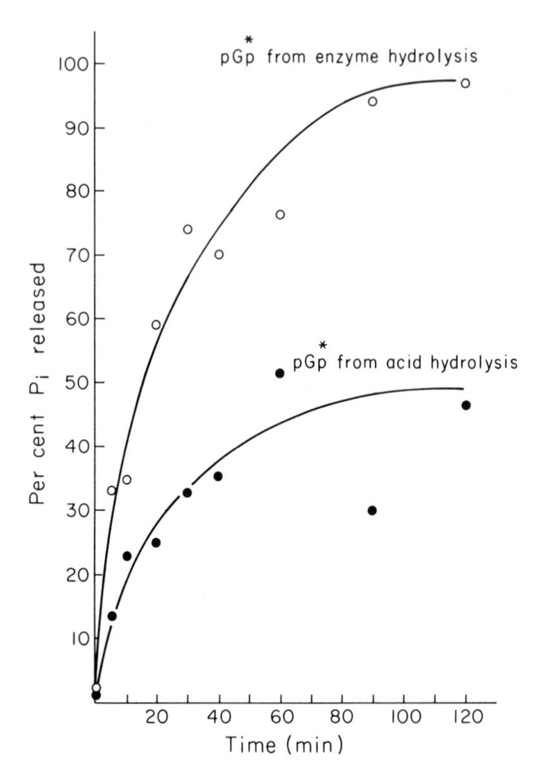

Fig. 6. Digestion by 3′-nucleotidase of pGp̊*, prepared enzymatically or by acid hydrolysis (Sy and Lipmann, 1973). pGp* obtained by inorganic pyrophosphatase digestion (4000 cpm) as described in Fig. 5A, and by hydrolysis in HCl for 16.5 hr at 37°C (5000 cpm) as described in Fig. 5B of the paper by Sy and Lipmann (1973), was incubated at 37°C in 100 μl of 20 m*M* Tris-HCl (pH 7.5) containing 0.002 unit of 3′-nucleotidase. At various time points, 10-μl aliquots were withdrawn and applied directly to polyethylenimine sheets; the spots were air-dried. Chromatograms were developed with 0.75 *M* KH_2PO_4 (pH 3.4) and were then radioautographed overnight. Areas corresponding to pGp and P_i were cut out and counted in a scintillation counter. The % PO_4 released corresponds to the % ^{32}P counts released from pGp*.

20-fold or greater stimulation through the attachment to the ribosomal system as compared to the best nonribosomal system is difficult to explain. However, a relaxed *E. coli* ribosomal mutant has now been found (Friesen *et al.,* 1974) which does not form the guanosine polyphosphates in spite of the presence of stringent factor on the ribosome. One should hope that with this mutant an explanation of the ribosomal effect will become available.

Dr. Sy (1974a) observed that pyrophosphate transfer from ATP to GTP or GDP is reversible. In the reverse direction, pppGpp is a better donor than

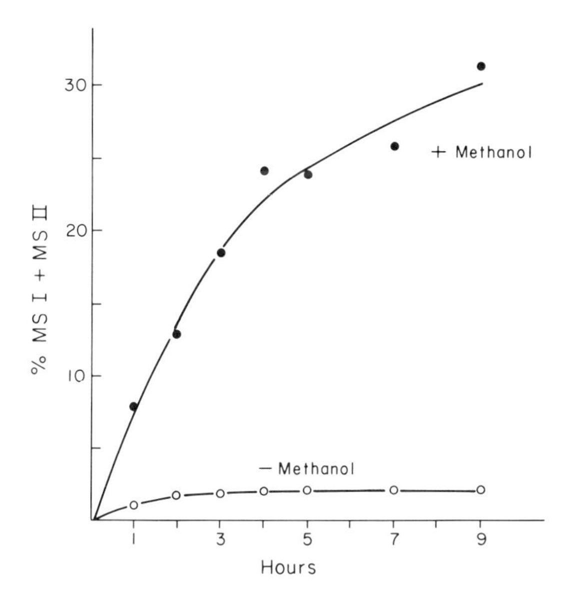

Fig. 7. Stringent factor-catalyzed ppGpp and pppGpp formation; methanol effect (Sy *et al.*, 1973). Reaction mixtures in 100 μl contained 20 m*M* Tris-acetate (pH 8.0), 20 m*M* $Mg(OAc)_2$, 4 m*M* dithiothreitol, 150 m*M* NH_4Cl, 0.4 m*M* [α-^{32}P]GTP (70 Ci/mol), 4 m*M* ATP, 4.5 μg of fraction I NH_4Cl wash, and 20% methanol (closed circles) or none (open circles). Incubations were at 28°C and, at various time points, aliquots of 5 μl were mixed with 5 μl of 1.76% formic acid. Precipitates were removed by centrifugation, and 2 μl of the supernatant were assayed for ppGpp and pppGpp formation. The results are expressed as % of added [α-^{32}P]GTP converted to ppGpp and pppGpp.

ppGpp, as shown in Fig. 9. For optimum reversal, a great excess of AMP has to be used; in the experiment shown, the AMP concentration was 10 m*M* in comparison to 10^{-5} *M* with pppGpp. Therefore, it is unlikely that the reverse reaction would play a role in the *in vivo* system.

Our interest has actually centered on the mechanism of synthesis and on pinpointing the enzymatic activity of the stringent factor. The stringency effect, i.e., the inhibition of RNA synthesis which, in some manner, is obviously caused by the production of pppGpp and ppGpp, still remains to be fully explained. However, recently a strong inhibition of ribosomal RNA synthesis by ppGpp has been found in an *in vitro* system (Reiness *et al.*, 1975).

The Blocking of the 5′-Terminal of Eukaryotic mRNA by Reaction with GTP

I would like to conclude by mentioning a recently discovered reaction studied by several groups (Adams and Cory, 1975; Muthukrishnan *et al.*, 1975; Abraham

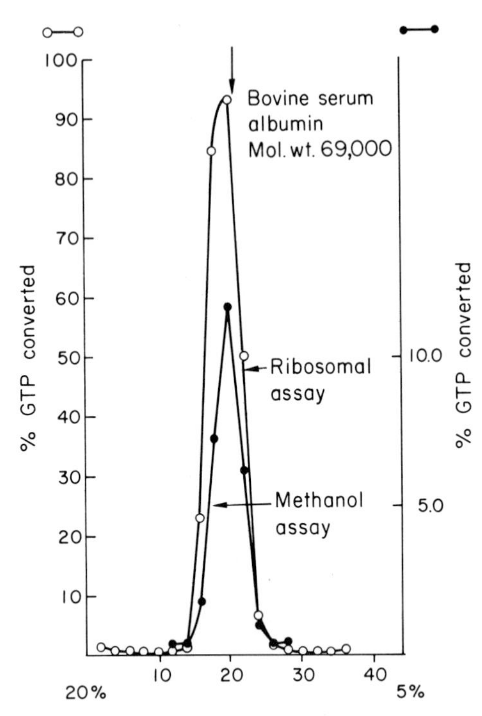

Fig. 8. Assay for ribosomal and nonribosomal synthesis after sucrose gradient centrifugation of stringent factor preparation (Sy, 1974a,b). Fraction II stringent factor (60 μg) was centrifuged in a SW 50 rotor at 45,000 rpm for 14 hr on a 5–20% linear sucrose gradient containing 20 m*M* Tris-Ac, pH 7.8, 20 m*M* $Mg(OAc)_2$, 100 m*M* NH_4Cl, and 1 m*M* dithiothreitol. Two-drop fractions were collected from the bottom and assayed for guanosine polyphosphate synthesis. Open circles, ribosomal system: 20-μl aliquots of the gradient fractions were incubated at 30°C for 90 min in a final volume of 50 μl that contained 40 m*M* Tris-Ac, pH 8.0, 4 m*M* dithiothreitol, 10 m*M* $Mg(OAc)_2$, 4 m*M* ATP, 0.4 m*M* [α-^{32}p]GTP (23 Ci/mol), 27 μg of tRNA, 10 μg of poly(A,U,G), and 80 μg of washed ribosomes. Incubation was stopped with HCOOH and guanosine polyphosphates were assayed as described (Sy and Lipmann, 1973). Closed circles, nonribosomal system: 20-μl aliquots of the gradient fractions were incubated at 30°C for 18 hr in a final volume of 50 μl that contained 40 m*M* Tris-Ac, pH 8.0, 4 m*M* dithiothreitol, 20 m*M* $Mg(OAc)_2$, 4 m*M* ATP, 0.4 m*M*[α-^{32}P]GTP (23 Ci/mol), 50 m*M* NH_4Cl, 15% methanol, 27 μg of tRNA, and 10 μg of poly(A,U,G). Incubation was stopped with HCOOH and guanosine polyphosphates were assayed as described (Sy and Lipmann, 1973).

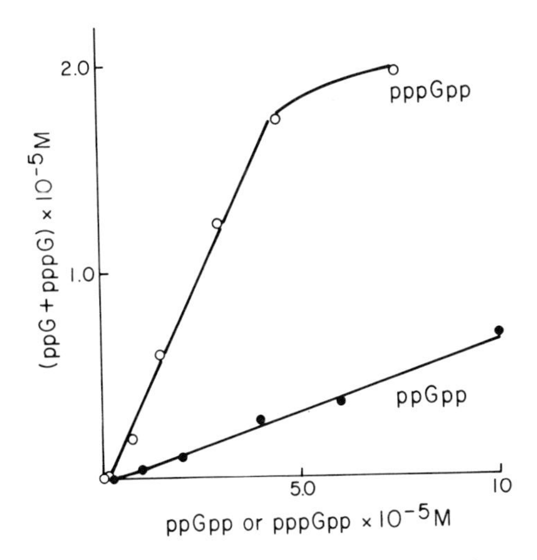

Fig. 9. Reverse reaction with pppGpp and ppGpp as donor to AMP. The reverse reactions were performed as described (Sy, 1974a,b). ppGpp as a ^{3}H-derivative (130,000 cpm/reaction) and pp$\overset{*}{p}$Gpp as the 5′-α-^{32}P-derivative (27,000 cpm/reaction) were added at the indicated concentrations; acceptor AMP concentration was 10 m*M*. Incubations were at 30°C for 90 min.

et al., 1975; Groner and Hurwitz, 1975). It appears that in eukaryote cells, including virus-induced tumor cells, mRNA's are blocked at the 5′-terminal by interaction with GTP which forms a 5′→5′ triphosphate bridge to the last nucleotide in mRNA. Thereby, as may be seen from Fig. 10, the 5′-terminal is blocked and may be protected in some manner against degradation. The mechanism of this GTP-linked reaction seems to be different in different systems, since it has been found that either the GTP marked in the α,β-phosphate retains these two phosphates in the bridge (Abraham *et al.*, 1975), or that two of the phosphates may be derived from the 5′-terminal nucleotide of mRNA (Ensinger *et al.*, 1975). It can be seen that, in addition, a methylation of the nucleotides has occurred on both sides of the phosphate bridge on the added guanosine in 7′-position and on the ribose of the terminal nucleotide of accepting mRNA in 3′-position. This methylation appears to be essential for the activity of the mRNA.

We have just begun some work on the mechanism of this blocking reaction. It interests us because of the new involvement of GTP in an important reaction in the protein synthesis system, in addition to the many roles of GTP which we have helped study through the years (Lucas-Lenard and Lipmann, 1971). Altogether, the role of the guanosine derivatives in an increasing number of key metabolic systems is quite remarkable.

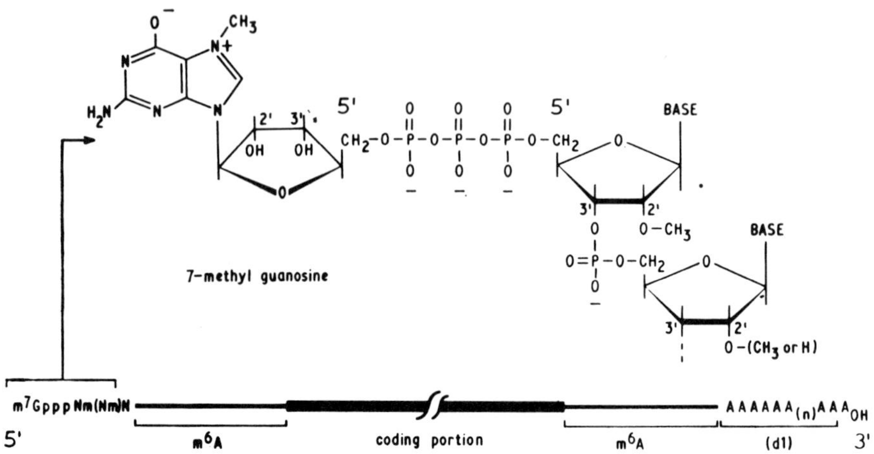

Fig. 10. 5′-Terminal structure and possible location of the major modified nucleosides in myeloma mRNA (Adams and Cory, 1975).

I offer here a discussion of the role of GTP in many metabolic reactions that have recently come to light. I hope that it will interest Albert Szent-Györgyi who has done so much to make us understand the role of ATP in the contraction of the muscle.

REFERENCES

Abraham, G., Rhodes, D. P., and Banerjee, A. K. (1975). *Nature (London)* **255**, 37–40.

Adams, J. M., and Cory, S. (1975). *Nature (London)* **255**, 28–33.

Cashel, M., and Gallant, J. (1969). *Nature (London)* **221**, 838–841.

Cashel, M., and Kalbacher, B. (1970). *J. Biol. Chem.* **245**, 2309–2318.

Chaykin, S., Law, J., Phillips, A. H., Tchen, T. T., and Bloch, K. (1958). *Proc. Natl. Acad. Sci. U.S.A.* **44**, 998–1004.

Ensinger, M. J., Martin, S. A., Paoletti, E., and Moss, B. (1975). *Proc. Natl. Acad. Sci. U.S.A.* **72**, 2525–2529.

Friesen, J. D., Fiil, N. P., Parker, J. M., and Haseltine, W. A. (1974). *Proc. Natl. Acad. Sci. U.S.A.* **71**, 3465–3469.

Gill, N., and Garren, L. D. (1970). *Biochem. Biophys. Res. Commun.* **39**, 335–343.

Greengard, P., Hayaishi, O., and Colowick, S. P. (1969). *Fed. Proc., Fed. Am. Soc. Exp. Biol.* **28**, 467.

Groner, Y., and Hurwitz, J. (1975). *Proc. Natl. Acad. Sci. U.S.A.* **72**, 2930–2934.

Grunberg-Manago, M., Ortiz, P. J., and Ochoa, S. (1955). *Science* **122**, 907–910.

Haseltine, W. A., Block, R., Gilbert, W., and Weber, K. (1972). *Nature (London)* **238**, 381–384.

Hayaishi, O., Greengard, P., and Colowick, S. P. (1971). *J. Biol. Chem.* **246**, 5840–5843.

Henning, U., Möslein, E. M., and Lynen, F. (1959). *Arch. Biochem. Biophys.* **83**, 259–267.

Ide, M. (1969). *Biochem. Biophys. Res. Commun.* **36,** 42–46.
Khorana, H. G., Fernandes, J. G., and Kornberg, A. (1958). *J. Biol. Chem.* **230,** 941–948.
Lipmann, F. (1971). *Adv. Enzyme Regul.* **9,** 5–16.
Lipmann, F., Tao, M., and Huberman, A. (1971). *Role Adenyl Cyclase Cyclic 3',5'–AMP (Adenosine 3',5'–Phosphate) Biol. Syst., Colloq., 1969* Fogarty Int. Cent. Proc., No. 4, pp. 29–39.
Lucas-Lenard, J., and Lipmann, F. (1971). *Annu. Rev. Biochem.* **40,** 409–448.
Makman, R. S., and Sutherland, E. W. (1965). *J. Biol. Chem.* **240,** 1309–1314.
Muthukrishnan, S., Both, G. W., Furuichi, Y., and Shatkin, A. J. (1975). *Nature (London)* **255,** 33–37.
Pedersen, F. S., Lund, E., and Kjeldgaard, N. O. (1973). *Nature (London) New Biol.* **243,** 13–15.
Reimann, E. M., Brostrom, C. O., Corbin, J. D., King, C. A., and Krebs, E. G. (1971). *Biochem. Biophys. Res. Commun.* **42,** 187–194.
Reiness, G., Yan, H.-L., Zubay, G., and Cashel, M. (1975). *Proc. Natl. Acad. Sci. U.S.A.* **72,** 2881–2885.
Robison, G. A., Butcher, R. W., and Sutherland, E. W. (1971). "Cyclic AMP." Academic Press, New York.
Schlesinger, M. J., and Coon, M. J. (1960). *Biochim. Biophys. Acta* **41,** 30–36.
Shuster, L., and Kaplan, N. O. (1953). *J. Biol. Chem.* **201,** 535–546.
Sy, J. (1974a). *Proc. Natl. Acad. Sci U.S.A.* **71,** 3470–3473.
Sy, J. (1974b). *In* "Lipmann Symposium: Energy, Regulation, and Biosynthesis in Molecular Biology" (D. Richter, ed.), pp. 599–609. de Gruyter, Berlin.
Sy, J., and Lipmann, F. (1973). *Proc. Natl. Acad. Sci. U.S.A.* **70,** 306–309.
Sy, J., Ogawa, Y., and Lipmann, F. (1973). *Proc. Natl. Acad. Sci. U.S.A.* **70,** 2145–2148.
Takai, K., Kurashina, Y., Suzuki, C., Okamoto, H., Ueki, A., and Hayaishi, O. (1971). *J. Biol. Chem.* **246,** 5843–5845.
Tao, M., and Huberman, A. (1970). *Arch. Biochem. Biophys.* **141,** 236–240.
Tao, M., and Lipmann, F. (1969). *Proc. Natl. Acad. Sci. U.S.A.* **63,** 86–92.
Tao, M., Salas, M. L., and Lipmann, F. (1970). *Proc. Natl. Acad. Sci. U.S.A.* **67,** 408–414.
Walsh, D. A., Perkins, J. P., and Krebs, E. G. (1968). *J. Biol. Chem.* **243,** 3763–3765.

3

Ektobiology, Transport of Nutrients, and Chemotaxis

HERMAN M. KALCKAR

I am being drawn into an important new field of cell biology and tumor biology which deals with membrane surface patterns and membrane functions. Since it deals with the surface of the plasma membrane which contains a number of proteins, among them transport carriers as well as glycoproteins and glycolipids (some of them displaying strong antigenic characters), I like to call this approach to biology "ektobiology"; I wrote an essay about it 10 years ago (Kalckar, 1965).

We know that both prokaryotic and eukaryotic cells carry a large number of surface antigens and that some of these antigens are simple sugars. This was first described by Heidelberger, Goebel, and Avery on pneumococci 50 years ago (see Heidelberger *et al.*, 1925).

The kind invitation to contribute to this volume induces me to try to write another essay, and since the volume is called "Search and Discovery," it calls for discussion of some new fundamental problems in biology, so why not ektobiology?

EKTOBIOLOGY AND SUGARS

So far as carbohydrates are concerned, an important new ektobiological approach in bacteriology was initiated around 1960 by my former co-workers

Hiroshi Nikaido and Annette Rapin (see reviews by Nikaido, 1965, and by Rapin and Kalckar, 1971). Nikaido stayed with us in our biochemistry laboratory at the Massachusetts General Hospital for 6–7 years and developed this approach further. He showed that *Escherichia coli* and *Salmonella* mutants unable to convert UDPG to UDPgalactose (epimeraseless mutants) contained highly defective lipopolysaccharides (LPS) with loss of several antigenic determinants. LPS from epimeraseless mutants were devoid not only of galactose but also of sugars located more terminally in the chain of the repeat units such as dideoxy sugars (Nikaido, 1962, 1965; Rapin and Kalckar, 1971). If UDPgalactose was added to broken cell preparations of the epimeraseless mutants, transfer of labeled galactose from UDPgalactose to the inner core of LPS took place. Mary Jane Osborn and co-workers have further demonstrated the successive transfer of glucose and *N*-acetylglucosamine to the galactosyl units of LPS in a particulate fraction from epimeraseless *Salmonella typhimurium* (Osborn *et al.*, 1962; Nikaido, 1962). Bruce Stocker and Nikaido found that epimeraseless *S. typhimurium* were not attacked by a semivirulent phage P22 (see Nikaido, 1965). Apparently galactose and the subsequent terminal sequence of the LPS are needed for absorption of the phage.

From an evolutionary point of view one could ask why any cell should waste its energy making UDPgalactose and galactosyl units. Not only are the galactose-free cells perfectly viable, they are even protected against certain "deadly" bacteriophages. Why "invent" these receptor sugars and preserve them when selection might go against such innovations? One possible answer to this question could be that the hazards are outweighed by the prospects of jumps in evolution brought about by "sneaking in" genes from bacterial viruses, at least the temperate viruses, to the fraction which survives (see Morse *et al.*, 1956; Weigle, 1966; Kalckar, 1966). We know that bacteriophages and plasmids can provide the host cells with new genetic programs.

TRANSPORT BIOLOGY AND CHEMOTAXIS

While isolating mutants defective in galactose metabolism, a talented graduate student in our department, Henry Wu (who also earned his Ph.D. degree at Harvard), isolated transport mutants, and our interest switched from metabolism and regulation and cell wall chemistry to transport mechanisms and the components involved. Anraku and Heppel (see Anraku, 1968) had described a periplasmic binding protein with high affinity for glucose and galactose and supposedly involved in transport of these sugars. When Winfried Boos from Freiburg University joined our laboratory in 1968, he pursued the studies of the high affinity galactose transport system with great skill, diligence, and scope (see Boos, 1974). He isolated a large number of mutants which had lost the highly active, high affinity galactose transport system and studied the periplasmic galactose-binding protein (GBP). GBP is presumably located on the outside of the membrane and released by a gentle nondenaturing technique described by Heppel (1969). Two types of defective galactose transport mutants were isolated. One of them was unable to synthesize the periplasmic galactose-binding protein. The other type showed normal biosynthesis of normal binding protein, yet no transport activity was discernible, presumbly due to a defect of a specific translocase in the inner membrane. We called the two types of galactose transport negatives (I) GBP^-, GTr^-, and (II) GBP^+, GTr^- (Boos, 1974; Silhavy *et al.,* 1974). However, the proof that GBP is involved in high-affinity transport remained admittedly incomplete.

At about this time my attention was drawn toward a new field developed by Julius Adler of the University of Wisconsin (see Adler, 1969). He revived an old forgotten biological field by using microbial genetics. This field, to which he gave such a dramatic renaissance, was bacterial chemotaxis, the ability of bacteria to detect a point source of a nutrient (or more general an "attractant") and act accordingly, i.e., respond to swimming toward the source. The type of modulation of their behavior by a gradient of an attractant has been described by Howard Berg (1974) and Dan Koshland (1974) as follows. In the absence of any gradient of attractant the bacteria move around by a so-called random walk, composed of straight swimming and sharp turns called twiddles or tumbling. The modulation of behavior brought about by a cell swimming up a gradient of an attractant and hence presumably filling some specific chemosensors results in a suppression of tumbling in favor of swimming (biased random walk up the gradient). "Memory" is part of the response (Koshland, 1974).

The modulations depend on the increase in the fractional amount of "chemoreceptor" attractant complex formed within a limited time span. It stood to reason that the chemoreceptors must be on the surface of the cell, either in the membrane or on the outside of the membrane. In search of specific chemotaxis mutants, Adler and his co-workers (see Adler, 1976) designed a new type of plating technique (so-called "sugar-swarm plates") and isolated a number of specific mutants highly defective in galactose chemotaxis ($GTax^-$), yet normal in their chemotactic responses to mannose, aspartate, or serine. Subsequently the strains with positive galactose chemotaxis were screened for galactose transport and metabolism (see Adler, 1969). I was most puzzled by the finding that several of Adler's mutants, which were unable to transport galactose, still retained their chemotaxis toward galactose (and glucose). Boos in our laboratory had been studying the two different types of galactose transport negative mutants, GTr^-, GBP^+, and the other transport defective type which had no functioning GBP, i.e., GTr^- and GBP^-. It now occurred to me that GBP may be the common first "receptor" for galactose chemotaxis as well as for galactose transport. Perhaps Adler's GTr^-, $GTax^+$ strain corresponded to our GTr^-, GBP^+

and his GTr⁻, GTax⁻ matched our GTr⁻, GBP⁻. The startling possibility evolved that GBP could therefore be involved in two functions. Further tests in the Adler laboratory in Wisconsin did indeed indicate that GBP is the common receptor for galactose chemotaxis and for galactose transport (Hazelbauer and Adler, 1971). This history I presented in my Weigle Memorial Lecture at California Institute of Technology in 1970 (Kalckar, 1971). A more popular article by Adler on chemotaxis appeared recently (Adler, 1976).

The galactose-receptor complex is supposed to release its substrate to the translocase of the membrane. For the operation of chemotaxis it must release a "signal." Boos was exploring the physical-chemical and biochemical properties of the purified GBP. Among the novel findings, I shall mention the conformational changes observable also by a distinct increase in tryptophan fluorescence emission elicited specifically by addition of galactose, and in concentrations as low as 10^{-7} *M* or lower, thiogalactosides being unable to elicit any detectable fluorescent change except at 1000-fold higher concentrations (see Fig. 1).

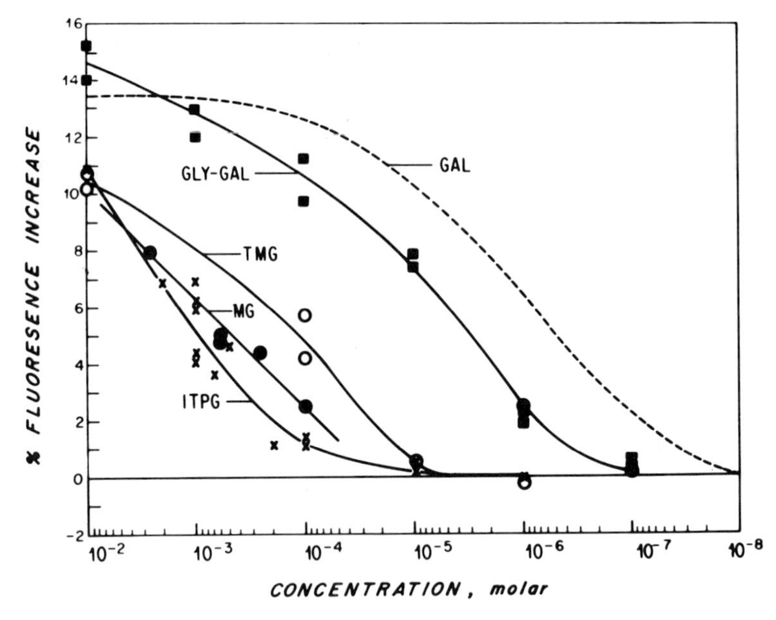

Fig. 1. Percent increase in fluorescence emission at 330 nm of galactose-binding protein versus total concentration of various sugars. Closed circles, MG; open circles, TMG; X, IPTG; closed squares, β-glycerol galactoside; broken line, D-galactose. Protein concentration, 16.4 μg/ml; temperature, 24°C. Abbreviations: TMG, methyl-1-thio-β-galactopyranoside; IPTG, isopropyl-1-thio-β-D-galactopyranoside; MG, methyl-1-*O*-β-D-galactopyranoside; β-glycerol galactoside, (D-glyceryl)-1-β-D-galactopyranoside. (From Boos *et al.*, 1972, with permission of the journal and author.)

Another unusual feature is the generation of an extra negative charge when the galactose complex is formed (Boos *et al.,* 1972; Boos, 1972; Silhavy *et al.,* 1974). We do not know at the present time which of the two "signals" is transmitted to the next step in the chemotactic system. However, since the charge difference was not discernible in an affinity mutant in the presence of 1000 times higher galactose concentration, whereas the fluorescence difference was expressed (Boos, 1972), the latter may be a more relevant signal for chemotaxis as well. The differential increase of fluorescence at 300–336 nm was found to be half maximum at 3 m*M* in the low affinity mutant whereas in a revertant the half point was reached at 3 μ*M* like the wild type (Boos, 1972).

The interpretation of the fluorescence change is probably far from simple and it is made more perplexing by the seemingly opposite response of the maltose chemotactic binding protein. That is, the maltose complex with its specific binding protein showed a decrease of the tryptophan emission fluorescence (Szmelcman *et al.,* 1976). Hopefully new techniques might eventually offer opportunities for a common interpretation.

The mutants which possess GBP with normal high affinity galactose receptor activity and hence, also normal galactose chemotaxis, yet defective galactose transport, is presumably programming a defective membrane protein, which normally specialized in active translocation of galactose or couples the GBP to the specific translocase of Rotman (Robbins and Rotman, 1975). The recent genetic studies by Ordal and Adler (1974) indicate once more that the only gene common for galactose chemotaxis and the high-affinity galactose transport is the one which programs for GBP, i.e., the periplasmic galactose receptor (see Adler, 1976; Silhavy *et al.,* 1974).

REGULATION OF TRANSPORT AND UPTAKE AND TUMOR BIOLOGY

In 1968 my colleague Donald Wallach suggested that neoplastic uncontrolled growth is part of the pleotropic changes which are to be found in the membranes of higher eukaryotic cells during oncogenesis. Wallach (1968) considered cancer a structural mutation in the programming of membrane proteins. This suggestion was of course a departure from the classical Warburg hypothesis of a regulatory failure of the Pasteur reaction. Some of the features can still be combined. In the 1970's much interest has been centered around oncogenesis and enhancement of uptake of nutrients (see Hatanaka's review, 1974). The enhancement of 2-deoxyglucose uptake or of facilitated exchange of 3-*O*-methylglucose (a nonphosphorylated hexose analogue) early after oncogenic transformation is not even dependent on cell division (Eckhardt and Weber, 1974). Addition of cytosine arabinoside, an effective inhibitor of DNA biosynthesis (Furth and Cohen, 1968), to the transformed cultures arrests DNA replication in the transformed culture, yet uptake rates remain high (Eckhardt and Weber, 1974).

Another nonmitogenic stimulus of uptake, deprivation of glucose (in the presence of the usual excess of glutamine) elicits a spectacular stimulation of hexose uptake in chick embryo fibroblast cultures (Martineau *et al.,* 1972) as well as in mouse and hamster cell cultures (Ullrey *et al.,* 1975). This type of stimulus can be repressed by actinomycin (Martineau *et al.,* 1972; Kletzien and Perdue, 1975) and the stimulation of uptake is classified as a "derepression." More *in vitro* studies are needed in order to settle this question. Glucose starvation does affect transport as well as uptake since facilitated influx as well as efflux of 3-*O*-methylglucose was found to be higher in glucose-starved cultures than in glucose-fed cultures (Christopher *et al.,* 1976a). Glucose starvation is said to permit derepression of hexose transport and uptake (Martineau *et al.,* 1972; Kletzien and Perdue, 1975; Christopher *et al.,* 1967a). Conversely glucose transport and metabolism bring about a repression of the hexose uptake system. Most intermediates of the Embden-Meyerhof pathway have been ruled out as co-repressors (Christopher *et al.,* 1976b). However, in view of the interest in methylglyoxal as an inhibitor of growth (Egyud and Szent-Györgyi, 1966) this alternate product of glycolysis deserves careful examination as a potential co-regulator. From our recent studies of protein turnover we are led to suspect that the glucose starvation enhancement (or derepression) may be largely due to a curbing in turnover rates of the hexose uptake systems–most likely the hexose carriers presumably located in the plasma membranes (Christopher *et al.,* 1976b).

In the same series of studies we also found that refeeding the cultures glucose gradually restored the repression of the hexose transport and uptake systems. In contrast, cycloheximide, which inhibits protein synthesis, prevented the return to the repressed state, even if excess glucose was present (Christopher *et al.,* 1976b). Perhaps regulator units, lost during the period of glucose stavation, are barred from being synthesized in the presence of clycloheximide.

Returning to oncogenesis and the enhancement of nutrient uptake, the oncogenically-induced increase in hexose transport and uptake seems to be affecting the V_{max} rather than the K_m of the transport system (see Kletzien and Perdue, 1975). In this case too, one should not only consider regulation of protein synthesis, but also regulation of processes like inactivation or degradation of hexose carriers. Since amino acid transport is also enhanced (Foster and Pardee, 1969; Isselbacher, 1972), the regulation of their carrier systems must also be considered.

Since intact cells contain varying levels of metabolic pools which affect transport rates, isolation of plasma membranes and their carriers will become necessaary in order to reach real insight into the nature of regulation. Before an isolation of transport carrier from cultured cells and oncogenically transformed cells can be achieved, isolation of plasma membrane preparations by nondenaturing techniques is a mandatory step. Such methods have been developed by Wallach (1967) and the first attempts to study the nature of the above-mentioned regulations using this method are underway (Quinlan and Hochstadt, 1974; Lever, 1976).

How do these studies of transport regulation converge toward ektobiology? I mentioned the periplasmic galactose (glucose) receptors of *E. coli* which serve two biological functions, transport and chemotaxis. I believe that one should be prepared for such surprises too in higher eukaryotes. Cytochalasin B is a strong inhibitor of cell movement and also a specific inhibitor of hexose transport; in fact the cytochalasin inhibition curves are strikingly similar (Brownstein *et al.*, 1975). Neither nucleoside transport nor amino acid transport are sensitive to the minute amounts of cytochalasin B which arrest hexose transport (Brownstein *et al.*, 1975; Christopher *et al.*, 1976b). Do the hexose carrier system or the cytochalasin binding system serve more than one function, i.e., cell migration as well as facilitated diffusion of hexose or its regulation?

ACKNOWLEDGMENT

This essay was formulated and written while the author was Scholar in Residence, Fogarty International Center, Bethesda, Maryland.

REFERENCES*

Adler, J. (1969). *Science* **166,** 1588–1597.

Adler, J. (1976). The sensing of chemicals by bacteria, *Sci. Am.,* (Apr) 40–47.

Anraku, Y. (1968). *J. Biol. Chem.* **243,**3116–3135.

Berg, H. C. (1974). *Nature* **249,** 77–79.

Boos, W. (1972). *J. Biol. Chem.* **247,** 5414–5424.

Boos, W. (1974). Pro and contra carrier proteins; Sugar transport via the periplasmic galactose-binding protein. *In* "Current Topics in Membranes and Transport" (A. Kleinzeller, ed.), Volume 5. Academic Press, New York.

Boos, W., Gordon, A. S., Hall, R. E., and Price, H.D. (1972). *J. Biol. Chem.* **247,** 917–924.

Brownstein, B. J., Rozengurt, E., De Asua, L. J., and Stoker, M. (1975). *J. Cell. Physiol.* **85,** 579–586.

Christopher, C. W., Kohlbacher, and Amos, H. (1976a). *Biochem. J.* **158,** 439–450.

Christopher, C. W., Ullrey, D., Colby, W., and Kalckar, H. M. (1976b). *Proc. Natl. Acad. Sci. U.S.A.* **73,** 2429–2433.

Eckhardt, W., and Weber, M. (1974). *Virology* **61,** 223–228.

Egyud, L. G., and Szent-Györgyi, A. (1966). *Proc. Natl. Acad. Sci. U.S.A.* **55,** 388–393.

*Article titles have been retained for reviews for clarity.

Foster, D. O., and Pardee, A. B. (1969). *J. Biol. Chem.* **244,** 2675–2681.
Furth, J. J., and Cohen, S. S. (1968). *Cancer Res.* **28,** 2061–2069.
Hatanaka, M. (1974). Transport of sugars in tumor cell membranes. Reviews on Cancer, *Biochim. Biophys. Acta* **355,** 77–104.
Hazelbauer, G. L., and Adler, J. (1971). *Nature (London) New Biol.* **30,** 101–104.
Heidelberger, J., Goebel, W. F., and Avery, O. T. (1925). *J. Exp. Med.* **42,** 727–745.
Heppel, L. A. (1969). *J. Gen. Physiol. Suppl.* **54,** 95–109.
Isselbacher, K. J. (1972). *Proc. Natl. Acad. Sci. U.S.A.* **69,** 585–589.
Kalckar, H. M. (1965). *Science* **150,** 305–313.
Kalckar, H. M. (1966). High energy phosphate bonds, optional or obligatory? *In* "Phage and the Origins of Molecular Biology" (J. Cairns, G. S. Stent, and J. D. Watson, eds.), pp. 43–49, Cold Spring Harbor Lab., Cold Spring Harbor, New York.
Kalckar, H. H. (1971). *Science* **174,** 557–565.
Kletzien, R. F., and Perdue, J. R. (1975). *J. Biol. Chem.* **250,** 593–600.
Koshland, D. E., Jr. (1974). The chemotactic response in bacteria. *In* "Biochemistry of Sensory Functions" (L. Jaenicke, ed.). Springer-Verlag, New York.
Lever, J. E. (1976). *J. Cell. Physiol., Suppl.* **81,** 779–787.
Martineau, R., Kohlbacher, M. S., Shaw, S. N., and Amos, H. (1972). *Proc. Natl. Acad. Sci. U.S.A.* **69** , 3407–3411.
Morse, M. L., Lederberg, E. M., and Lederberg, J. (1956). *Genetics* **41,** 142–156.
Nikaido, H. (1962). *Proc. Natl. Acad. Sci. U.S.A.* **48,** 1337–1341.
Nikaido, H. (1965). Bacterial cell wall–deep layers. *In* "The Specificity of Cell Surfaces" (B. D. Davis and L. Warren, eds.), pp. 3–30. Prentice-Hall, Englewood Cliffs, New Jersey.
Ordal, G. W., and Adler, J. (1974). *J. Bacteriol.* **117,** 517–526.
Osborn, M. J., Rosen, M. Rothfield, L., and Horecker, B. L. (1962). *Proc. Natl. Acad. Sci. U.S.A.* **48,** 1831–1838.
Quinlan, D. C., and Hochstadt, J. (1974). *Proc. Natl. Acad. Sci. U.S.A.* **71,** 5000–5003.
Rapin, A. M. C., and Kalckar, H. M. (1971). The relation of bacteriophage attachment to lipopolysaccharides structure. *In* "Microbial Toxins" Vol. IV, (G. Weinbaum *et al.,* eds.), pp. 267–307. Academic Press, New York.
Robbins, A. R., and Rotman, B. (1975). *Proc. Natl. Acad. Sci. U.S.A.* **72,** 423–427.
Silhavy, T., Boos, W., and Kalckar, H. M. (1974). The role of *Escherichia coli* galactose binding protein in galactose transport and chemotaxis. *In* "Biochemistry of Sensory Functions" (L. Jaenicke, ed.), pp. 165–205. Springer-Verlag, Berlin and New York.
Szmelcman, S., Schwartz, M., Silhavy, T. J., and Boos, W. (1976). *Eur. J. Biochem.* **65,** 13–19.
Ullrey, D., Gammon, M. T., and Kalckar, H. M. (1975). *Arch. Biochem. Biophys.* **167,** 410–427.
Wallach, D. F. H. (1967). Isolation of plasma membranes of animal cells. *In* "The Specificity of Cell Surfaces" (B. D. Davis, and L. Warren, eds.), pp. 129–164. Prentice-Hall, Englewood Cliffs, New Jersey.
Wallach, D. F. H. (1968). *Proc. Natl. Acad. Sci. U.S.A.* **61,** 868–874.
Weigle, J. (1966). Story and structure of the λ transducing phage. *In* "Phage and the Origins of Molecular Biology" (J. Cairns, G. S. Stent, and J. D. Watson, eds.), pp. 226–236. Cold Spring Harbor Lab., Cold Spring Harbor, New York.

VITAMIN C AND HEALTH

4

Albert Szent-Györgyi and Vitamin C

LINUS PAULING

In the period around 1900 it was recognized that certain diseases, including scurvy and beriberi, are deficiency diseases, caused by the lack of certain substances in the diet. A number of efforts were made to separate pure vitamin C, the substance that was thought to prevent scurvy, from lemon juice and other foods. Pure vitamin C, L-ascorbic acid, was obtained for the first time by Albert Szent-Györgyi in 1928. Szent-Györgyi was working to isolate a reducing substance that he had observed to be present in various plant and animal tissues. The name hexuronic acid was given to the substance that he obtained, and by 1932 it had been recognized that his substance was vitamin C, and the name was changed to L-ascorbic acid. Szent-Györgyi received the Nobel Prize for Physiology and Medicine for the year 1937 in recognition of his discoveries concerning the biological oxidation processes, with special reference to vitamin C and to the role of fumaric acid in these processes.

In 1939 Szent-Györgyi wrote that he believed that living organisms were very well adapted to their surroundings, but that recently the surroundings have been changing in such a way as to produce disharmony between man and his environment, causing poor health. The destruction of the natural environment, pollution of air and water, noise pollution, and poor nutrition are among the factors that cause poor health. He wrote the following statement: "I have a strong faith in the perfection of the human body, and I think that vitamins are an important factor in its coordination with its surroundings. Vitamins, if properly understood and applied, will help us to reduce human suffering to an extent which the most fantastic mind would fail to imagine" (Szent-Györgyi, 1939).

For a number of years around 1937 much effort was expended in checking the value of vitamin C and other vitamins in preventing and treating disease. Some diseases were found that responded in a striking way to a high intake of certain vitamins. With other diseases some patients seemed to be benefited by an increased intake of vitamins, whereas others were not. During this period the sulfa drugs, penicillin, and other agents effective against infections were discovered, and perhaps as a consequence of these discoveries interest in the vitamins decreased. The Food and Nutrition Board of the United States National Academy of Sciences–National Research Council (1974) in setting the Recommended Dietary Allowances (RDA) for vitamin C (45 mg/day for an adult) and other vitamins emphasizes that these amounts suffice to prevent manifestations of deficiency diseases in most persons. The possibility that a large intake would lead to improved health is ignored. The biochemist G. H. Bourne, now Director of the Yerkes Primate Laboratory, pointed out in 1949 that the food ingested by the gorilla consists largely of fresh vegetation in quantity such as to give the gorilla about 4.5 g of ascorbic acid/day. He stated also that before the development of agriculture man existed largely on green plants, supplemented with some meat, and concluded that "It may be possible, therefore, that when we are arguing whether 7 or 30 mg of vitamin C is an adequate intake we may be very wide of the mark. Perhaps we should be arguing whether 1 g or 2 g a day is the correct amount." Irwin Stone (1965, 1966a,b, 1967) formulated a number of arguments to support the thesis that the optimum intake for man, leading to the best of health, probably lies in the region between about 1 g and 5 g per day. About 10 years ago (April 1966) he wrote to me, sending copies of his articles, and I became interested in the question of the value of ascorbic acid in improving health and preventing disease. I wrote to Szent-Györgyi to ask his opinion. He gave me permission to quote part of his answering letter as follows:

> As to ascorbic acid, right from the beginning I felt that the medical profession misled the public. If you don't take ascorbic acid with your food you get scurvy, so the medical profession said that if you don't get scurvy you are all right. I think that this is a very grave error. Scurvy is not the first sign of the deficiency but a premortal syndrome, and for full health you need much more, very much more. I am taking, myself, about 1 g a day. This does not mean that this is really the optimum dose because we do not know what full health really means and how much ascorbic acid you need for it. What I can tell you is that one can take any amount of ascorbic acid without the least danger.

In my book "Vitamin C and the Common Cold" (Pauling, 1970a) and in a later paper (Pauling, 1974) and book (1976), I published discussions of the various arguments indicating that a high intake of vitamin C, as much as 100 times the RDA, leads to improvement in health and to protection against disease, and also may be valuable in treating disease. Some of the arguments presented below are taken from these sources. These arguments include those advanced by Bourne and Stone, and also another one, based upon consideration of the evolutionary processes that have led to our dependence on vitamins (Pauling, 1970b, 1976).

THE OPTIMUM INTAKE OF VITAMIN C

The Recommended Dietary Allowance (RDA) of ascorbic acid (vitamin C) has been set at 45 mg/day for adults (35 for infants, 60 for pregnant women, and 80 for lactating mothers) by the Food and Nutrition Board of the National Academy of Sciences–National Research Council (1974). Ascorbic acid differs from other vitamins in that an exogenous source is required by only a few animal species. This fact indicates that the amount contained in a diet of raw natural plant food is less than the optimum intake (Bourne, 1949; Pauling, 1970b). If the amount of an essential substance contained in the available food is greater than the optimum requirement, a mutant that had lost the machinery for manufacturing the essential substance would have been relieved of the burden of developing and operating the machinery, and would thus have an advantage over the wild type of the species. [This effect has been verified by experiments involving competition between different strains of microorganisms (Zamenhof and Eichhorn, 1967).] For ascorbic acid, the average value for 110 raw natural plant foods is 2.3 g/day per amount of food for 70 kg body weight (Pauling, 1970b). Animals of many species must surely have lived for long periods of time on a diet providing 2.3 g/day or more of ascorbic acid (per 70 kg of body weight), yet only a few species have lost the ability to synthesize it. Hence this value represents a lower limit to the optimum intake. This loss of the ability to manufacture ascorbic acid by a few species of animals, including the primates, presumably occurred while the primitive ancester was living in an environment that provided an especially large amount of the substance. Because of the burden of the machinery, it is unlikely that animals would synthesize more ascorbic acid that the amount required for optimum health (Pauling, 1968). The amounts made by mammals when calculated for 70 kg body weight are 3 to 19 g/day and similar amounts may be near the optimum for man. The mammals studied range in weight from the mouse (20 g body weight) with a rate of synthesis of 19 g/day per 70 kg to the goat (50 kg) with a rate of synthesis of

13 g/day per 70 kg. Values for the rat under various conditions of stress range from 4 to 15 g/day per 70 kg.

Yew (1973) reported that growth rate and other measures of good health indicate an optimum intake of 3.5 g/day per 70 kg body weight for guinea pigs, which require exogenous ascorbic acid. Taking the optimum intake proportional to body weight is indicated to be reasonable by the observed proportionality in the amounts manufactured by other animals, as mentioned above. Yew's observations and similar observations for monkeys and other animals provide additional support for the conclusion that the optimum rate of intake for man is in the range of a few grams per day.

THE COMMON COLD

There is much evidence that an increased intake of vitamin C strengthens the natural protective mechanisms of the body and decreases both the incidence and the severity of the common cold. So far there have been 14 controlled trials carried out in which some of the subjects received vitamin C (100 mg or more per day) and others received a placebo over a period of time during which they were exposed to cold viruses in the ordinary way, by contact with other people. Seven of these studies are in my opinion good studies. In these seven studies the values, shown in the following tabulation, of decrease in the average amount of illness for the vitamin C subjects relative to the controls were observed:

	Decrease in amount of illness (%)
Cowan, Diehl, Baker: Minnesota, 1942	31
Ritzel: Switzerland, 1961	63
Anderson, Reid, Beaton: Canada, 1972	32
Coulehan, Reisinger, Rogers, Bradley: Arizona, 1974	30
Charleston, Clegg: Scotland, 1972	58
Sabiston, Radomski: Canada, 1974	68
Anderson, Beaton, Corey, Spero: Canada, 1975	25
Average reduction in illness by vitamin C	44

The seven other studies, which are less reliable, gave an average protection of 26%. A discussion of all 14 studies has been published (Pauling, 1976).

There is accordingly strong evidence that the regular ingestion of supplementary vitamin C provides some protection against the common cold, decreasing the amount of illness by nearly half.

VIRUSES AND VIRAL DISEASES

Inactivation of viruses by ascorbic acid *in vitro* was first reported by Jungeblut (1935). He found that concentrations of ascorbic acid comparable to those achievable in the blood stream by a high intake of the vitamin inactivated poliomyelitis virus within 30 minutes, as shown by decreased incidence of paralysis in monkeys injected intracranially with the virus suspension. He also reported (1937) a smaller amount of paralysis in monkeys who had been receiving large doses of ascorbic acid than in those receiving small doses, when the active virus was injected into the brain. Sabin (1939) then reported that he had not found a protective effect against polio virus in monkeys who received a suspension of the virus applied to the tissues of the upper respiratory tract. Jungeblut (1939) repeated this work, both with the dose used by Sabin and with a smaller dose applied to the respiratory tract, and found that when the smaller dose was used the high intake of ascorbic acid was able to protect the monkeys from paralysis (although the control monkeys were paralyzed), but that they were not protected against a larger dose of the virus. There is accordingly indication that the protective effect of ascorbic acid against viral infection is limited: ascorbic acid may be effective when the number of virus particles is small, and not when the number is large.

Other early investigators also reported inactivation of viruses by ascorbic acid *in vitro*, including herpes simplex, vaccinia, foot and mouth, rabies, and bacterial viruses. The references to the early papers are in the monograph by Stone (1972). The most recent work is that of Murata and Kitagawa and their collaborators (1971, 1972, 1973). They found that all kinds of bacterial viruses tested by them could be inactivated by ascorbic acid *in vitro*, the rates of inactivation being different for different viruses. They found also that the inactivation does not occur except in the presence of oxygen, that it is prevented by free-radical quenchers, and that strands of the nucleic acid of the virus are split during the process of inactivation. They attribute the inactivation to radicals formed by ascorbic acid and atmospheric oxygen.

Early papers in which some control of viral diseases by ascorbic acid was observed, involving viral hepatitis, chicken pox, measles, virus encephalitis, mononucleosis, mumps, shingles, and influenza, are referred to by Stone (1972). The most recent work in this field is on ascorbic acid and viral hepatitis in surgical patients given transfusions of blood. In 1975 Murata reported on the observations of Dr. Fukumi Morishige in Torikai Hospital, Fukuoka, Saga, Japan. They found that the incidence of serum hepatitis in patients given transfusions of blood in connection with surgical operations was 7% under ordinary conditions. When they studied a series of 1250 thoracic-surgery patients receiving different amounts of ascorbic acid, they found that there were

11 cases of serum hepatitis in 150 patients who received little or no ascorbic acid (less than 1.5 g/day). This frequency is similar to that reported elsewhere. For example, in a recent study in the United States of 108 prospectively followed, multiply transfused, open-heart surgery patients, 12 (11%) developed hepatitis (Alter *et al.*, 1975). They had received only volunteer donor blood tested by counterelectrophoresis for hepatitis-B surface antigen prior to transfusion. In contrast, in the Murata-Morishige study there were no cases of serum hepatitis among 1100 patients with thoracic surgery who received blood transfusions and also received at least 2 g of ascorbic acid/day. Dr. Morishige (personal communication) now recommends that surgical patients receive 10 g of ascorbic acid/day for at least 1 month, beginning, if possible, before the operation, and 6 g of ascorbic acid/day for another 6 months. Murata and Morishige have reported also that a high intake of ascorbic acid has therapeutic effect against other viral diseases, including infectious hepatitis, measles, mumps, viral orchitis, viral pneumonia, pleuritis, herpes zoster, herpes facialis, stomatitis aphthosa, encephalomyelitis, and certain types of meningitis, but they have not yet carried out a statistical analysis of their observations (Murata, 1975, and personal communication).

Their observations strongly suggest that extensive clinical trials of the value of ascorbic acid in preventing and treating hepatitis and other viral diseases should be carried out.

BACTERIA AND BACTERIAL DISEASES

The inactivation *in vitro* of diphtheria toxin by ascorbic acid was reported by Jungeblut and Zwemer (1935). Jungeblut and others later reported inactivation of tetanus toxin, *Staphylococcus* toxin, and dysentery toxin (references in Stone, 1972). Bacteriostatic and bactericidal action of ascorbic acid against *Staphylococcus aureus* and several other bacteria was observed by Gupta and Guha (1941) and others. Some success in controlling various bacterial infections in man by an increased intake of ascorbic acid has been reported (references in Stone, 1972).

It has been known for 30 years that ascorbic acid is needed for effective phagocytic activity of leukocytes in concentration about 20 $\mu g/10^8$ cells. It is known also that wounds, infections, and other stresses lead to a decrease in the serum and leukocyte concentrations of ascorbic acid. Hume and Weyers (1973) found that the average concentration of ascorbic acid for subjects receiving an ordinary Scottish diet was 20.0 (SD ±3.3) $\mu g/10^8$ cells. On the first day of a cold, the concentration dropped to 10.3 (SD ±0.3) μg, and remained below the phagocytically effective level for 3 days. A regular intake of 1 g of ascorbic acid/day plus 6 g/day for 3 days when a cold is contracted sufficed to keep the concentration high, above 23.9 μg/10^8 cells, but with an intake of 200 mg/day the concentration on the first 3 days of a cold fell into the range 8–14 μg/10^8 cells. This evidence shows that a large intake of ascorbic acid is needed to provide protection against the secondary bacterial infections that often accompany the common cold.

The bactericidal effect of ascorbic acid probably involves free radicals formed during oxidation of ascorbic acid. Hydrogen peroxide is formed during the reaction of ascorbic acid and oxygen (Udenfriend *et al.*, 1952), and macrophages lack peroxidase. It has been shown that ascorbic acid and hydrogen peroxide together have a pronounced bactericidal effect, which is increased by a small concentration of copper ions (Ericsson and Lundbeck, 1955; Miller, 1969). The presence of free radicals has been demonstrated by electron-spin resonance spectroscopy (Yamazaki *et al.*, 1960), and the bactericidal activity is completely inhibited by free-radical inhibitors (Miller, 1969).

VITAMIN C AND CARDIOVASCULAR DISEASE

Cardiovascular disease now causes about 50% of all deaths. It is accordingly very important to search for its causes and for ways of controlling it. A valuable contribution was made in 1974 by Krumdieck and Butterworth in their paper "Ascorbate-cholesterol-lecithin interactions: factors of potential importance in the pathogenesis of atherosclerosis." They state that "A rather large volume of literature supports the hypothesis that vitamin C decreases susceptibility to vascular injury. There is also evidence that both vitamin C and certain unsaturated lecithins participate in the mobilization and excretion of cholesterol," and they conclude that "Vitamin C seems to occupy a position of unique importance by virtue of its involvement in two systems: the maintenance of vascular integrity and the metabolism of cholesterol to bile acids. It is our belief that the available scientific evidence clearly justifies, and indeed calls for, a carefully controlled evaluation of the effects of vitamin C and the lecithins on atherosclerosis." They point out that injury and repair of the arterial wall are processes that go on continuously during life, and that disease will occur if more injury is inflicted than can be repaired or if the normal processes of repair are slowed down.

McCormick (1957) discussed vitamin C deficiency in relation to coronary thrombosis, and wrote that "Thrombosis is not in itself a pernicious development but rather a protective response of the organism designed normally to effect repair of damaged blood vessels by cicatrization. High blood pressure, excessive stretching of blood vessels, and deficiency of vitamin C, resulting in rupture and bleeding of the intima at the site of such stress, initiate the development of the thrombosis by means of the clotting of the blood, which is also a protective reaction. This multiple protective mechanism should be sustained and controlled by physiologic means (vitamin C therapy) rather than suppressed by anticoagulants with their dangerous side effects."

The similarity between the vascular injury in scurvy and that in atherosclerosis was studied by Willis and his collaborators (Willis, 1953, 1957; Willis *et al.*, 1954; Willis and Fishman, 1955). Krebs (1953) quotes the Scottish physician James Lind, who in 1747 carried out his famous experiment with twelve patients severely ill with scurvy in which the two who received two oranges and one lemon each day were well in 6 days, as having written in 1757 that "persons that appear to be but slightly scorbutic are apt to be suddenly and unexpectedly seized with some of its worst symptoms. Their dropping down dead upon an exertion of their strength, or change of air is not easily foretold." Krebs (1953) reported that in the experiment at the Sorby Research Institute in Sheffield, England, with ten healthy young men (age 21 to 34 years) who were given a scorbutogenic diet, two became suddenly severely ill with obvious cardiac emergencies requiring hospitalization.

These observations indicate that a deficiency in vitamin C may lead to cardiovascular disease, even in young people.

The evidence for a negative correlation between intake of ascorbic acid and concentration of cholesterol in the blood for both guinea pigs and humans was reviewed by Krumdieck and Butterworth (1974) and by Ginter (1975). The mechanism seems to be that an increase in concentration of ascorbic acid leads to an increase in the rate of conversion of cholesterol to bile acids (Ginter, 1973). The recognized correlation between the serum cholesterol level and the incidence of coronary heart disease thus also indicates that an increased intake of vitamin C should be of value in providing some protection against heart disease.

Some indication of the amount of protection against cardiovascular disease that would result from an increased intake of vitamin C is provided by an epidemiological study of 577 older persons (over 50 years old at the beginning of the study) reported by Chope and Breslow (1956). During the 7 years of the study 88 of the subjects died. The mortality rate for this period was 27.7% for those who ingested less than 50 mg of ascorbic acid per day (36/130) and 11.0% for those who ingested 50 mg or more (average about 100 mg) per day (44/399). Ingestion of about twice the RDA of vitamin C is thus correlated with a decrease in death rate by 60%. Heart disease was the cause of 55% of the deaths. A smaller apparent protective effect was reported also for increased intakes of vitamin A (42%) and niacin (28%). It is interesting that these effects are associated with an increase in average intake from about half the RDA to about twice the RDA. No epidemiological studies have been reported for larger intakes.

VITAMIN C AND CANCER

Part of the extensive literature on vitamin C and cancer has been discussed by Stone (1972), who pointed out that rats and mice when exposed to carcinogens increase their rate of synthesis of ascorbic acid, whereas in guinea pigs, which cannot synthesize the substance, its concentration in the blood decreases. A decrease in the concentration of ascorbic acid in the blood has also been observed in human beings with malignancies. Several physicians in the period 1940 to 1956 reported a favorable effect of large doses of vitamin C, sometimes together with vitamin A, on the general condition of cancer patients (references given by Stone, 1972). The Canadian physician W. J. McCormick (1954, 1959, 1963), on the basis of the literature and his own observations, developed the hypothesis that cancer is a preventable collagen disease that results from a deficiency of ascorbic acid. He wrote that "We maintain that the degree of malignancy is determined inversely by the degree of connective tissue resistance, which in turn is dependent upon the adequacy of vitamin C status." Cameron and Pauling (1973, 1974) emphasized the value of ascorbic acid in maintaining the integrity of the intercellular cement and therefore increasing its resistance to malignant invasive growth, and pointed out that it might operate to strengthen several of the natural protective mechanisms of the body.

There is some direct evidence that an increase in intake of vitamin C provides an increased degree of resistance to cancer. Stocks and Karn (1933) in a study of the diet of 462 patients with cancer and 435 control patients in England found a consistent negative correlation between the occurrence of cancer and the increased ingestion of carrots, turnips, cauliflower, cabbage, onions, watercress, and beetroot, and concluded that an increased intake of vitamin C, riboflavin, and vitamin D decreases the incidence of cancer. Other similar studies have been discussed by Bjelke (1974). Bjelke himself in extensive epidemiological studies in Norway and Minnesota of cancer of the stomach, colon, and rectum in relation to diet, involving about 40,000 persons, has reported finding a negative correlation between these types of cancer and the intake of fruits, vegetables, and vitamin C.

The most extensive clinical trial of ascorbic acid in human cancer is that of Cameron and Campbell (1974), who reported on 50 patients with advanced human cancer who received no treatment other than ascorbic acid, usually 10 g/day. They concluded that "this simple and safe form of medication is of definite value in the palliation of terminal cancer." The findings suggest that it should be employed as a standard supportive measure to reinforce established methods of treatment in the general management of earlier and more favorable cases. Five of the patients showed striking improvement. One patient, with reticulum cell sarcoma, showed two complete remissions clearly induced by the high-dose ascorbic acid therapy (Cameron *et al.,* 1975). The disease was widely disseminated at the time of first diagnosis. The patient improved rapidly on 10 g/day of ascorbic acid, and within 2 months was back at work, fit and well in all respects, and with a normal chest x-ray. After another 3 months his supplemental vitamin C was stopped, and in 4 weeks the cancer had returned. He responded with a second remission to 20 g/day of intravenous ascorbic acid for 14 days, followed by 12.5 g/day taken by mouth, and remains fit and well.

A report on supplemental ascorbate (usually 10g/day) in the supportive treatment of cancer has recently been published by Cameron and Pauling, 1976. Their observations are summarized in the abstract of their paper.

> Ascorbic acid metabolism is associated with a number of mechanisms known to be involved in host resistance to malignant disease. Cancer patients are significantly depleted of ascorbic acid, and in our opinion this demonstrable biochemical characteristic indicates a substantially increased requirement and utilization of this substance to potentiate these various host resistance factors.
>
> The results of a clinical trial are presented in which 100 terminal cancer patients were given supplemental ascorbate as part of their routine management. Their progress is compared to that of 1000 similar patients treated identically, but who received no supplemental ascorbate. The mean survival time is more than 4.2 times as great for the ascorbate subjects (more than 210 days) as for the controls (50 days). analysis of the survival-time curves indicates that deaths occur for about 90% of the ascorbate-treated patients at one-third the rate for the controls and that the other 10% have a much greater survival time, averaging more than 20 times that for the controls.
>
> The results clearly indicate that this simple and safe form of medication is of definite value in the treatment of patients with advanced cancer.

It has also been reported that the cancers that often appear in the bladders of cigar smokers and other users of tobacco regress if the patient ingests 1 g/day of more of ascorbic acid (Schlegel *et al.*, 1970), and that 3 g/day is effective in controlling the genesis of cancer of the colon in some patients with active adenomatous polyp formation (DeCosse *et al.*, 1975). Many other studies bearing on vitamin C in relation to cancer in man and other animals have been reported. There is no doubt that vitamin C has prophylactic and therapeutic value for cancer, but extensive carefully controlled trials that could determine the optimum intakes for these purposes and the amount of protection that would result have not been carried out.

CONCLUSION

A certain amount of vitamin C, usually 5 or 10 mg/day but different for different people, is needed to prevent overt manifestations of scurvy. As was pointed out by Szent-Györgyi many years ago, a larger amount leads to improvement of health and greater resistance to disease. It is likely that an increased intake of ascorbic acid increases the effectiveness of the protective mechanisms in the human body, leading to some degree of control of essentially all diseases. The optimum intake of vitamin C, leading to the best of health, is not reliably known, but may well be in the range from 1 to 10 g for most people. The study by Breslow and Chope and other epidemiological studies indicate that with an increased intake of ascorbic acid the age-specific incidence of disease and mortality might well be decreased to less than 50% of the value on the ordinary intake (the RDA), and the decrease might in fact be considerably greater.

The evidence now available strongly suggests that thorough, carefully controlled trials should be made of the value of ascorbic acid for improving the health and decreasing the incidence of disease, and also for treatment of disease.

REFERENCES

Alter, H. J., Holland, P. V., Morrow, A. G., Purcell, R. H., Feinstone, S. M., and Moritsugu, Y. (1975). *Lancet* **2,** 838.

Anderson, T. W., Reid, D. B. W., and Beaton, G. H. (1972). *Can. Med. Assoc. J.* **107,** 503; correction **108,** 133 (1973).

Anderson, T. W., Beaton, G. H., Corey, P. N., and Spero, L. (1975). *Can. Med. Assoc. J.* **112,** 823.

Bjelke, E. (1974). *Scand. J. Gastroenterol.* **9,** Suppl. 31, 1–235.

Bourne, G. H. (1949). *Br. J. Nutr.* **2,** 346.

Cameron, E., and Campbell, A. (1974). *Chem.-Biol. Interact.* **9,** 285.

Cameron, E., and Pauling, L. (1973). *Oncology* **27,** 181.

Cameron, E., and Pauling, L. (1974). *Chem.-Biol. Interact.* **9,** 273.

Cameron, E., and Pauling, L. (1976). *Proc. Natl. Acad. Sci, USA* **73,** 3685.

Cameron, E., Campbell, A., and Jack, T. (1975). *Chem.-Biol. Interact.* **11,** 387.

Charleston, S. S., and Clegg, K. M. (1972). *Lancet* **1,** 1401.

Chope, H. D., and Breslow, L. (1956). *Am. J. Public Health* **46,** 61.

Coulehan, J. L., Reisinger, K. S., Rogers, K. D., and Bradley, D. W. (1974). *N. Engl. J. Med.* **290,** 6.

Cowan, D. W., Diehl, H. D., and Baker, A. B. (1942). *J. Am. Med. Assoc.* **120,** 1268.

DeCosse, J. J., Adams, M. B., Kuzma, J. F., LoGerfo, P., and Condon, R. E. (1975). *Surgery* **78,** 608.

Ericsson, Y., and Lundbeck, H. (1955). *Acta Pathol. Microbiol. Scand.* **37,** 493.

Food and Nutrition Board of the United States National Academy of Sciences–Research Council. (1974). "Recommended Dietary Allowances." 8th rev. ed. Nat. Acad. Sci., Washington, D.C.

Ginter, E. (1973). *Science* **179,** 702.

Ginter, E. (1975). *Biol. Pr.* **21,** 33.

Gupta, G. C. D., and Guha, B. C. (1941). *Ann. Biochem. Exp. Med.* **1,** 14.

Hume, R., and Weyers, E. (1973). *Scott. Med. J.* **18,** 3.

Jungeblut, C. W. (1937). *J. Exp. Med.* **66,** 459.

Jungblut, C. W. (1939). *J. Exp. Med.* **70,** 315.

Jungeblut, C. W., and Zwemer, R. L. (1935). *Proc. Soc. Exp. Biol. Med.* **52,** 1229.

Krebs, H. A. (1953). *Proc. Nutr. Soc.* **12.** 237.

Krumdieck, C., and Butterworth, C. E., Jr. (1974). *Am. J. Clin. Nutr.* **27,** 866.
McCormick, W. J. (1954). *Arch. Pediatr.* **71,** 313.
McCormick, W. J. (1957). *Clin. Med.* **4,** 839.
McCormick, W. J. (1959). *Arch. Pediatr.* **76,** 166.
McCormick, W. J. (1963). *Clin. Physiol.* **5,** 198.
Miller, T. E. (1969). *J. Bacteriol.* **98,** 949.
Murata, A. (1975). *Proc. Intersect. Congr. Int. Assoc. Microbiol. Soc., 1st, 1970* Vol. 3, 432.
Murata, A., and Kitagawa, K. (1973). *Agric. Biol. Chem.* **37,** 1145.
Murata, A., Kitagawa, K., and Saruno, R. (1971). *Agric. Biol. Chem.* **35,** 294.
Murata, A., Kitagawa, K., Inmark, H., and Saruno, R. (1972). *Agric. Biol. Chem.* **36,** 2597.
Pauling, L. (1968). *Science* **160,** 265.
Pauling, L. (1970a). "Vitamin C and the Common Cold." Freeman, San Francisco, California.
Pauling, L. (1970b). *Proc. Natl. Acad. Sci. U.S.A.* **67,** 1643.
Pauling, L. (1974). *Proc. Natl. Acad. Sci. U.S.A.* **71,** 4442.
Pauling, L. (1976). "Vitamin C,, the Common Cold, and the Flu." Freeman, San Francisco, California.
Ritzel, G. (1961). *Helv. Med. Acta* **27,** 63.
Sabin, A. B. (1939). *J. Exp. Med.* **69,** 507.
Sabiston, B. H., and Radomski, M. W. (1974). "Health Problems and Vitamin C in Canadian Northern Military Operations." DCIEM Report No. 74-R-1012. Defense Research Board, Department of National Defense, Ontario, Canada.
Schlegel, J. U., Pitkin, G. E., Mishimura, R., and Schultz, G. N. (1970). *J. Urol.* **103,** 155.
Stocks, P., and Karn, M. K. (1933). *Ann. Eugen. (London)* **5,** 237.
Stone, I. (1965). *Am. J. Phys. Anthropol.* **23,** 83.
Stone, I. (1966a). *Acta Genet. Med. Gemellol.* **15,** 345.
Stone, I. (1966b). *Perspect. Biol. Med.* **10,** 133.
Stone, I. (1967). *Acta Genet. Med. Gemellol.* **16,** 52.
Stone, I. (1972). "The Healing Factor: Vitamin C Against Disease." Grosset, New York.
Szent-Györgyi, A. (1939). "On Oxidation, Fermentation, Vitamins, Health and Disease." Williams & Wilkins, Baltimore, Maryland.
Udenfriend, S., Clark, C. T., Axelrod, J., and Brodie, B. B. (1952). *J. Biol. Chem.* **208,** 731.
Willis, G. C. (1953). *Can. Med. Assoc. J.* **69,** 17.
Willis, G. C. (1957). *Can. Med. Assoc. J.* **77,** 106.
Willis, G. C., and Fishman, S. (1955). *Can. Med. Assoc. J.* **72,** 500.
Willis, G. C., Light, A. W., and Gow, W. S. (1954). *Can. Med. Assoc. J.* **71,** 562.
Yamazaki, I., Mason, H. D., and Piette, L. (1960). *J. Biol. Chem.* **235,** 2444.
Yew, M.-L. S. (1973). *Proc. Natl. Acad. Sci. U.S.A.* **70,** 969.
Zamenhof, S., and Eichhorn, M. M. (1967). *Nature (London)* **216,** 456.

5

What's Ahead for Medical Research and Health Care?*

JOHN PLATT

The role of the biomedical profession is the prevention, cure, relief, or comfort of human illness or malfunction. Relief and comfort are often limited to the alleviation of symptoms and human sympathy and support. Prevention and cure, however, require knowledge of a practical therapy, such as vaccination, and the social organization and implementation to supply it; but they can be most powerful when there is also a good biological knowledge of the malfunction or disease entity, such as the life-cycle of an infectious organism, and its feedbacks and controls and the body's responses to it. On the other hand, there can sometimes be considerable biological knowledge without a practical therapy, as in the case of schistosomiasis.

The mission of biomedical research is to provide biological knowledge and, eventually, therapy for diseases and malfunction. This task has already been substantially completed in the last century, since Pasteur, for most of the classic killers and life-shortening diseases of mankind. This has added considerably to health and longevity and prosperity in the Western World, but for that half of

*Reprinted from "Medical Care and Society" (M. Belchior, ed.), pp. 10–15. World Health Organization, Geneva, Switzerland, 1975.

the human race still afflicted by poverty and illiteracy, there is frequently not the social organization necessary to apply this therapy or the basic social structure that itself can prevent many of the more debilitating diseases. As René Dubos has emphasized in "Mirage of Health" (Harper and Row, New York, 1971), "The great advances in health in the 18th and 19th centuries were largely the result of social reforms that alleviated some of the *pollution, dirt, poor housing* and *crowding,* and *malnutrition* . . . . 'Indeed many of the most terrifying microbial diseases—leprosy, plague, typhus, and the sweating sickness—had all but disappeared from Europe long before the advent of the germ theory.' " (Quote from R. J. Haggerty, The boundaries of health care, in *The Pharos,* July 1972. Italics added.)

The result is that for most of the world today, the most rapid gains in health and longevity will come, not from new research, but from application of existing knowledge and from the changes in education or social organization necessary to make this knowledge effective. In considering practical policy and implementation for improved health, it is therefore helpful to distinguish four different levels of biomedical intervention in society:

1. Individualized medical and surgical care and treatment of illness, with one doctor serving hundreds of patients per year
2. Mass treatment, prevention, and cure, as with vaccines, contraceptives, and mass drugs, with one doctor and several paramedicals serving tens of thousands per year
3. Public health and epidemiology, as with improved sewage disposal, chlorinated water, mosquito control, or highway design, with each doctor, biologist, or engineer contributing to the health of millions
4. General nutrition and self-help knowledge, involving better diets, home sanitation, birth control, infant care, and family and self-help treatments, where the doctor or biologist or official, using the mass media, can redirect the available resources and education of society for improved health for tens of millions

The costs of these different levels of intervention are roughly in inverse proportion to the number of people served by each doctor or biologist. Thus the per capita cost of Level 1 intervention and associated services will be in the range of several hundred dollars per year, or several percent of personal and national income. The cost of Level 2 will be in the range of a few dollars per capita per year, and Levels 3 and 4 may produce great improvements in health for a few cents per capita per year.

Level 1 is therefore expensive, ameliorative (concerned with "illness care") rather than preventive, does not much improve average health or length of life, and is not applicable for most of the world's population. It will never disappear where doctors and hospitals are available. But even in the richer countries there is some evidence that it is reaching the limits of its applicability, while in the poorer countries, with scarce resources and few doctors, it may never be available, at least on a mass basis, for the foreseeable future.

Levels 2, 3, and 4 are far more powerful for society but, being generally preventive methods, they require more knowledge and application of knowledge of the biological origins of disease and of the subtle natural methods of feedback control and response in complex systems of organisms. (The side effects and long-run secondary effects and dangers of mass drugs require particularly careful understanding and management.) It will help us understand the power—and the social difficulty—of intervention at these Levels if we look at some major components of health, longevity, and quality of life that have little to do with the traditional medical bedside or hospital role in the care of illness. These components include:

Food supply, protein supply (against kwashiorkor), productive and sanitary farming, fishing, meat supply, and food distribution (Levels 3 and 4 and general social organization)

Birth control, contraceptives, sterilization, and abortion (Levels 2 and 4 and socio-cultural pattern; occasionally Level 1)

Home knowledge, sanitation, diet, birth control, infant care, and self-help with first-aid and minor drugs (Level 4)

Slowing of aging, affected by life-style, diet, hormones (Level 4, occasionally Levels 1 and 2)

Reduction of drug abuse, including alcohol, tobacco, and tranquilizers (Levels 3 and 4 and socio-cultural pattern)

Reduction of environmental pollution and carcinogens (Levels 3 and 4)

Reduction of accidents, home, industrial, and highway (Levels 3 and 4; the fourth leading cause of death in the United States)

Reduction of suicides and homicides (Socio-cultural pattern, with some psychiatric help at Level 1; the ninth and tenth causes of death in the United States)

In addition, there are the nonbedside components of health care which are of central biomedical concern but still involve Level 1 intervention only to a minimal degree. These include:

Mild neuropsychiatric help, increasingly being returned to family and community care and counseling, with drug help (Levels 1 and 4)

Genetic counseling and prenatal analysis, with abortion to prevent birth defects (Levels 1 and 2)

Dental care, including diet and hygiene (Levels 1 and 4)

Prostheses, glasses, hearing aids, pacemakers (Level 1 plus Level 4)

Reduction of upper respiratory infections (Levels 3 and 4, occasionally Level 1; the second highest cause of bed disability in the United States)

Reduction of iatrogenic diseases (Level 1 malfunction; requiring improved education of doctors, midwife care, patient records, and reduced hospitalization)

For many countries, the approach to better health through intervention at Levels 2, 3 and 4 in these areas will be difficult but rewarding. When this is combined with application of existing biomedical knowledge, it may do more for the health of the world than any new research findings can be expected to do. On the other hand, these considerations strongly suggest the need for much new research to be oriented toward social and mass prevention and therapy at Levels 2, 3, and 4, for maximum speed and effectiveness. Thus, for diseases such as leprosy and even schistosomiasis, it is more important to expand research on the possibilities of a vaccine than on palliatives and hopeful cures. For schizophrenia, it is urgent to follow up on the evidence of a strong genetic component, and to look for biochemical diagnostic indicators, so that the mental disorder might be prevented by a therapeutic diet, as with phenylketonuria (PKU), or controlled by injections, as diabetes is controlled with insulin.

These are but some of the still intractable diseases. Today in the United States, the leading diseases in terms of death rates are heart disease, cancer, stroke, influenza-pneumonia, and diabetes, while the leading diseases in terms of days of bed disability are influenza-pneumonia, colds, heart disease, arthritis, and mental disorders (especially schizophrenia). It is interesting to note how these lists are now dominated by functional and genetic disorders along with labile viruses and viro-genetic diseases, such as cancer may be. It is clear that the reason why they are still intractable is that they are not susceptible to the "one-infectious-organism" approach of the Pasteurian age, and no early "break-throughs" can be assured. What is needed here is to understand complex feedback controls, regenerative and immune mechanisms and their breakdown, and enzyme and hormone systems, and how they can be modified on a steady-state basis. Success may depend on such things as the computer-assisted interpretation of the complex and individualized biochemical fingerprints of the different stages of a disease, or on the discovery of better animal models or animal hosts.

The last 10 years have opened up a number of new approaches, which may greatly increase the power of research methods as well as applications. Several directions of great promise can be listed.

1. *A theory of biochemistry,* of the relation between structure and function of macromolecules, seems closer than ever. It would permit a prediction of reactions, the design of synthetic molecules to do particular jobs, and the design of biochemical cures.

2. *Somatic-cell technology,* of secondary relation to health but perhaps still very important, produces whole plants from single cells after DNA injection or hybridization. If the mass side effects and dangers can be foreseen and con-

trolled, this could produce multitudes of new plants with new properties, and perhaps someday new meat animals for better diets as well as for more powerful research tools.

3. *Genetic transfer,* with altered cells grafted back to a host or perhaps with directed alterations of his genes, may cure enzyme and functional deficiencies. This could have great dangers, but it also has great potentialities.

4. *Prenatal genetic analysis,* with abortion to prevent birth defects, is already being practiced.

5. *Understanding the genetic orgin* of many diseases (hundreds of them now known) could lead to simple biochemical diagnostic indicators, often permitting dietary prevention or chemical amelioration or cure.

6. *Regeneration of tissues,* in addition to our familiar regeneration of skin and bone to heal a break, is suggested by several experiments. Regeneration of fingers or hands or eyes (common in vertebrates as high as the newts) might replace clumsy mechanical prostheses.

7. *Artificial organs* now include thousands of kidney dialysis machines and the prospects of artificial hearts. In the long run, they should be much cheaper than transplants, even cheaper than an automobile, although they will never be as cheap or effective as biochemical reversal of the original disease.

8. *Research on the aging process* may show us how to reverse or delay genetically programmed aging, or how to replace broken-down repair mechanisms. Concentrated research in this field might add 20 vigorous years to our life expectations.

9. *Behavior therapy and bio-feedback,* with acquired control of internal states, is able to produce rapid cures of stuttering and bed-wetting, obesity, and anorexia, as well as control of blood pressure, and it may be extended rapidly to many other behavior-affected disorders.

10. *Automated biochemical diagnosis,* even with lab tests at home, is spreading rapidly and is cheaper, faster, and more accurate than present laboratory methods. Quicker, simpler tests and computer analysis of complex biochemical fingerprints could greatly increase the reliability of diagnosis and the genetic and epidemiological screening of populations.

11. *Data-bank records of patient histories,* with suitable safeguards, could simplify physicians' files and hospital admissions, reduce errors, preserve continuity of health care while traveling, and warn of epidemiologically significant fluctuations. When coupled with automated diagnosis, in the long run it may be one of the cheaper aids to improving individual health care (Level 1) for the world.

MUSCLE

6

Past and Present Studies on the Interaction of Actin and Myosin

H. E. HUXLEY

During the years 1941–1943, Dr. Albert Szent-Györgyi and his collaborators at Szeged (notably F. B. Straub, I. Banga, and M. Gerendàs) described two interrelated discoveries which revolutionized muscle research and which, I think it's fair to say, have formed the basis of virtually all subsequent work (of which there has been a great deal!) on the molecular mechanism of contraction. They found, first of all, that the protein then known as myosin and long recognized as a major structural component of muscle, was in fact a complex of two proteins. One of these (the major component) they continued to call myosin. The other component they named actin (Straub, 1942). The complex formed by actin and myosin was called actomyosin.

The second discovery, made by Szent-Györgyi in 1942, was that threads of actomyosin, in the presence of magnesium and potassium ions, contracted rapidly upon the addition of ATP. Dr. Szent-Györgyi has written that this observation made the deepest impression on him of his whole research career, and the effect on the study of muscle has been equally profound. It made it possible to think about the problem in definite and manageable molecular terms, since it showed that the basis of contractility resided simply (perhaps not quite the right word!) in the interaction of actin, myosin and ATP, i.e., a relatively small number of components. Szent-Györgyi and his group showed how these proteins could be prepared and purified, so the stage was set for all the analysis of their properties and interactions, and of their arrangement in muscle, which has, during the last 25–30 years, led to our present concept of the mechanism of contraction.

The first beginnings of what one might call the modern era of muscle research had, of course, begun a little earlier in 1939 when Engelhardt and Ljubimova reported the finding of an ATPase activity in myosin preparations that did not seem to be separable from the protein (complex) "myosin" itself. They later reported (Engelhardt *et al.*, 1941) that the elasticity of myosin threads was increased by ATP; and Needham and his co-workers (1941a,b; Dainty *et al.*, 1944) showed that the viscosity and flow birefringence of "myosin" solutions were decreased by the presence of ATP. However, the most crucial discovery was the contraction of actomyosin.

Even in these early days, a remarkable number of important properties of these proteins were recognized by the Szeged group. Thus it was observed that purified myosin would "crystallize" into filamentous aggregates, and that fibers made from myosin alone (or from actin alone) would *not* contract in ATP. It was found that actin could exist in two forms, globular and fibrous; and it was found that the magnesium ATPase activity of myosin alone was low but was activated by combination with actin. To quote from Szent-Györgyi's own writings "Actin thus turns Mg ion from an inhibitor into a promotor of enzyme activity. Myosin as ATPase thus becomes activated by excitation, which makes actin and myosin unite to form actomyosin". Those who work on muscle will know how exactly correct this has turned out to be.

After World War II, the results obtained by Szent-Györgyi's group were summarized in a more accessible form (Szent-Györgyi, 1945) and also became widely known through Dr. Szent-Györgyi's writings and lectures about muscle (1947, 1948, 1951, 1953). These books were extremely influential, not only because they described what were clearly important results, but because they were written in a marvellously enthusiastic and invigorating style and were full of stimulating ideas and speculations. Dr. Szent-Györgyi was, I believe, unique among biochemists working at that time in the amount of emphasis he placed upon structure and in the way he encouraged and exhorted people to be aware of the ramifications and implications of a problem at all levels of organization. For example, I'd like to quote a short passage from one of his books on muscle, in which he says the following "It is equally difficult to understand the engine of a car without opening the hood, as it is to understand this mechanism from the heap of the smallest parts into which it can be disintegrated. Fortunately, there is a third way: to start with the whole, keep it together as long as possible, learning all we can about it. Then start pulling it to pieces bit by bit, trying after every step to put it together again, to correlate the new partial system with the whole. This is what we will try to do". He was almost unique in not only wanting to gradually disassemble some cellular function to its most elementary components, but in trying to put some of them together again to reconstruct model systems having some elements of the organized functioning system again. Nowadays, people think nothing of reconstituting cell membranes, or diffusing tubulin monomers into dividing cells to have them attach to chromosomes, or putting ribosomal subunits together again and having them synthesize protein, or even injecting alien messengers into living cells. But biochemists were much less daring 30 years ago, and I think many were quite sceptical of the relevance of Szent-Györgyi's reconstituted systems to the original contractile apparatus. Szent-Györgyi differed from them in being much more adventurous, more optimistic, and also, I believe, in his instincts in realizing ahead of other people that the processes he was studying were so important and so basic in Nature that they would have to have a basic simplicity and sturdiness about them; so that they would have, as it were, a strong inner life of their own, the ability to survive and emerge through our clumsy attempts to handle them—what we should now call the built-in capacity for self-assembly into functional structures. Szent-Györgyi not only exhorted others to pay attention to structure at all levels—"using our eyes"—but did in fact carry out work in this way himself, with Rozsa and Wyckoff (Rozsa *et al.*, 1949). They produced some very stimulating pictures of actin in the process of polymerization, showing an approximate 400 Å period (which we know now corresponds to the pitch of the actin helix). He pointed out its similarity to the fine periodicity seen in early micrographs of muscle, and concluded that actin filaments probably formed a continuous longitudinal structure in the muscle.

He also showed great foresight about the universality of biological mechanisms. Again to quote from the same introductory chapter, he said this "If muscle is really such an extraordinary material, and its mechanical function so unique, we may ask whether it will tell us anything about life in general, anything that will help us to understand other living systems. Four decades of research have left no doubt in the author's mind that there is only one life and one living matter, however different its structures and varied its appearance. We are all but recent leaves on the same old tree of life, and even if this life has adapted itself to new functions and new conditions, it uses the same old principles over and over again. In principle, it does not matter which material we chose for our study of life, whether it be grass or muscle, virus or brain. If only we dig deep enough we always arrive at the centre, the basic principles on which life was built and due to which it still goes on."

At this time, very little was known about the structure of the myosin molecule, save that it had a very high molecular weight (estimated then in the region of 1 million) and was several thousand angstroms in length and about 25 Å in diameter. Also, electron microscope studies on muscle structure were only just beginning. Nevertheless, in this extremely uncharted field, Dr. Szent-Györgyi's reasoning and intuition were remarkably accurate. He visualized, as I've mentioned, that a muscle fibril consisted essentially of a bundle of longitudinally oriented actin filaments, which transmitted the tension developed by myosin molecules or rodlets lying longitudinally between them. Actin and myosin were thought probably to be dissociated in relaxed muscle, and to combine during contraction. In one discussion (Szent-Györgyi, 1953), he even suggested (though did not follow up the idea) that individual myosin molecules (which were presumed to shorten during contraction) might be attached to a central noncontracting chain, so that the extent of movement of each myosin molecule need only be about 500 Å. In other work with Rozsa and Wyckoff (Rozsa *et al.*, 1949) he obtained evidence that the filaments were packed in an hexagonal lattice within the fibrils, and argued (Szent-Györgyi, 1953) that this suggested a hexagonal packing of myosin molecules around each actin filament.

Undoubtedly the most important and far reaching innovation introduced by Dr. Szent-Györgyi during these years was the use of the glycerol-extracted rabbit psoas muscle as a model system to study the events during contraction. He recognized very early that one needed a system intermediate between the artificial actomyosin fiber and the intact muscle, in which the much higher degree of order and high concentration of protein present in muscle would be preserved, but in which the complications introduced by the presence of the activation and relaxation mechanisms could be avoided, and in which added ions would have free access to the contractile material. He showed that these glycerinated preparations, which could be preserved in the deep freeze almost indefinitely, would, in the presence of ATP or ITP, generate tensions comparable with those produced *in vivo*; and he argued that contraction was occurring by essentially the same mechanism, as indeed all later studies confirmed. This type of preparation has been of inestimable value, as everyone who works on muscle will agree, and its usefulness was appreciated and exploited very rapidly.

The single fiber preparation derived from it was put to excellent use by the school of H. H. Weber (Weber and Portzehl, 1952, 1954), especially in their studies of the dual role of ATP—in producing relaxation when ATP-splitting was prevented, and in producing contraction when splitting took place—and which they contrasted with the behavior of other polyphosphates which could dissociate actomyosin but were not themselves hydrolyzable, and which could produce only relaxation. At that time, it was very important conceptually to establish, as the Weber school did, that ATP splitting occurred concurrently with contraction, not with relaxation, and the clear formulation of the concept that actin and myosin were dissociated in relaxation was of great value too, especially to my own work.

There were two other very important early applications of the glycerinated psoas preparation. My late colleague, Dr. Jean Hanson, realized in about 1951 that the uncontracted myofibrils which could be prepared very readily from glycerinated muscle offered ideal experimental material for observation of the band-pattern in the phase contrast light microscope, and she began the studies (Hanson, 1952) which we eventually continued together at MIT in 1953 and 1954 (Hanson and Huxley, 1953; Huxley and Hanson, 1954). At about the same time, working as a Ph.D. student in the MRC unit in Cambridge, I realized that the glycerinated psoas provided very convenient additional material for x-ray diffraction observations of muscle in rigor, to compare with the patterns I had obtained from frog muscle. The psoas muscles gave much superior patterns to those I was obtaining at that time from frog muscle in rigor. These patterns were crucial and enabled me to deduce that there must be a double hexagonal array of filaments present, with actin in one set of filaments and myosin in the other, with cross-links between them which became attached in rigor and probably during contraction too.

After completing my Ph.D. in the MRC Unit in Cambridge, I went to work in Professor F. O. Schmitts' laboratory at MIT as a post-doctoral fellow, and it was shortly after arriving there, in the autumn of 1952, that I first visited Dr. Szent-Györgyi, in his house in Penzance Point in Woods Hole—an area that has come to play a very important part in my life! It was a very memorable occasion for me to be describing the x-ray results, which indicated how actin and myosin were organized into separate but interacting filaments, to the person who had shown that these proteins constituted the contractile machine of muscle. Dr. Szent-Györgyi gave me a great deal of encouragement, and as a consequence of that and of the close links I also formed then with the other members of his group, especially Andrew and Eva Szent-Györgyi, I think an awareness of at least some muscle biochemistry has continued to exercise a moderating influence over the more unrealistic structural fantasies which I might otherwise have developed!

A few months later I was able to obtain electron micrographs of thin cross-sections of glycerinated psoas muscle which showed the filaments of actin and myosin in the double hexagonal array; and again I felt very proud and privileged to be able to show these to Dr. Szent-Györgyi. Very soon after, Jean Hanson and I started working together on myofibrils from glycerinated muscle, discovered that the myosin filaments were present only in the A-bands, and were able to put forward the interdigitating filament model of striated muscle.

The subsequent development of the sliding filament, moving cross-bridge model by Jean Hanson and myself, and by A. F. Huxley and R. Niedergerke, does not need to be described again today, save to say that it embodied in concrete terms a mechanism by which structural changes in myosin molecules caused by their interaction with actin were integrated together to exert a sliding force on the actin filaments, which transmitted that force along the length of the muscle (see Fig. 1). While I do not wish to imply that Dr. Szent-Györgyi necessarily agrees with all the features of the present models of the mechanism, I think it is clear that the general concept owes a great deal to the work and the ideas for which he was responsible. And, of course, the discovery by his pupils (Mihalyi, Gergely, and Andrew Szent-Györgyi) that the myosin molecule could be cleaved into two types of fragments, one of which possessed the solubility properties of the parent myosin, while the other possessed the ATPase and actin-binding ability but was soluble at physiological ionic strength, gave considerable support to my picture of myosin cross-bridges projecting sideways from a thick filament backbone and interacting with actin filaments alongside.

There is one point, however, where I might go into a little more detail about the earlier work on the actin-myosin interaction. This concerns the question of directionality. In an intact muscle working by a sliding filament mechanism, the actin filaments have to be moved in a specific direction, i.e., toward each other in the center of each A-band by the myosin cross-bridges. On a moving cross-bridge model, this implies that the whole sarcomere is built with a precise structural polarity, so that the identical relative steric arrangement of actin and myosin molecules is maintained everywhere, but with the absolute orientation of each reversing in the center of the sarcomere. While it was possible to believe that a real muscle was constructed this way, it was difficult at first to imagine how an artificial fiber of actomyosin could possess sufficient of the same features to function in a similar manner. Yet it was a fact, shown by Dr. Szent-Györgyi, that such threads did contract!

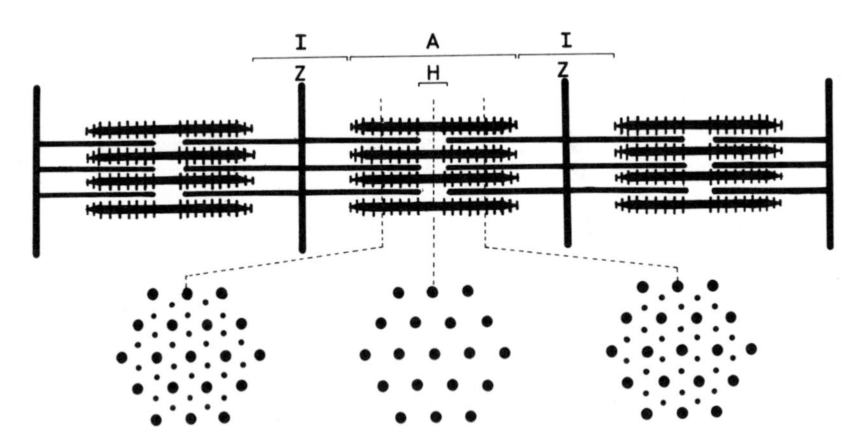

Fig. 1. Diagrammatic representation of the structure of striated muscle, showing overlapping arrays of actin- and myosin-containing filaments, the latter with projecting cross-bridges on them. For convenience of representation, the structure is drawn with considerable longitudinal foreshortening; with filament diameters and side-spacings as shown, the filament lengths should be about five times the lengths shown. For rabbit muscle, sarcomere length would be approximately 2.5 μm.

A possible resolution of this difficulty became apparent when it was noticed (Huxley, 1963) that the filaments formed by the precipitation of purified myosin (essentially similar to the individual fibrils in the myosin "crystals" originally described by Szent-Györgyi) frequently exhibited a characteristic bipolar appearance. It was shown that this arises because the myosin molecules themselves have the capacity for self-assembly into filaments with the required reversal of polarity at the center. That is, the linear tail portions (LMM) of the myosin molecules, which pack together to form the backbone of the thick filaments, do so in such a way that along each half of the length of the filament they are oriented in the same sense as their neighbors, but so that the direction reverses halfway along the filament. Thus the orientation of the myosin cross-bridges follows the same pattern. Since such bipolar filaments will form spontaneously, they are undoubtedly present in oriented threads of actomyosin, and will be capable of drawing together any adjacent pairs of actin filaments which happen to be oriented in the appropriate sense. Such a mechanism will not be a very efficient contractile system, for it lacks the other component present in the intact muscle, namely a specific structure (the *Z*-line) which joins together oppositely polarized sets of actin filaments. However, it is not unreasonable to suppose that there is sufficient internal friction within such threads so that the shortening of individual elements of the structure will give rise to the contractions actually observed.

SOME CURRENT WORK ON ACTOMYOSIN

I would like to turn now to much more recent work on the detailed nature of the myosin-actin interactions, as it occurs at the cross-bridges in muscle, and also to the findings that closely similar proteins, and a closely similar interaction, are involved in many different kinds of motility in nonmuscle cells. A convenient point to begin is the particular type of cross-bridge mechanism known as the "swinging cross-bridge" model (Fig. 2) (Huxley, 1969). In this model, as in previous ones, the heads of the myosin molecules (the S_1 subunits) project out sideways from the backbone of the myosin filaments and interact with actin, in a cyclic manner. In the resting state it is supposed that the myosin heads, which are thought to be somewhat asymmetrical structures, are oriented approximately perpendicular to the filaments. Upon activation, the heads begin to attach to neighboring actin monomers, and, when attached, undergo some structural change which causes them to forcibly alter their angle of attachment, tilting over and pulling the actin filament along, by a distance of the order of 50–100 Å, in the appropriate direction. At the end of the working stroke, the S_1 subunit detaches from that particular actin monomer, and returns toward its original orientation, ready to attach to another actin monomer further out along the

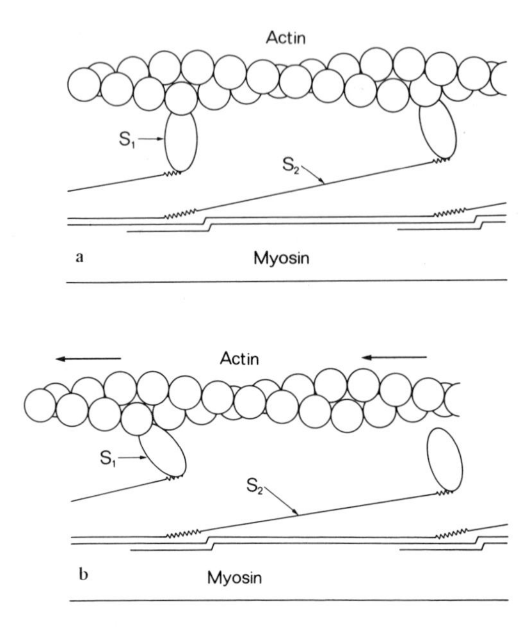

Fig. 2. Active change in angle of attachment of cross-bridges (S_1 subunits) to actin filaments could produce relative sliding movement between filaments maintained at constant lateral separation (for small changes in muscle length) by long range force balance. Bridges can act asynchronously since subunit and helical periodicities differ in the actin and myosin filaments. (a) Left hand bridge has just attached; other bridge is already partly tilted. (b) Left hand bridge has just come to end of its working stroke; other bridge has already detached, and will probably not be able to attach to this actin filament again until further sliding brings helically arranged sites on actin into favorable orientation.

actin filament. Similar cycles of interaction are taking place at all the other cross-bridges in the overlap area, so that a steady sliding motion is produced. From the work of Lymm and Taylor and others, it is believed that the myosin heads split the ATP attached to them *before* they attach to actin, but that the split products–ADP and P_i–remain bound to them and are released during the force-generating stroke when actin and myosin are combined. The binding of a fresh molecule of ATP then dissociates the cross-bridge at the end of the working stroke. This particular cross-bridge model differs from earlier ones in that it supposes that the myosin heads are attached relatively flexibly to the myosin filament backbone by the S_2 linkage, a section of the linear portion of the myosin molecule, which is not strongly bonded into the assembly of the terminal light meromyosin (LMM) portions of the myosin molecules which make up the shaft of the filaments. The function of the S_2 linkage is to transmit tension to the myosin filaments, but to allow a considerable latitude in the exact lateral and radial position of the heads of the molecules, relative to the myosin backbone, while still enabling them to make a precisely defined attachment to actin. In this way, the mechanism could operate over a range of side spacings between actin and myosin filaments (which do occur in practice at different sarcomere lengths), and would also allow some flexibility in the orientation of the actin helix relative to the myosin filament at which attachment could occur, and even in the selection of which particular neighboring actin filament a cross-bridge attached to at a particular moment.

In this mechanism the driving force which causes the bridge to tilt on its active stroke is *not* produced at the junctions between the myosin heads and the myosin filament backbones, but at or near the interacting surface between the myosin head and actin. The return stroke in the unattached state is envisaged as one requiring only weak forces. Thus while the myosin head-actin filament interaction has to be the invariant feature of all such mechanisms—as indeed appears to be the case from the remarkable constancy of the appearance of HMM-decorated actin, not just from muscle but from a wide variety of cellular systems throughout nature—the model places much looser restrictions on the precise mode of assembly of the myosin filaments, providing their structural polarity is maintained. Considerable differences are in fact found in the diameters and lengths and exact packing arrangements of the myosin containing filaments between different animal species and even within the same animal. What form the myosin assemblies take in nonmuscle motile systems is still very uncertain, but it is noteworthy that in every case which has been examined so far except one [*Acanthamoeba* myosin, Pollard and Korn (1973)], the myosin molecules have the familiar "tadpolelike" structure (i.e., a globular region at one end of an approximately 1400 Å long tail), and in many cases will assemble to form bipolar filaments, often somewhat smaller than those from muscle myosin. Thus there is a strong presumption that a mechanism similar to the actin-myosin interaction in muscle is likely to be involved in the operation of these nonmuscle motile systems too. This is a development which should bring Dr. Szent-Györgyi—and indeed everyone who works on muscle—a considerable amount of interest and satisfaction.

The model which I have described has a considerable amount of experimental support, both from the point of view of the substructure of the myosin molecule (especially as determined by Lowey and her collaborators, 1969), and from a number of electron microscope observations (Huxley, 1963; Reedy *et al.*, 1965; Moore *et al.*, 1970). In our most recent work, however, we have been particularly concerned with advancing the scope of x-ray diffraction observations, and so I will summarize briefly some aspects of the present state of this field.

The x-ray technique provides one of the best, though by no means straightforward methods of investigating the nature and behavior of the cross-bridges, since a large part of the low angle diagrams from striated muscle comes from the cross-bridges themselves, and since it is possible to study a muscle by this technique under almost normal working conditions. The disadvantage of this technique—besides the inherent and well-known ambiguities of x-ray diagrams!—is that the reflections from muscle are rather weak (about one millionth as strong as the direct beam), and it has therefore been necessary to invest a considerable amount of time and effort into the technical innovations required to record the patterns sufficiently rapidly. This is not an appropriate occasion to discuss these developments. Suffice it to say that we can now record changes in the strongest reflections with a time resolution of 100 milliseconds; but we could still make use of gains of x-ray intensity by several orders of magnitude in order to measure weaker reflections and to avoid having to use very long series of contractions. However, what of the results? In the diagram from a resting muscle there is a well developed system of layer lines with a 429 Å axial repeat and a strong third order meridional repeat at 143 Å (Huxley and Brown, 1967; Elliott *et al.*, 1967). This pattern arises from a regular helical arrangement of cross-bridges on the thick filaments, with groups of cross-bridges occurring at intervals of 143 Å along the length of the filaments with a helical repeat of 429 Å. The number of cross-bridges in each group is not yet absolutely certain, but it is more likely to be three than two (Squire, 1973). When a muscle goes into rigor, the axial x-ray diagram loses nearly all the features associated with the myosin filament helix and shows instead a pattern of reflections which can be indexed on the actin helix. This shows that the cross-bridges, the S_1 heads of myosin, are relatively flexibly attached to the myosin filament backbone, but form a relatively rigid attachment (in rigor anyway) to actin. The same conclusion can be drawn from a comparison of the very regular and highly ordered appearance in the electron microscope of negatively stained specimens of actin filaments decorated with S_1 and the very disordered appearance of the cross-bridges on myosin filaments examined by the same technique (Huxley, 1963, 1969). The characteristics of the two types of attachment suggest very strongly that the site at which force originates is at the interacting surface of the myosin head and the monomer on the actin filament.

Diagrams from contracting muscle may be recorded on film using a shutter so as to transmit the x-ray beam only when the muscle is being stimulated. A long series of tetani is necessary, with intervals in between them for recovery—usually one second tetani and two minute intervals. Such diagrams show that the whole pattern becomes very much weaker during contraction, indicating that the cross-bridges are much less regularly arranged, as would be expected if they were undergoing asynchronous longitudinal or tilting movements (and possibly lateral ones too) during their tension-generating cycles of attachment to actin. If this is indeed the case, then the very rapid development of the active state of a muscle following stimulation should be accompanied by an equally rapid decrease in intensity of the layer line pattern and it is most important to find out if this is really the case. Recent technical developments have now made it possible to record the x-ray diagram sufficiently rapidly to investigate this question. This involves the use of higher power x-ray sources and of various types of x-ray photon counters, and the accumulation of electronically gated data in multichannel scalars, which will be described in detail elsewhere. Studying single twitches of frog sartorius muscle, Dr. John Haselgrove and I have found that the change in pattern (at 10°C) begins about 10–15 milliseconds after stimulation and is half complete by about 20 milliseconds, a time when the externally-measured tension (which in a normal isometric contraction is delayed behind the onset of the active state by the necessity to stretch the series elastic elements before the internal activity can manifest itself) has hardly begun to rise.

The equatorial part of the x-ray diagrams from muscles is also very informative. It is generated by the regular side-by-side hexagonal lattice in which the filaments are arranged, and the relative intensity of the two principle reflections is strongly influenced by the lateral position of the cross-bridges. Large changes occur as between resting muscle and muscle in rigor (Huxley, 1952, 1953, 1968) when a high proportion, if not all, of the cross-bridges will be attached to actin. These have been interpreted as indicating that in a resting muscle, the cross-bridges lie relatively closer to the backbone of the thick filaments, whereas when they attach to actin they hinge further out and lie with their centers of mass nearer to the axes of the actin filaments at the trigonal positions of the hexagonal lattice.

Similar changes have been observed in contracting muscles (Haselgrove and Huxley, 1973), though the extent of change is less and would correspond to about half the cross-bridges being in the vicinity of the actin filaments at any one time. Again, the observations are consistent with a model in which the cross-bridges are undergoing a mechanical cycle of attachment to and detachment from the actin filaments during contraction. However, it should be appreciated that the parameter that is being measured is the average lateral position of the cross-bridges and this does not provide conclusive evidence about the proportion actually attached. Nevertheless, the changes do indicate that a substantial lateral movement of the cross-bridges takes place during contraction and, as in the case of the layer-line changes, it is important to establish whether the movement occurs at a sufficiently rapid rate for it to arise from a force-generating attachment of cross-bridges. Again, this is a quantity that we can now measure, and Dr. Haselgrove and I have found that the expected changes in the equatorial x-ray diagram (decrease in intensity of [10] reflection, increase in intensity of [11] reflection) do indeed occur with great rapidity after stimulation. At 10°C, for example, the change is half complete within about 20–30 milliseconds, according well with the expected temporal characteristics of the active state.

These types of observations (and analogous ones on insect flight muscle, see, for example, Armitage *et al.*, 1975) are thus beginning to provide moment-to-moment information about cross-bridge behavior which strongly supports the general features of the swinging cross-bridge model, and which we are beginning to be able to compare with other kinetic measurements on muscle and muscle proteins. This should make it possible eventually to establish whether muscle behavior can be completely described in these terms. The x-ray technique is not one, however, that readily lends itself to the study of the much smaller and less well oriented motile systems *in situ* in nonmuscle cells, even with very high-power x-ray sources, and radiation damage is likely to be a severely limiting factor. The study of isolated components would appear to offer more promise, however, especially if more of the muscle and nonmuscle proteins can be crystallized. Indeed, it may well turn out that the cellular motile systems will lead the way in this respect!

In general then, there are good prospects for further advances in the structural field in muscle, and the wide application this information may have throughout the whole of cell biology will make it all the more interesting to obtain.

I hope I have been able to demonstrate a small part at least of how much has followed from Dr. Szent-Györgyi's fundamental observations in this field.

REFERENCES

Armitage, P. M., Tregear, R. T., and Miller, A. (1975). *J. Mol. Biol.* **92**, 39.

Dainty, M., Kleinzeller, A., Lawrence, A. S. C., Needham, J., Needham, D. M., and Shen, S. C. (1944). *J. Gen. Physiol.* **27**, 355.

Eliott, G. F., Lowy, J., and Millman, B. M. (1967). *J. Mol. Biol.* **25**, 31.

Engelhardt, W. A., and Ljubimova, M. N. (1939). *Nature (London)* **144**, 668.

Engelhardt, W. A., Ljubimova, M. N., and Meitina, R. A. (1941). *Dokl. Akad. Nauk SSSR* **30**, 644.

Hanson, J. (1952). *Nature* **169**, 530.

Hanson, J., and Huxley, H. E. (1953). *Nature* **172**, 530.

Haselgrove, J. C., and Huxley, H. E. (1973). *J. Mol. Biol.* 77, 549.

Huxley, H. E. (1952). Ph.D. Thesis, University of Cambridge, England.

Huxley, H. E. (1953). *Proc. R. Soc. London, Ser. B* **141**, 59.

Huxley, H. E. (1963). *J. Mol. Biol.* 7, 281.

Huxley, H. E. (1968). *J. Mol. Biol.* **37**, 507.

Huxley, H. E. (1969). *Science* **164**, 1356.

Huxley, H. E., and Brown, W. (1967). *J. Mol. Biol.* **30**, 383.

Huxley, H. E., and Hanson, J. (1954). *Nature* **173**, 973.

Lowey, S., Slater, H. S., Weeds, A. G., and Baker, H. (1969). *J. Mol. Biol.* **42**, 1.

Moore, P. B., Huxley, H. E., and DeRosier, D. J. (1970). *J. Mol. Biol.* **50**, 279.

Needham, J., Shen, S. C., Needham, D. M., and Lawrence, A. S. C. (1941a). *Nature (London)* **147**, 466.

Needham, J., Kleinzeller, A., Miall, M., Needham, D. M., and Lawrence, A. S. C. (1941b). *Nature (London)* **147**, 766.

Pollard, T. D., and Korn, E. D. (1973). *J. Biol. Chem.* **248**, 4682.
Reedy, M. K., Holmes, K. C., and Tregear, R. T. (1965). *Nature (London)* **207**, 1276.
Rozsa, G., Szent-Györgyi, A., and Wyckoff, R. W. G. (1949). *Biochim. Biophys. Acta* **3**, 561.
Squire, J. (1973). *J. Mol. Biol.* **72**, 291.
Straub, F. B. (1942). *Stud. Inst. Med. Chem. Univ. Szeged* **2**, 3.
Szent-Györgyi, A. (1942). *Stud. Inst. Med. Chem. Univ. Szeged* **1**, 17.
Szent-Györgyi, A. (1945). *Acta Physiol. Scand.* **9**, Suppl. 25, p. 25.
Szent-Györgyi, A. (1947). "Chemistry of Muscular Contraction," 1st ed. Academic Press, New York.
Szent-Györgyi, A. (1948). "Nature of Life." Academic Press, New York.
Szent-Györgyi, A. (1951). "Chemistry of Muscular Contraction," 2nd ed. Academic Press, New York.
Szent-Györgyi, A. (1953). "Chemical Physiology of Contraction in Body and Heart Muscle." Academic Press, New York.
Weber, H. H., and Portzehl, H. (1952). *Adv. Protein Chem.* **7**, 162.
Weber, H. H., and Portzehl, H. (1954). *Prog. Biophys. Biophys. Chem.* **4**, 60.

7

Troponin and Its Function

SETSURO EBASHI

INTRODUCTION

Calcium is now widely recognized as a universal mediator between the signal taking place at the surface membrane and the intracellular processes such as contraction, secretion and some metabolic processes (cf. Carafoli *et al.*, 1975). This role of calcium was recognized in muscle by Heilbrunn (1940) when he observed more pronounced shrinking of injured muscle fibers in the presence of Ca^{2+} than in its absence and he suggested the contraction-inducing role of the Ca^{2+}. This was confirmed by the elegant work of Kamada and Kinosita (1943) using a microinjection technique.

Unfortunately, these observations, which were made at the same time as the epoch-making studies on muscle proteins in Szeged (see Szent-Györgyi, 1944), were not followed up by muscle biochemists or physiologists, and the role of calcium had been almost forgotten until the late 1950's. In the 1950's interesting work on the "relaxing factor" was energetically pursued by muscle biochemists. "Relaxing factor" was first described by Marsh (1951) using muscle homogenates, and, later, Bendall (1953) and Fujita (1954) extended the studies by using glycerinated muscle bundles, another great invention of Szent-Györgyi (1949). In 1955, we concluded that the relaxing factor was a microsomal fraction, identical with Kielley-Meyerhof's granular ATPase (Kielley and Meyerhof, 1948; Kumagai *et al.*, 1955). This important conclusion, however, did not immediately contribute to the clarification of the relaxation mechanism but rather added some confusion for a while. The situation was pertinently described by Szent-Györgyi (1960) as follows:

"The relaxing factor presents a number of most fascinating problems. Its function has been made still more enigmatic by the discovery of Kumagai *et al.* (1955) according to which this factor can be spun down at relatively low speeds and thus can be located in granules as definite morphological entities. Thus far, all attempts have failed to show that a substance is produced by these particles which diffuses out of them and makes actomyosin relax. But, if nothing diffuses out, how can changes be induced in actomyosin by something which happens in (or on) microsomes? This is one of the most fascinating puzzles at present."

The solution of the mechanism of the "relaxing factor" was reached in Lipmann's laboratory. Yet another example of the essential role of the energy rich bond in biological reactions (Lipmann, 1941) was found when ATP activated the accumulation of calcium by the microsomal fraction (Ebashi, 1961a; Ebashi and Lipmann, 1962*). In addition, the unique effect of Ca^{2+} on actomyosin ATPase (Weber, 1959; Ebashi, 1961a; Weber and Winicur, 1961), on superprecipitation (Ebashi, 1961a; Weber and Winicur, 1961) and on glycerinated fibers (Ebashi, 1960, 1961a) was clearly demonstrated. These results together with the fact that the "relaxing factor" is identical with the sarcoplasmic reticulum (Ebashi and Lipmann, 1962) has led us to the now widely accepted concept on excitation-contraction coupling (Ebashi, 1961b; cf. Weber, 1966; cf. Ebashi and Endo, 1968).

The next question raised was naturally how Ca^{2+} affected the contractile system. Confirming the suggestive work of Perry and Grey (1956), made before the awareness of the role of Ca^{2+}, Weber and Winicur (1961) showed that synthetic, or reconstituted actomyosin as compared to natural actomyosin is only weakly sensitive to Ca^{2+} or completely insensitive; the degree of sensitivity varied with the actin preparation. An attempt was then made without success to isolate "native" actin that would provide reconstituted actomyosin with full sensitivity to Ca^{2+}. Subsequently we showed the existence of a third factor which conferred calcium sensitivity to actomyosin (Ebashi, 1963). This factor,

*According to Ebashi and Lipmann, the mechanism of calcium concentration was the transport of Ca^{2+} into the vesicles, but the emphasis was later laid on the rapid Ca-binding, suggesting the binding of calcium to the surface of vesicular membrane (Ebashi, 1961a, 1965). There is no doubt that the calcium accumulation at a slow steady rate in the presence of oxalate (Hasselbach and Makinose, 1961) is carried out by the Ca-transport mechanism (cf. Hasselbach, 1964), but this slow process does not explain the rapid rate of muscle relaxation (cf. Ebashi and Endo, 1968). Recently, Y. Ogawa (personal communication) has succeeded in preparing fragmented sarcoplasmic reticulum which shows 30 to 50 fold faster rate of calcium accumulation than the slow steady rate. This rate may be enough to meet the physiological requirement, but its mechanism of the rapid uptake has remained to be solved.

tentatively named "native tropomyosin" because of its similarity to tropomyosin found by Bailey (1946), was later shown to be composed of classic tropomyosin and a new protein, named troponin (Ebashi and Kodama, 1965).

In this connection it should be pointed out that Szent-Györgyi had intuitively sensed the presence of such a system and described a method to isolate a protein fraction (Szent-Györgyi and Kaminer, 1963). This fraction was later worked on by Azuma and Watanabe (1965); its major and minor components corresponded to tropomyosin and troponin, respectively.

Troponin is located on the actin filament with a periodicity of about 400 Å (more precisely, 380 Å) and this periodicity is undoubtedly due to the distribution along the actin filament of tropomyosin the length of which is approximately 400 Å (Ohtsuki *et al.*, 1967). From these and other observations, a model for the thin filament has been presented (Ebashi *et al.*, 1969; Fig. 1) in which a tropomyosin molecule lies in the groove formed by the double actin strands, covering seven actin molecules, and troponin is located at a specified part of a tropomyosin molecule. The assumed molar ratio of 7 actin to 1 tropomyosin and troponin molecules, was later confirmed by optical diffraction studies (Ohtsuki and Wakabayashi, 1972; Spudich *et al.*, 1972).

The function of troponin was summarized as follows (Ebashi and Endo, 1968; Ebashi *et al.*, 1969): (i) The effect of troponin is transmitted to actin through tropomyosin. (ii) Troponin depresses the actomyosin–ATP system in the absence of Ca^{2+}. (iii) Ca^{2+} removes this depression. Although (i) might be debatable, (ii) and (iii) have been accepted as the basis of the troponin mechanism.

Troponin has been shown to be a complex of three subunits (Greaser and Gergely, 1971), troponin T (tropomyosin-binding factor; TN T), troponin I (inhibitory factor; TN I) and troponin C (Ca-binding factor; TN C). It is interesting that functions (i), (ii), and (iii) correspond to the function of these

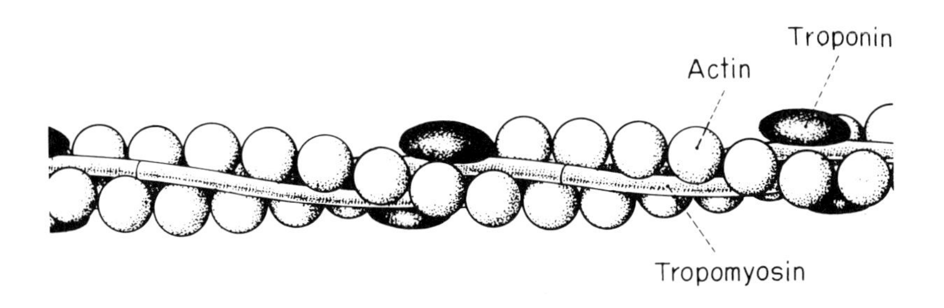

Fig. 1. Model for the fine structure of the thin filament (from Ebashi, 1972; but redrawn for this article).

subunits (see Ebashi, 1974a). This may be an example supporting Oosawa's concept (1972) of "one protein for one function" (cf. Ebashi, 1974a).

THE MECHANISM OF TROPONIN ACTION

The mechanisms of troponin subunit interaction with one another, and also with tropomyosin (and actin) in regulating the interaction of actin with myosin has been discussed in detail (Ebashi, 1974a). Here, we briefly summarize the present ideas from the observations of several workers (see Drabikowski *et al.*, 1974; Hitchcock *et al.*, 1973). Since the affinity between TN I and TN T is very weak, the function of troponin is essentially dependent on the two interactions, TN C–TN I and TN C–TN T (Ebashi, 1974a). The properties of these two interactions are summarized below.

TN C–TN I

1. The affinity between TN C and TN I is entirely calcium-dependent; the mode of its Ca dependence resembles that of contraction.
2. The affinity in the presence of Ca^{2+} is not completely abolished even by 6 *M* urea, though it is considerably weakened.
3. The affinity is weakened by the removal of Ca^{2+}, but not abolished. The remaining interaction in the absence of Ca^{2+} is eliminated by 6 *M* urea.
4. In the presence of Ca^{2+}, the TN C–TN I complex has practically no affinity for tropomyosin or the tropomyosin–F–actin complex (this term denotes the complex in which the molar ratio of actin to tropomyosin is 7:1), but in the absence of Ca^{2+} it has some affinity for it, though weak.
5. A relatively large amount of TN C–TN I (equimolar to actin or more) sensitizes the tropomyosin–F–actin complex to Ca^{2+} to some extent. The activities of various combinations of TN T–TN C–TN I from two sources in hybridization experiments (Table I) are somewhat parallel with those of corresponding combinations of only TN C–TN I (Fig. 2).

TN T–TN C

1. The affinity between TN T and TN C is mainly dependent on calcium, but in higher concentrations than those required for contraction.
2. There is also a weak but a definite Ca-independent affinity between TN T and TN C.
3. The Ca-dependent affinities exhibited by hybrid forms of TN C–TN T are inversely related to the regulatory activities of hybrids containing corresponding

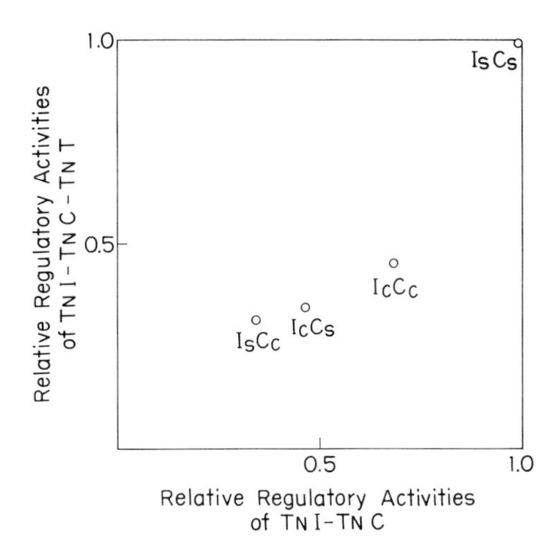

Fig. 2. Relationship between regulatory activities of various kinds of TN I–TN C of a large amount and those of troponins containing respective TN I–TN C. I, TN I; C, TN C; c, cardiac origin; s, skeletal origin. The value on the ordinate is the average of the regulatory activities of two troponins which contain the same kind of TN I–TN C, in combination with TN T from 2 sources (see Table I), e.g., $I_cC_cT_s$ and $I_cC_cT_c$ for I_cC_c. For regulatory activities of TN I–TN C see the text (for the details, see Ebashi, 1974b).

TN C–TN T in combination with TN I (Fig. 3). The most effective hybridized troponin (see Table I) has no Ca-dependent affinity between TN T and TN C at low Ca^{2+} concentrations at which a definite regulation is observed.

In view of the above observations, it is assumed that the crucial process of troponin function is the binding of TN I, relatively free from TN C, to the tropomyosin in the absence of Ca^{2+} and that the dissociation of TN I from TN C is catalyzed by TN T through its Ca-independent affinity for TN C. A result conforming with this assumption has been obtained by spin label studies (Ohnishi *et al.*, 1975). It is, however, quite possible that TN T requires the collaboration of tropomyosin to play its mediating role.

It is questionable whether TN I binds to actin. The fact that a large amount of TN I inhibits the actomyosin ATPase in the absence of tropomyosin might appear to support the existence of such binding (Perry *et al.*, 1973). However, this inhibition does not take place if a freshly prepared desensitized myosin B (myosin B, free of troponin and tropomyosin) is used. Furthermore, basic proteins such as protamine and cupreine form co-precipitates with F–actin and intensely inhibit the myosin-actin interaction as does TN I (Ebashi *et al.*, 1974). Therefore, the inhibition of actomyosin ATPase by a large amount of TN I may not be considered as evidence of physiological binding of TN I to actin. On the other hand, TN I has practically no affinity for F–actin, if the molar ratio of TN

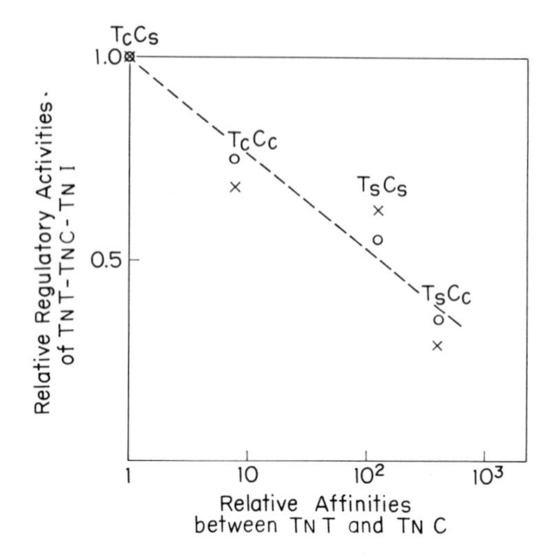

Fig. 3. Relationship between affinities of various kinds of TN C–TN T and regulatory activities of corresponding TN T–TN C in combination with TN I. The value on the ordinate is the average of regulatory activities of two similar TN T–TN C–TN I complexes. x, the simple average; o, corrected for the regulatory activities of TN I–TN C. For other abbreviations see the legend to Fig. 2.

TABLE I
Calcium-Sensitizing Activities of Various Kinds of Combinations of Troponin Subunits[a]

Combinations			
TN T	TN C	TN I	Ca sensitivities
c	c	c	15.4 ± 2.1 (6)
c	c	s	7.4 ± 1.2 (4)
c	s	c	8.6 ± 0.6 (4)
c	s	s	32.4 ± 3.4 (5)
s	c	c	3.4 ± 0.7 (4)
s	c	s	4.6 ± 0.6 (4)
s	s	c	2.6 ± 0.3 (4)
s	s	s	20.8 ± 1.7 (8)

[a]c or s indicates the cardiac or skeletal origin of the subunit, respectively. Figures in parentheses show the number of experiments. For further details, see Ebashi (1974b).

I to actin is kept around 1:7; under the same conditions, TN I certainly binds to tropomyosin. Since the affinity of TN I for tropomyosin is too weak to be the basis for fairly strong binding of TN I to the tropomyosin–F–actin complex, it does not exclude the possibility that TN I binds also to actin when it binds to the tropomyosin–F–actin complex (Ebashi *et al.*, 1974).

In view of above observations and considerations, the mechanism of how troponin works in the regulation mechanism is schematically illustrated in Fig. 4.

Comparing the three dimensional reconstructed image of the rigor complex of F–actin and subfragment-1 (S-1) (Moore *et al.*, 1970) with the x-ray diffraction pattern of living muscle, which indicates the displacement of tropomyosin, Huxley (1973) and Haselgrove (1973) have reached the conclusion that the position of tropomyosin on the actin filament in the resting state overlaps the site for S-1 interaction. They ascribe this steric hindrance as the mechanism of inhibition of the interaction of actin with myosin in the resting state. This view is further verified by the finely reconstructed three dimensional images of tropomyosin–F–actin complex with or without TN I and TN T (Wakabayashi *et al.*, 1975).

The steric blocking hypothesis has so far been the most elegant idea to explain the inhibition conferred by TN I on the tropomyosin–F–actin complex. However, any hypothesis cannot escape an element of doubt. Such doubt arises in smooth muscle (chicken gizzard and bovine stomach). Its myosin and actin form the arrowhead structure, but the ATPase activity is very low. Tropomyosin alone does not markedly activate the ATPase but does so in the presence of troponin and Ca (see the next section).

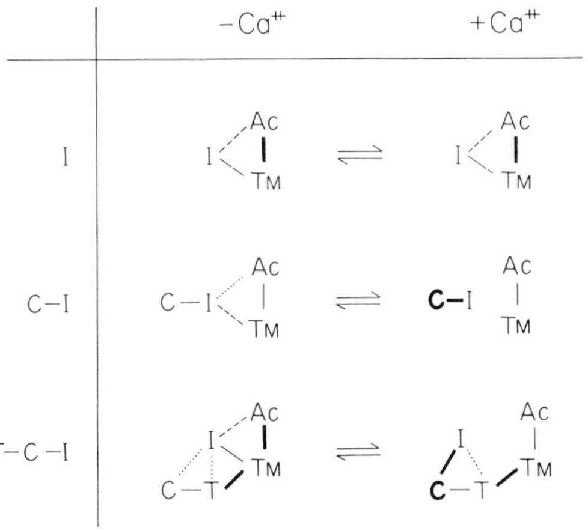

	$-Ca^{++}$		$+Ca^{++}$
I	I, Ac, TM	⇌	I, Ac, TM
C–I	C–I, Ac, TM	⇌	C–I, Ac, TM
T–C–I	I, C–T, Ac, TM	⇌	I, C–T, Ac, TM

Fig. 4. Schematic illustration of the interactions of troponin subunits underlying the troponin function.

Another hypothesis on the action of tropomyosin-troponin may be related to the flexible nature of the actin filament which may be crucial for its reactivity with myosin (Oosawa *et al.,* 1977). One could assume that the movement of actin molecules in the relaxed or inactivated state is restricted by the firm attachment of tropomyosin to the actin strand at the side of the groove. During contraction tropomyosin shifts to the center of the groove (Wakabayashi *et al.*, 1975) and binds the actin more loosely, thus enabling it to move relatively freely and to react with the myosin molecule.

TROPONIN AND TROPOMYOSIN IN SMOOTH MUSCLE

In 1966 Ebashi *et al.* isolated "native tropomyosin" from chicken gizzard muscle by essentially the same procedure as used for skeletal muscle. This naturally gave the impression that the events in smooth muscle might be essentially the same as in skeletal muscle.

Recently, we started working again on the regulatory system, confirming that "native tropomyosin"-like protein of gizzard restored the Ca-sensitivity of desensitized gizzard myosin B. Furthermore, the "native tropomyosin" was separated into tropomyosin and troponin-like protein by almost the same method used for skeletal native tropomyosin. If the separated components were recombined, they could sensitize not only desensitized gizzard myosin B but also reconstituted gizzard actomyosin (Ebashi *et al.,* 1975a,b, 1976). In these respects smooth muscle is apparently similar to striated muscle.

We realize, however, that the mode of action of gizzard troponin–tropomyosin system differs considerably from that of striated muscle in the following ways:

1. Reconstituted actomyosin as well as desensitized myosin B shows only a weak ATPase and practically no superprecipitation irrespective of the presence or absence of Ca^{2+}. (If aged myosin preparation is used, an appreciable activation is observed.)
2. Addition of tropomyosin to actomyosin does not significantly alter it, although a slight activation can often be observed.
3. In the absence of Ca^{2+}, addition of troponin to the tropomyosin-actin-myosin system does not increase the level of ATPase or superprecipitation.
4. Marked ATPase and superprecipitation are observed only when troponin is present together with Ca^{2+}.

Thus Ca^{2+} acts as a real activator in smooth muscle in sharp contrast to striated muscle, in which Ca^{2+} acts as a kind of derepressor (Ebashi *et al.*, 1969) (Fig. 5). It should also be noted that the contractile system of smooth muscle does not show a biphasic response, or substrate-inhibition in response to increasing Mg-ATP concentrations. The system is also insensitive to ionic strength; it

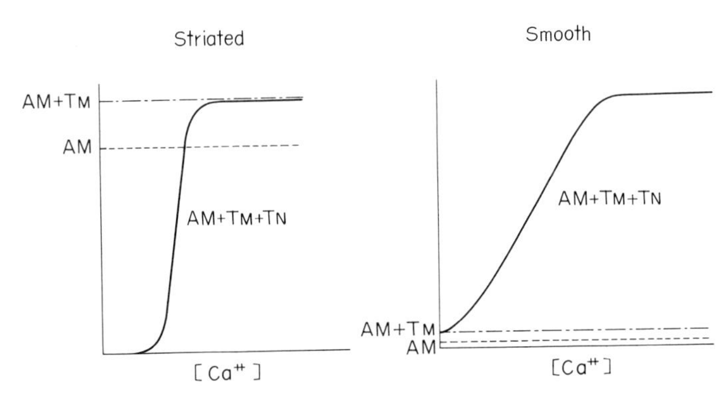

Fig. 5. Schematic illustration of the myosin–actin–ATP interaction as a function of Ca^{2+} concentration. Left, for skeletal muscle. Right, for smooth muscle. Ordinate, the degree of the interaction in arbitrary unit (from Ebashi *et al.*, 1975a, but redrawn for this article).

shows almost constant activities at ionic strength from 0.02 to 0.12, and in the absence of Ca^{2+} these activities are much lower (Fig. 6).

In *in vitro* experiments a fairly high concentration of Mg^{2+}, more than 4 mM, was necessary for the activation of smooth muscle contractile system. Similar concentrations of magnesium were required for optimal myosin filament formation; this may be one of the important reasons for the high magnesium requirement for the contraction of smooth muscle (see Plate 3 in Ebashi *et al.*, 1976). Since the intercellular free Mg^{2+} concentration is not much higher than 1 mM, the necessity for a high Mg^{2+} concentration in *in vitro* system is rather puzzling. Perhaps the high concentration of myosin *in vivo*, more than 10 mg/ml, does not require such a high Mg^{2+} concentration. This may be related to the controversial problems on the stability of the myosin filament in smooth muscle.

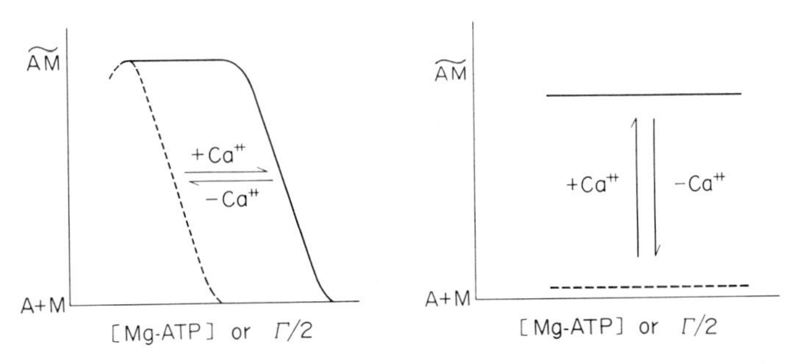

Fig. 6. Schematic illustration of the myosin–actin–ATP interaction as a function of Mg-ATP concentration or ionic strength. Left, for skeletal muscle. Right, for smooth muscle. $\widetilde{AM}$, contracting state, in which myosin and actin are interacting in the presence of ATP. A+M, relaxed state, in which myosin and actin are dissociated by ATP.

Although myosin and troponin are considered to be tissue-specific, actin is thought to be present in a highly conserved form in different types of muscle and in nonmuscle tissues. However, this may be a misleading concept. For example, skeletal actin cannot react properly with gizzard myosin. Tropomyosin is also said to be similar in different types of muscle. This is also not true. Skeletal tropomyosin cannot replace gizzard tropomyosin in a gizzard contractile system. In Table II, the cross reactions between contractile proteins of skeletal and smooth muscle are listed.

Although not directly related to the title of this paper, reference must be made to the role of α-actinin (Ebashi *et al.,* 1964; Ebashi and Ebashi, 1965). This protein was isolated from the native tropomyosin-containing "extract" of muscle; in a sense, it was a by-product of troponin. If we use very pure preparations of myosin, actin, tropomyosin and troponin, only a very weak activation is observed even in the presence of Ca^{2+}. Addition of α-actinin to the system restores the ordinary activation (Ebashi *et al.,* 1976). This reminds us of the old work on α-actinin in the skeletal actomyosin system. α-Actinin was first found as the factor which bound to F–actin and promoted the myosin-actin interaction in the presence of ATP. In this respect, α-actinin seemed to play an equivalent role to "protin" which was once considered by Szent-Györgyi (1947) to be an activator of the actomyosin-ATP system, although protin was thought to be bound to myosin.

TABLE II
Calcium Sensitivities of Various Combinations of Skeletal and Gizzard Proteins[a]

M	A	TM	TN	ATPase	Superprecipitation	Ca sensitivity
G	G	G	G	++	++	+++
G	S	G	G	+	–	–
S	G	G	G	+~++	+	+
G	G	S	G	+	±	–
G	G	S	S	+	±	±
G	G	G	S	+	±	–
S	S	G	G	+++	+++	(±)[b]
S	S	G	S	++++	++++	(+++)[b]
S	S	S	G	+++	+++	–
S	S	S	S	+++	++++	+++

[a]S, rabbit skeletal proteins; G, chicken gizzard (bovine stomach proteins show essentially the same results as chicken gizzard proteins; T. Toyo-oka, T. Mikawa, K. Maruyama and S. Ebashi, unpublished).

[b]Specified conditions are required to show the Ca sensitivity.

The remarkable effect observed with α-actinin, however, was later considered to be a mere *in vitro* phenomenon, because the presence of α-actinin was shown to be confined to the *Z*-band (Masaki *et al.*, 1967) and its effect was abolished by native tropomyosin (Ebashi, 1967). However, this does not necessarily mean that the effect of α-actinin in smooth muscle is also physiologically insignificant. It is quite possible that minor proteins, such as α-actinin or β-actinin, involved in the fine structural arrangement of myofilaments in skeletal muscle would play more important roles in smooth muscle and also in nonmuscle tissues.

CONCLUDING REMARKS

It was once believed by many muscle scientists that Szent-Györgyi's actomyosin–ATP system fully represented the molecular mechanism underlying muscle contraction. To those with such a belief, the discovery of regulating proteins such as troponin-tropomyosin appeared as a new concept in the molecular mechanisms of contraction. But Szent-Györgyi himself had no such belief. In 1947 he proposed the presence of small molecular weight proteins in muscle, which he termed "protins," and considered them as regulators of the myosin–actin–ATP interaction. Thus Szent-Györgyi, the discoverer of the actomyosin–ATP system, recognized its limitations and in his visionary way foresaw the future developments in muscle research.

REFERENCES

Azuma, N., and Watanabe, S. (1965). *J. Biol. Chem.* **240**, 3852.
Bailey, K. (1946). *Nature (London)* **157**, 368.
Bendall, J. R. (1953). *J. Physiol. (London)* **121**, 232.
Carafoli, E., Clementi, F., Drabikowski, W., and Margreth, A., eds. (1975). "Calcium Transport in Contraction." North-Holland Publ., Amsterdam.
Drabikowski, W., Strzelecka-Golaszewska, H., and Carafoli, E., eds. (1974). "Calcium Binding Proteins." Elsevier, Amsterdam (cf. Drabikowski *et al.*, Winter *et al.*, and van Eerd and Kawasaki).
Ebashi, S. (1960). *J. Biochem. (Tokyo)* **48**, 150.
Ebashi, S. (1961a). *J. Biochem. (Tokyo)* **50**, 236.
Ebashi, S. (1961b). *Prog. Theor. Phys., Suppl.* **17**, 35.
Ebashi, S. (1963). *Nature (London)* **200**, 1010.
Ebashi, S. (1965). *Mol. Biol. Muscular Contract.* BBA Libr. Vol. 9, p. 197.
Ebashi, S. (1967). *In* "Symposium on Muscle" (E. Ernst and F. B. Straub, eds.), p. 77. Akadémiai Kiadó, Budapest.
Ebashi, S. (1972). *Nature (London)* **240**, 217.
Ebashi, S. (1974a). *Essays Biochem.* **10**, 1.

Ebashi, S. (1974b). *In* "Lipmann Symposium: Energy, Biosynthesis and Regulation in Molecular Biology" (D. Richter, ed.), p. 165. de Gruyter, Berlin.
Ebashi, S., and Ebashi, F. (1965). *J. Biochem. (Tokyo)* **58**, 7.
Ebashi, S., and Endo, M. (1968). *Prog. Biophys. Mol. Biol.* **18**, 123.
Ebashi, S., and Kodama, A. (1965). *J. Biochem. (Tokyo)* **58** 207.
Ebashi, S., and Lipmann, F. (1962). *Cell Biol.* **14**, 389.
Ebashi, S., Ebashi, F., and Maruyama, K. (1964). *Nature (London)* **203**, 645.
Ebashi, S., Iwakura, H., Nakajima, H., Nakamura, R., and Ooi, Y. (1966). *Biochem. Z.* **345**, 201.
Ebashi, S., Endo, M., and Ohtsuki, I. (1969). *Q. Rev. Biophys.* **2**, 351.
Ebashi, S., Ohnishi, S., Maruyama, K., and Fujii, T. (1974). *Fed. Eur. Biochem. Soc. Meet., 9th, 19xx*, p. 71.
Ebashi, S., Nonomura, Y., Kitazawa, T., and Toyo-oka, T. (1975a). *In* "Calcium Transport in Contraction and Secretion" (E. Carafoli *et al.*, eds.), p. 405. North-Holland, Amsterdam.
Ebashi, S., Toyo-oka, T., and Nonomura, T. (1975b). *J. Biochem. (Tokyo)* **78**, 859.
Ebashi, S., Nonomura, Y., Toyo-oka, T., and Katayama, E. (1976). *In* "Calcium in Biological Systems" (C. J. Duncun, ed.), p. 349. Cambridge University Press, London.
Fujita, K. (1954). *Folia Pharmacol. Jpn.* **50**, 183.
Greaser, M. L., and Gergely, I. (1971). *J. Biol. Chem.* **246**, 4226.
Haselgrove, J. C. (1973). *Cold Spring Harbor Symp. Quant. Biol.* **37**, 341.
Hasselbach, W. (1964). *Prog. Biophys. Chem.* **14**, 167.
Hasselbach, W., and Makinose, M. (1961). *Biochem. Z.* **333**, 518.
Heilbrunn, L. V. (1940). *Physiol. Zool.* **13**, 88.
Hitchcock, S. E., Huxley, H. E., and Szent-Györgyi, A. G. (1973). *J. Mol. Biol.* **80**, 825.
Huxley, H. E. (1972). *Cold Spring Harbor Symp. Quant. Biol.* **37**, 361.
Kamada, T., and Kinosita, H. (1943). *Jpn. J. Zool.* **10**, 469.
Kielley, W. W., and Mayerhof, O. (1948). *J. Biol. Chem.* **174**, 387.
Kumagai, H., Ebashi, S., and Takeda, F. (1955). *Nature (London)* **176**, 166.
Lipmann, F. (1941). *Adv. Enzymol.* **1**, 99.
Marsh, B. B. (1951). *Nature (London)* **167**, 1065.
Masaki, T., Endo, M., and Ebashi, S. (1967). *J. Biochem. (Tokyo)* **62**, 630.
Moore, P. B., Huxley, H. E., and DeRosier, D. J. (1970). *J. Mol. Biol.* **50**, 279.
Ohnishi, S., Maruyama, K., and Ebashi, S. (1975). *J. Biochem. (Tokyo)* **78**, 73.
Ohtsuki, I., and Wakabayashi, T. (1972). *J. Biochem. (Tokyo)* **72**, 369.
Ohtsuki, I., Masaki, T., Nonomura, Y., and Ebashi, S. (1967). *J. Biochem. (Tokyo)* **61**, 817.
Oosawa, F. (1972). *In* "The Basis of Modern Physics" (F. Oosawa and E. Teramoto, eds.), Vol. 8, p. 1. Iwanami Publ. Co., Tokyo (in Japanese).
Oosawa, F., Maeda, Y., Fujime, S., Ishiwata, S., Yanagida, T., and Taniguchi, M. (1977). *J. Mechano-chem.* **4**, 63.
Perry, S. V., and Grey, T. C. (1956). *Biochem. J.* **64**, 5.
Perry, S. V., Cole, H. A., Head, J. F., and Wilson, F. J. (1973). *Cold Spring Harbor Symp. Quant. Biol.* **37**, 251.
Spudich, J. A., Huxley, H. E., and Finch, J. T. (1972). *J. Mol. Biol.* **72**, 619.
Szent-Györgyi, A. (1944). "Studies on Muscle. Inst. Med. Chem., Univ. Szeged.
Szent-Györgyi, A. (1947). "Chemistry of Muscular Contraction," 1st ed. Academic Press, New York.
Szent-Györgyi, A. (1949). *Biol. Bull.* **96**, 140.
Szent-Györgyi, A. (1960). *In* "The Structure and Function of Muscle" (G. H. Bourne, ed.), 1st ed., Vol. 3, p. 445. Academic Press, New York.

Szent-Györgyi, A., and Kaminer, B. (1963). *Proc. Natl. Acad. Sci. U.S.A.* **50**, 1033.
Wakabayashi, T., Huxley, H. E., Amos, L. A., and Klug, A. (1975). *J. Mol. Biol.* **93**, 477.
Weber, A. (1959). *J. Biol. Chem.* **234**, 2764.
Weber, A. (1966). *In* "Current Topics in Bioenergetics" (D. R. Sanadi, ed.), p. 203. Academic Press, New York.
Weber, A., and Winicur, S. (1961). *J. Biol. Chem.* **236**, 3198.

8

Molecular Movements and Conformational Changes in Muscle Contraction and Regulation*

J. GERGELY

It is indeed a great honor and pleasure to be part of this tribute to Albert Szent-Györgyi, whose influence on modern biochemistry and biophysics is being continuously reflected in numerous papers of his former pupils and admirers scattered all over the world and, in a special way, attested to by this volume.

To begin on a personal note, I first heard about Albert Szent-Györgyi's work on intermediate metabolism and on vitamin C as a medical student at the University of Budapest in the late 1930's. No scientist, physician or student at that time failed to be thrilled by the news of the award of the Nobel prize to him. It was, however, the publication of the Studies of the Institute of Biochemistry in Szeged in the early 1940's (Szent-Györgyi, 1941–1944) breaking

*The preparation of this manuscript was supported by grants from the National Institutes of Health (HL-5949), the National Science Foundation and the Muscular Dystrophy Associations of America, Inc.

new ground in muscle biochemistry that opened up to me completely new vistas. These publications differed from the usual articles of that era in their conciseness and strong dependence on experimental facts. At the same time, their bold imagination and beauty had a tremendous impact on me. Having just graduated from medical school I realized how little I knew. Another of Szent-Györgyi's interests at that time was the application of quantum mechanics to biological systems. The idea that biological phenomena had to be understood not only in terms of molecules but also in terms of the interactions of electrons within molecules and indeed in the interaction of electrons in one molecule with those in another molecule using the analogy of conductors and semiconductors was fascinating. I was fortunate to be able to join Albert Szent-Györgyi's institute on his move to Budapest after the war. I studied quantum chemistry with the late M. G. Evans, at Szent-Györgyi's suggestion, and later rejoined him in this country. Ever since I have remained "hooked" on muscle; and although I have turned from the theoretical quantum mechanical approach to more experimental pursuits, I feel that the early ideas of Albert Szent-Györgyi's quantum chemical thinking, and his strong emphasis on linking thermodynamics, physical chemistry and biochemistry have been powerful forces in shaping my own approach.

There is one more personal note that I should like to add. While Albert Szent-Györgyi, or Prof as many still fondly call him, has always been in the forefront in introducing new ideas and in applying sophisticated concepts to biology, those of us who had a chance to observe him closely have always been impressed with the simplicity of his approach and by his insistence on close observation of phenomena. I still have a vivid image of him, going back more than 25 years, as he stood before a window observing a test tube in which the superprecipitation of actomyosin was proceeding. Eventually one had to use more and more complicated pieces of equipment, but the need to stay in personal contact with the material which was, as it were, living in the test tube, remained one of the fundamental precepts impressed on our minds by the man whom we honor with this book.

INTRODUCTION

This paper is a discussion of some current problems in muscle research which in many of their aspects can be traced back to the pioneering work of Albert Szent-Györgyi. Muscle biochemistry and muscle physiology, as recently emphasized by Dorothy Needham (1971), have a history that extends back over 2000 years. But during the last 30 years, two of these scientific revolutions, a concept introduced by Thomas Kuhn (1962), that happen once in a while in various disciplines took place in the field of muscle. The first of these was the discovery of actin and myosin, as separate entities, by Szent-Györgyi and his colleagues in

Szeged. The second involved the realization that these two proteins are organized in two distinct sets of filaments and that muscle contraction is not based on the shortening of microscopic elements but rather on the relative sliding of the two sets of filaments in such a way that individual elements do not shorten on the microscopic scale but undergo cyclic changes (cf. Huxley, in this volume). These discoveries changed many of the original problems that emerged after the discovery of actin and myosin. Yet the rereading of the papers in the "Studies" (Szent-Györgyi, 1941–1944) or a later review (Szent-Györgyi, 1958) of the then current problems of muscle research shows that, with changes in emphasis, some of the unsolved problems are still with us.

INTERNAL MOVEMENTS IN MYOSIN

The interaction of ATP with myosin is clearly a key process in muscle contraction, and one of the outstanding problems of muscle biochemistry is to achieve an understanding of how this interaction results in the conversion of chemical into mechanical energy. Intimately connected with this fundamental problem of bioenergetics is the notion that certain movements within and between molecules making up the contractile apparatus underlie both muscle contraction and its regulation. These ideas are described in considerable detail in various contributions in this volume. I shall restrict my own contribution to the discussion of some work of my colleagues and myself that deals with changes that occur in the myosin molecule on interaction with ATP, with the internal flexibility in myosin, and with some aspects of the movement of tropomyosin in the thin filaments which in turn is involved in the regulation of contraction (see papers by Ebashi and by Weber, this volume).

The essential feature of the driving mechanism in the sliding filament theory is movement of the cross-bridges corresponding to the heads of the myosin molecules (see Huxley, this volume). While x-ray data have for some time indicated that the cross-bridges move, the *in vitro* demonstration of the internal flexibility of the myosin molecule has only been achieved recently. The first successful demonstration was obtained by Morales and his colleagues (Mendelson *et al.*, 1973) in their studies of the depolarization of fluorescence emission by a fluorescent chromophore attached to the head portion of the myosin molecule. This technique permits the calculation of the correlation time of the protein that carries the fluorescent dye. The correlation time of the smallest active proteolytic fragment, subfragment-1, (S-1), was not too different from the correlation time of a larger active fragment, heavy meromyosin, (HMM), and the intact molecule. These data clearly indicate that the myosin heads possess motion of their own. The drawback of the fluorescence method is that slower motions, which are exhibited by myosin heads organized into filaments and particularly myosin heads interacting with actin, can be less accurately studied because of the short fluorescence lifetime of the dye.

Another technique employed recently by my colleagues and myself (Thomas *et al.*, 1975a,b) utilizes a new variant of the electron spin resonance (ESR) technique developed by Hyde and his colleagues (Hyde and Dalton, 1972; Hyde and Thomas, 1973). Spin labels have now been used for some years to study biological systems and they have been particularly useful for studying changes in the internal structure of proteins (McConnell and McFarland, 1970). The newer techniques referred to as saturation transfer spectroscopy permit the study of the motion of the molecule, or part of the molecule, to which the spin labels are attached and extend the useful time range to 10^{-3} sec. This contrasts with the time domain of fluorescence polarization and conventional electron spin resonance techniques, which are of the order of 10^{-7} sec.

The correlation times characteristic of the motion of subfragment-1 obtained by the fluorescence technique and the saturation transfer ESR technique are in good agreement, viz. about 200 nsec. Comparison of the data on the isolated subfragment-1 with intact myosin and the two-headed proteolytic fragment heavy meromyosin suggests that the two heads in myosin retain their independent motion. The ESR technique shows even less difference between the correlation times of S-1, HMM and myosin; this strongly suggests the conclusion, arrived at from fluorescence studies concerning segmental flexibility of the myosin molecule, of at least one hinge being located in the HMM portion at the junction of S-1 and the so-called subfragment-2 (S-2). Saturation transfer spectroscopy permits a more detailed investigation of the motion of the myosin heads when myosin molecules are organized into filaments or when myosin or its fragments are attached to actin. The motion of the heads is only slightly slowed upon formation of myosin filaments, allowing for the change in size of the particles, but the attachment of subfragment-1 to actin changes the correlation time to about 200 nsec to about 0.2 msec. (Table I).

TABLE I
Correlation Times of Spin Label Attached to Myosin or Subfragment-1 Obtained by Saturation Transfer ESR Spectroscopy[a]

	Correlation times (sec)
Myosin, solution	3.7×10^{-7}
Heavy meromyosin	3.2×10^{-7}
Subfragment-1	2.2×10^{-7}
Myosin, filament	3×10^{-6}
Myosin attached to F-actin	2×10^{-4}
Subfragment-1 attached to F-actin	2×10^{-4}

[a]Based on Thomas *et al.* (1975a,b).

The most interesting question from the point of view of the mechanism of muscle contraction is the elucidation of the changes brought about in myosin by the interaction with ATP, whose hydrolysis furnishes the energy for contraction. So far we have not been able to utilize the technique discussed above for investigating the effect of ATP on myosin-head motion, because with the labels used ATP produces another type of change in myosin which could be described as an internal or conformational one. This is revealed in increased motion of the spin label attached to the active center of the myosin head, specifically to one of

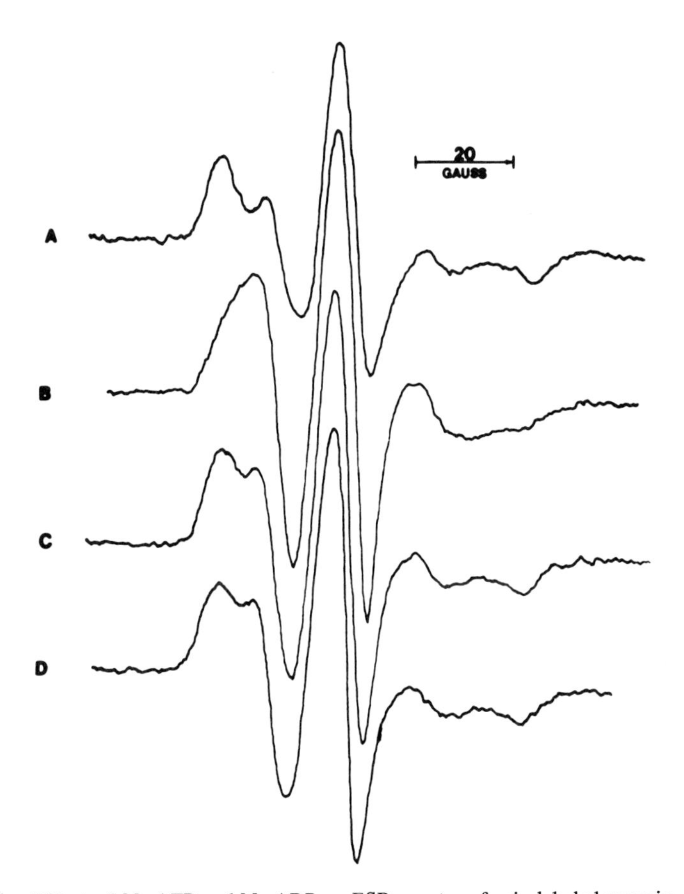

Fig. 1. Effect of Mg-ATP and Mg-ADP on ESR spectra of spin labeled myosin. Selective spin labeling at the S-1 thiol groups with *N*-(1-oxyl-2,2,6,6-tetramethyl-4-piperidinyl) was carried out. Spectra of solutions containing S-1 labeled myosin, 14 mg/ml, 0.4 *M* KCl and 0.04 *M* Tris, pH 7.5 were recorded at room temperature. (A) No further addition; (B) 5 m*M* $MgCl_2$ and 5 m*M* ATP, recorded 2 min after addition of ATP; (C) as (B) recorded 10 min after addition of ATP; (D) 5 m*M* $MgCl_2$ and 5 m*M* ADP. (Reproduced with permission from Seidel and Gergely, 1971.)

the reactive sulfhydryl groups. This increased mobility of the spin label is characteristic of the form of myosin present in the steady state during ATP hydrolysis (Fig. 1) and has been identified with the complex formed with tightly bound products (Seidel and Gergely, 1971). The same intermediate complex, also identified by its increased fluorescence in unlabeled myosin (Werber *et al.*, 1972; Bagshaw *et al.*, 1974) is presumably the one which is then able to interact with actin. Thus the success of utilizing the spin labeled myosin for the demonstration of the conformational change that occurs upon interaction of myosin with ATP has so far been hindering us in utilizing the saturation transfer technique for possible changes in the motion of the ATP head. The reason for this is the motion of the spin label, caused by structural changes occurring in the myosin head upon interaction with ATP, which increases independently of the motion of the protein to which it is attached. Experiments are now under way to obtain spin labeled myosin which does not respond with an internal conformational change to ATP, so that any change occurring in the slower range domain can be interpreted in terms of the motion of the protein. These studies will permit a comparison of the ATP-induced changes in myosin and actin *in vitro* with the motional states deduced from transients observed in physiological experiments (e.g., Huxley and Simmons, 1972). Another possibility opened up by this technique is to study the dynamics of actin filaments as they interact with the cross-bridges (Thomas *et al.*, 1976).

MOLECULAR CHANGES IN REGULATION

A new type of molecular motion has been discovered in recent years in the translocation of tropomyosin within the thin filaments depending on whether the regulatory protein troponin is in combination with calcium (see, e.g., Wakabayashi *et al.*, 1975). The regulation of the movement of tropomyosin is an example of the type of phenomenon that has always fascinated Szent-Györgyi, viz. how long-range effects result from localized interactions between proteins and small molecules. For it is the combination of Ca^{2+} with the troponin C (TN C) subunit, resulting in conformational changes within that subunit (see Ebashi, this volume), which according to the concepts developed by several investigators results in the release of another subunit, troponin I (TN I), of troponin from the thin filament, thereby permitting the movement of tropomyosin. This conclusion is based, among other things, on the release of the complex of TN C and TN I from the thin filament upon combination with Ca^{2+} (for background, see Potter *et al.*, 1974).

Another approach to the analysis of this process can be made by studying changes in the properties of TN I when Ca^{2+} combines with TN C. Spin label studies of this sort have been reported by Ebashi *et al.;* and recently my colleague, Leavis (1976) has been able to demonstrate changes in a fluorescent label attached to TN I on combination of Ca^{2+} with TN C. This system is particularly interesting in that it permits a functional distinction among the various kinds of calcium binding sites recently found in TN C. Two of these are able to bind both Ca^{2+} and Mg^{2+}, while two others, having a somewhat lower Ca^{2+}-affinity, are specific for Ca^{2+} (Potter and Gergely, 1975). Ca^{2+}-binding to the former (Fig. 2) produces a large conformational change in TN C, presumably preparing this protein for its regulatory role, while combination with the Ca^{2+}-specific sites results in a change in TN I reflected in the altered fluorescence of the label which might be related to the interaction of tropomyosin movement. This result is in harmony with our earlier work showing that the Ca^{2+}-dependence of actomyosin ATPase activity involves binding to the Ca^{2+}-specific site (Potter and Gergely, 1975).

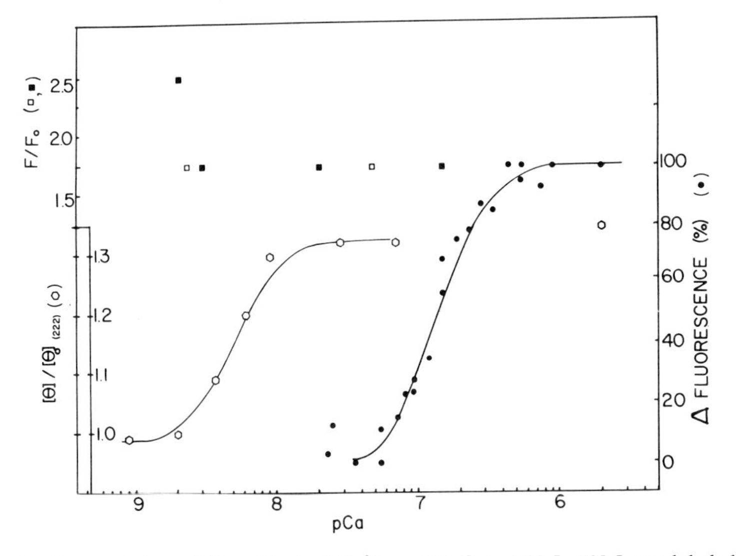

Fig. 2. Transmission of the effect of Ca^{2+} on TN C to TN I. TN I was labeled with *N*-dansyl-*S*-mercuric cysteine (see Leavis and Lehrer, 1974) and to its 1:1 complex with TN C. Calcium was added to attain the pCa indicated on the abscissa. The fluorescence enhancement λ_{excit} = 33 nm, $\lambda_{emission}$ = 500 nm is shown on the right hand side of the ordinate (circles). For comparison, changes of circular dichroism of 222 nm are also shown (hexagons), as well as changes in fluorescence when the label is attached to TN C with (open squares) and without (closed squares) 2 m*M* Mg^{2+}. The solutions contained 20 m*M* KCl, 25 m*M* HEPES buffer, pH 7.5, 2 m*M* EGTA and protein, 0.1 mg/ml, 25°C. $CaCl_2$ was added in amounts calculated to reach the desired pCa. For other details of protein preparation, calculations of pCa and techniques of measurement, see Potter *et al.* (1974). (P. Leavis, unpublished.)

CONCLUSIONS

The relation of protein motions or conformational changes to the mechanism of muscle contraction represents an extension of the early ideas of Albert Szent-Györgyi and his colleagues, seeking to explain muscle contraction in terms of changes in the molecular structure of constituents of the contractile apparatus. Although the details have evolved with an increase in our knowledge of the details of the proteins of the myofibrils, the basic concepts originated by the man to whom this volume is dedicated have remained a powerful driving force for all those who have come into contact with him.

REFERENCES

Bagshaw, C. R., Eccleston, J. F., Eckstein, F., Goody, R. S., Gutfreund, H., and Trentham, D. R. (1974). *Biochem. J.* **141**, 351–364.
Huxley, A. F., and Simmons, R. M. (1972). *Nature (London)* **237**, 281–283.
Hyde, J. S., and Dalton, L. R. (1972). *Chem. Phys. Lett.* **16**, 568–572.
Hyde, J. S., and Thomas, D. D. (1973). *Ann. N.Y. Acad. Sci.* **222**, 680–692.
Kuhn, T. (1962). "The Structure of Scientific Revolutions." University of Chicago Press, Chicago.
Leavis, P. (1976). *Fed. Proc., Fed. Am. Soc. Exp. Biol.* **35**, 1746.
Leavis, P., and Lehrer, S. S. (1974). *Biochemistry* **13**, 3042–3048.
McConnell, H. M., and McFarland, B. G. (1970). *Q. Rev. Biophys.* **3**, 91–136.
Mendelson, R. A., Morales, M. F., and Botts, J. (1973). *Biochemistry* **12**, 2250–2255.
Needham, D. M. (1971). "Machina Carnis." Cambridge Univ. Press, London and New York.
Potter, J. D., and Gergely, J. (1975). *J. Biol. Chem.* **250**, 4628–4633.
Potter, J. D., Seidel, J. C., Leavis, P. C., Lehrer, S. S., and Gergely, J. (1974). *In* "Calcium Binding Proteins" (W. Drabikowski, H. Strzeelecka, and E. Carafoli, eds.), pp. 129–152. Elsevier, Amsterdam.
Seidel, J. C., and Gergely, J. (1971). *Biochem. Biophys. Res. Commun.* **44**, 826–830.
Szent-Györgyi, A. (1941–1944). *Stud. Inst. Med. Chem. Univ. Szeged,* Vols. I–IV. [Volume IV also was published in *Acta Physiol. Scand., Suppl.* XXV, **9**, (1945).]
Szent-Györgyi, A. (1958). *Science* **128**, 699–702.
Thomas, D. D., Seidel, J. C., Gergely, J., and Hyde, J. S. (1975a). *J. Supramol. Struct.* **3**, 376–390.
Thomas, D. D., Seidel, J. C., Hyde, J. S., and Gergely, J. (1975b). *Proc. Natl. Acad. Sci. U.S.A.* **72**, 1729–1733.
Thomas, D. D., Seidel, J. C., and Gergely, J. (1976). *Fed. Proc., Fed. Am. Soc. Exp. Biol.* **35**, 1745.
Wakabayashi, T., Huxley, H. E., Amos, L. A., and Klug, A. (1975). *J. Mol. Biol.* **93**, 477–497.
Werber, M. M., Szent-Györgyi, A. G., and Fasman, G. D. (1972). *Biochemistry* **11**, 2872–2883.

9

How Active Is Actin? Propagated Conformational Changes in the Actin Filament

JOSEPH LOSCALZO, GEORGE H. REED, and ANNEMARIE WEBER

I (A.W.) encountered Albert Szent-Györgyi first through my father's description. My father visited Albert Szent-Györgyi during World War II, I think in the early 1940's. Whatever my father said about the science then is mingled with my later memories, but I do remember him talking about the man. It was not often that my father was so fascinated and impressed by someone. There was the image of courage and strength: Albert Szent-Györgyi was an acknowledged enemy of the regime as the authorities in Germany were quick to inform my father. Then there was the man of unusual outgoing warmth, filled with that incredible vitality of which the infectious enthusiasm for science is a part. As a scientist he is unique, for not many are also artists, as Carl Cori described him, or Dionysians, as Albert Szent-Györgyi called himself. Under whatever name, the intuitive genius is a very rare phenomenon of nature.

During those years, the most striking example of his intuitive insight was the statement that superprecipitation of actomyosin represents the fundamental reaction of muscle contraction. It outraged physiologists for years afterwards.

The intuitive insight together with systematic studies from the Szeged laboratory resulted in a remarkable advance in the understanding of muscle within a few years. It was found that myosin consisted of two fractions, one capable of superprecipitation, the other not. Using a method of differential extraction and denaturation, Straub achieved the isolation of actin, the protein which forms a complex with myosin and is indispensable for superprecipitation. Invention of glycerination of fibers made it possible to combine mechanical and enzymatic measurements. As a result, the basic outline of the biochemistry of the interaction between actin and myosin was obtained, although soon, in part, to be lost again in the confusion created by large numbers of studies of myosin ATPase activity. Nevertheless, in the early 1940's it emerged from the work of Albert Szent-Györgyi's group (1942) that actin and myosin cooperated in contraction as well as in hydrolysis of ATP, and this confirmed the belief that ATP provides the energy for contraction.

In this cooperation between the two proteins Albert Szent-Györgyi visualized actin as an active participant, quite different from the passive role it was later assigned after Huxley's and Hanson's studies led to an understanding of the mechanical events responsible for shortening. At that time actin was viewed as a rope that is pulled along by myosin.

Attention again centered on the actin filament when Ebashi discovered that in vertebrate skeletal muscle the whole apparatus regulating contraction is located on that filament (Ebashi and Endo, 1968). Our interest in the dynamic behavior of actin was aroused by the cooperative phenomena displayed by regulated actin filaments containing troponin and tropomyosin (Weber and Murray, 1973). For instance, if a fraction of the actin molecules in a regulated filament is combined with nucleotide-free myosin, forming rigor complexes, interaction between the remaining free actin molecules and ATP-activated myosin takes place in the absence of calcium; contraction thus becomes independent of calcium. This effect may occur without active participation by actin molecules; it is quite possible that a few rigor complexes (rigor complexes are known to be formed under all conditions) just displace tropomyosin from a position on the actin filament where it blocks the binding of ATP-activated myosin. By contrast, conformational changes in actin may contribute to potentiation, a phenomenon which cannot be attributed as easily to changes in tropomyosin position. Potentiation is caused by rigor complexes which, in the presence of calcium, increase activation of ATP hydrolysis by regulated actin well above the level of that achieved by pure actin or regulated actin free of rigor complexes. One could assume that rigor complexes cause sufficient transposition of tropomyosin enabling it to contribute some of its amino acid sequences to the active site of actin, thereby improving the cofactor effect of actin. The difficulty in this assumption is that tropomyosin does not have a sevenfold repeat of the same amino acid sequence so that more than one sequence would have to be capable of providing the same potentiating effect. Furthermore, neither x-ray diffraction (Haselgrove, 1972) nor spin label attachment to tropomyosin (as recently observed by J. Loscalzo) gives positive evidence of a change of tropomyosin position during potentiation, although it cannot be ruled out. The lack of an adequate explanation for potentiation led us to look for propagated conformational changes along the actin molecules induced by myosin binding.

We decided to look at a part of the actin molecule for which there is a probe capable of distinguishing conformational changes within actin from those due to binding of another protein. Such a probe exists for the region of the unexchangeable bound divalent cation in actin, since Burley *et al.* (1972) had shown that the spin of manganous ion placed in this position interacts with the spin of a nitroxide label attached to a fast reacting sulfhydryl group.

We determined the effect of binding, of myosin subfragment-1 (S-1), on the spin-spin interaction by measuring the extent of quenching of the central peak amplitude of the spectrum of the nitroxide label. Adding S-1 to give 30% saturation of the filament (the binding was stoichiometric since our protein concentrations were greater than 10^{-5} *M* and the dissociation constant for the complex is about 2×10^{-8} *M* [Marston and Weber, 1975]) resulted first in a large increase in peak amplitude, then a decrease returning to the original value at about 70% saturation with S-1 (Fig. 1) (Loscalzo *et al.*, 1975). These changes in spin interaction show that actin responds to S-1 binding by a change in

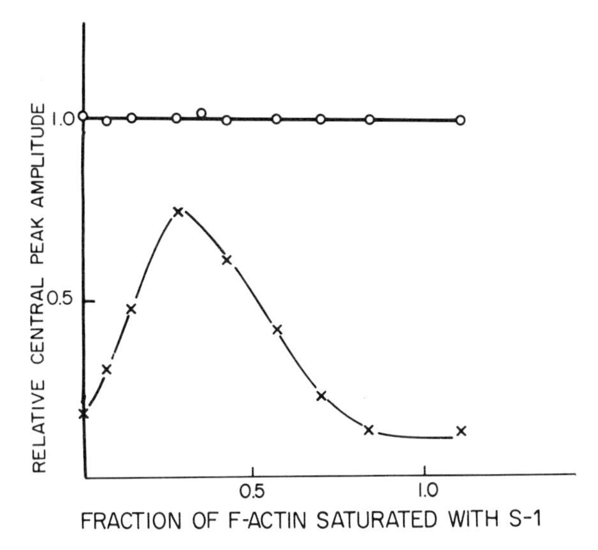

Fig. 1. Relative central peak amplitude of the spin label ESR spectrum as a function of F-actin saturation with S-1: (o), spin labeled-actin; (x), spin labeled-Mn^{2+}–actin. Central peak amplitude in the absence of Mn^{2+} equals 1.0.

conformation. We do not know where the unexchangeable cation or the cysteine residue are situated with respect to the myosin binding site. However, we can say that the response must involve protein-protein interactions other than those between S-1 and a single monomer since the conformational response changes direction with increasing S-1.

What sort of conformational change are we observing? A quantitative interpretation of the signal requires knowledge of the extent and purity of the label. We established that, after incubation of monomeric actin with spin label, more than 95% of the actin monomers in the filament carried a single nitroxide label (only actin incapable of polymerization had two) bound to cysteine 373. The cysteine was identified by using the procedures of digestion and chromatography described by Bridgen (1971). Ninety-nine % of the unexchangeable cation sites were occupied by manganous ions. For a completely immobilized nitroxide label, quenching of the spin label central peak amplitude depends on the distance of the manganous ion from the spin label, the orientation of the manganous ion with respect to the spin label, and the correlation time for the interaction between the manganous ion and the spin label. Control experiments (measurements of the paramagnetic contribution to the proton relaxation time of water as a function of bound S-1) showed the correlation time to be unchanged by S-1 binding. Simulated spectra modeled after the equations of Leigh (1970) indicated that the observed extent of change in peak amplitude was too large to be explained by changes in orientation alone and that the distance between spins must have increased by at least 6–8 Å.

The next question concerns the location of the two interacting spins: are they on the same actin molecule or do the spins of neighboring molecules interact? The general experimental design to answer this question is obvious. First, make a mixed polymer from two different singly labeled actin molecules, interact. Second, surround doubly labeled actin molecules (manganous ion and nitroxide label on the same molecule) with unlabeled, native monomers. The execution of the experiment depends on certain conditions. First, since manganous ion bound to actin in the monomeric state is exchangeable with other ions (it only becomes unexchangeable after polymerization), one must be able to complete mixing and polymerization before a substantial fraction of the bound manganous ion has exchanged with calcium and magnesium ions of the other monomers. Fortunately, manganous ion exchanges very slowly, with a half time of about 25 sec, whereas mixing and polymerization can be completed within 5 sec. Second, labeled and unlabeled monomers must form truly random copolymers and not sort themselves out according to species. In other words, labeling must not alter the reactivity of the sites involved in polymerization (measurements of actin-activated ATPase activity under a variety of conditions suggested that the myosin binding site was not disturbed). Polymerization seemed to remain unaffected according to measurements of rates of polymerization determined by viscosity and release of inorganic phosphate (liberated when an ATP-carrying monomer polymerizes and converts its structural ATP to ADP). The formation of random copolymers was proved by the following experiment suggested to us by Steven Marston. It is based on the fact that polymerization behaves like a phase change: polymerized actin is in equilibrium with a critical concentration of monomer which remains constant, independent of the polymer concentration (Oosawa and Kasai, 1971). Below the critical concentration no polymers are formed (Table I). If one mixed two species of actin, each below the critical concentration for that species, formation of two kinds of polymers, each consisting of one species only, is impossible by definition. Only a mixed copolymer could be formed if the combined concentration of both species is above critical. On mixing singly labeled with native actin, each species below critical concentration, we found that polymerization occurred and that the critical concentration of the combined species was equal to that of native actin (Table I). Polymerization was measured by sedimentation at 100,000 *g* over a period of 2 hr or by the activation of myosin ATPase, a function monomeric actin lacks. This experiment proved that random copolymers were formed between all species of actin used in the experiment (Table I). No spin-spin interaction with or without S-1 was observed when manganous ion and nitroxide label were placed on different actin molecules. On the other hand, when we copolymerized doubly labeled actin with a six-fold excess of unlabeled actin, the signal change for each labeled actin was the same as that in a homogeneously doubly labeled polymer. These experiments prove the spin-spin interaction to be intramolecular, i.e., the distance between the divalent ion and the nitroxide label on the cysteine of the same molecule had changed by about 6–8 Å. This does not necessarily suggest that the 40 Å diameter molecule undergoes a drastic

TABLE I
Critical Concentrations of Actin Copolymers[a]

Copolymer	Actin (μM)		
	Total added	Supernatant	Sedimented
Spin labeled-actin + native (1:1)	0.75	0.73	0.02
	1.00	0.97	0.03
	1.25	0.97	0.28
	1.50	0.95	0.55
Mn^{2+}–Actin + native (1:1)	0.75	0.72	0.03
	1.00	0.95	0.05
	1.26	1.01	0.24
	1.50	0.93	0.57

[a]Concentration of actin remaining in supernatant equals critical concentration.

change in tertiary structure since the rather long spin label arm (2 rings in sequence) allows considerable amplification of a much smaller actin side chain movement.

What can one say about the cooperativity of the response? The increase in peak amplitude at 30% saturation with S-1 is too great to be restricted to the actin molecules complexed with S-1. The reasoning is as follows. Spin-spin interaction is lost and peak amplitude maximal when it has become equal to that in the absence of manganous ion; spin-spin interaction between manganous ion and nitroxide label ceases when their distance exceeds 25 Å. If the conformational change in each S-1 complexed-actin gave rise to the maximal peak amplitude, the resulting peak amplitude for a filament 30% saturated with S-1 would be

$$0.3 + \frac{\text{peak}_{\text{S-1 absent}}}{\text{peak}_{\text{Mn}^{2+}\text{ absent}}} (1 - 0.3)$$

(S-1 binding causes no signal change in the absence of manganous ion.) We find, however, twice that value indicating that the conformational change occurred in at least two actin molecules for each bound S-1, and hence was propagated (Fig. 1). If S-1 binding only reduced but did not abolish the spin-spin interaction, more actin molecules would have contributed to the increase in peak amplitude; i.e., all of the actin molecules in the filament may have undergone a conformational change.

As mentioned above, beyond 30% saturation with S-1, the peak amplitude declined again. Since at greater concentrations more and more S-1 molecules are occupying positions next to each other, one could attribute the effect to steric crowding, forcing an adjustment in each actin-S-1 bond. We do not consider this a likely explanation because a titration curve of actin with HMM (the proteolytic myosin fragment that contains two active sites on the same molecule) measuring peak amplitude as a function of active sites (not molecules), superimposed on a titration curve with S-1. If placing myosin active sites next to each other prevented the separating movement of the spins, one would not expect an identical response to S-1 and HMM since HMM provides two sites next to each other at all degrees of saturation. One may argue that only one of the two sites binds to the filament even though a filament is saturated with HMM at a ratio of one active site/actin or 1 molecule/2 actin molecules (Eisenberg *et al.*, 1971). Again, this is not compatible with a superimposable plot. Therefore, the biphasic nature of the response probably is not related to steric interference between myosin heads. Although we ruled out steric interference, the effect must be related to the increasing density of myosin molecules on the actin filament. We have no data that provide further information, but we can offer a speculation. If, for instance, actin molecules complexed with myosin retain their own spin-spin interaction, i.e., make no contribution to increased peak amplitude but impose a conformational change on two neighboring actin molecules as long as they are free of bound myosin, the findings in Fig. 1 would result. At S-1 saturations greater than 30% the number of acto–S-1 complexes surrounded by free actin molecules begin to decline as more and more S-1 molecules are placed next and opposite to each other.

Biphasic changes in response to HMM binding to actin have been observed previously (Fujime and Ishiwata, 1971; Yanagida *et al.*, 1974; Tawada, 1969). However, the methods did not permit identification as propagated conformational changes in actin. These data, obtained by laser light scattering, linear ultraviolet dichroism, and ultraviolet absorption, were usually attributed to changes in filament rigidity. In light of our data it appears that the absorption measurements are caused by a conformational change in the actin molecule, suggesting that, in addition to the manganous ion site and cysteine 373, it also involves a tryptophan. This conformational change would seem to be responsible for the change in rigidity indicated by laser light scattering.

We do not know whether the propagated conformational changes involve the binding site for myosin. No definite answer is obtained from existing measurements of ATPase activity in the presence of varying concentrations of rigor complexes (i.e., complexes between actin and nucleotide-free myosin under conditions of S-1/actin equal to or greater than 1 (Bremel *et al.*, 1972)). While such curves do not show any gross abnormality, such as a biphasic shape, they have not been precisely analyzed kinetically because it is very difficult, for technical reasons, to obtain values for free ATP under these conditions.

Placing tropomyosin on the actin filament completely alters the biochemical responsiveness of the actin molecule. Although in the absence of S-1 the spin-spin interaction is not altered by the addition of tropomyosin, on addition of S-1 we no longer see any change in the spin-spin interaction (Table II). The same is true if the filament contains, in addition, calcium-free troponin. Addition of calcium in the absence of S-1 results in a slight decrease of the spin-spin interaction which returns to the original value on addition of S-1. However, once

TABLE II
Effect of Tropomyosin on Spin-Spin Interaction

S-1 saturation	Relative central peak amplitude[a]
0	0.22
28	0.20
56	0.22
84	0.23

[a]The relative central peak amplitude of spin labeled-actin in the absence of bound manganous ion was taken as 1.00 and did not change with increasing S-1 saturation.

the actin molecules are no longer uniform as a result of binding troponin to every seventh actin and of exposing them to different parts of the tropomyosin molecule, the data become more difficult to interpret. The conformational change following calcium addition may be restricted to the actin molecule complexed to troponin. The observation that tropomyosin-actin, which in contrast to pure actin displays no conformational changes in the region of our probe when myosin is bound, is compatible with the view that tropomyosin causes potentiation of ATPase activity (see above) by direct participation in the reaction between actin and ATP-activated myosin. However, we shall postpone the final verdict until a probe is found that reports conformational changes directly from the active site.

A very important conclusion from these experiments pertains to the function of tropomyosin. In addition to providing steric blockage of the myosin binding site during relaxation in troponin-containing actin filaments, tropomyosin alters the biochemical response of the actin molecule to myosin. This may be one of its major functions in muscles without troponin where contraction is regulated by a light chain on myosin.

In conclusion, we have found that actin is capable of being a very active participant in interaction with myosin, regulating its response to myosin binding by propagated conformational changes.

REFERENCES

Bremel, R. D., Murray, J. M., and Weber, A. (1972). *Cold Spring Harbor Symp. Quant. Biol.* **37**, 267.
Bridgen, J. (1971). *Biochem. J.* **123**, 591.
Burley, R. W., Seidel, J. C., and Gergely, J. (1972). *Arch. Biochem. Biophys.* **150**, 792.
Ebashi, S., and Endo, M. (1968). *Prog. Biophys. Mol. Biol.* **18**, 123.
Eisenberg, E., Dobkin, L., and Kielley, W. W. (1971). *Fed. Proc., Fed. Am. Soc. Exp. Biol.* **30**, 1310.
Fujime, S., and Ishiwata, S. (1971). *J. Mol. Biol.* **62**, 251.
Haselgrove, J. C. (1972). *Cold Spring Harbor Symp. Quant. Biol.* **37**, 341.
Hodges, R. S., and Smillie, L. B. (1973). *Can. J. Biochem.* **51**, 56.
Leigh, J. S., Jr. (1970). *J. Chem. Phys.* **52**, 2608.
Loscalzo, J., Reed, G. H., and Weber, A. (1975). *Proc. Natl. Acad. Sci. U.S.A.* **72**, 3412.
Marston, S. A., and Weber, A. (1975). *Biochemistry* **14**, 3868.
Oosawa, F., and Kasai, M. (1971). *Biol. Macromol.* **5**, 261.
Szent-Györgyi, A. (1942). *Stud. Inst. Med. Chem. Univ. Szeged* **1**, 17.
Tawada, K. (1969). *Biochim. Biophys. Acta* **172**, 311.
Weber, A., and Murray, J. M. (1973). *Physiol. Rev.* **53**, 612.
Yanagida, T., Taniguchi, M., and Oosawa, F. (1974). *J. Mol. Biol.* **90**, 509.

10

Myosin-Linked Regulation of Muscle Contraction

ANDREW G. SZENT-GYÖRGYI

PERSONAL REMARKS

Albert Szent-Györgyi is totally responsible for my entry into science and for my particular interest in muscle research. If not for him it is doubtful that I would be writing this paper in his honor. I call him "Prof" as do all his associates and other friends. But as a Prof he has never been involved in a systematic fashion in education, never having actually taught a graduate or undergraduate course. Nevertheless, the term comes quite naturally, and the question is: What has one learned from him? Not necessarily techniques or instrumentation, for he has always considered one's fingers, eyes, and even one's brain the most important instruments for conducting scientific research. Neither did we learn systematic scholarship. His reading habits have always been very selective as is evident from one of his stories: Crossing the border from Canada several decades ago, he was asked by the immigration officer: "Oh, you are invited here to be a Professor. Can you read and write?" Prof's answer was: "I occasionally write but never read." Much of the trade we had to learn on our own; what he taught us were the essentials. There was the prevailing attitude that to do research was tremendous fun, and the most exciting thing one could do with one's life. This was possible only if one attempted to solve important problems, and, to achieve it, one's approach had to be direct and simple. Prof's personality, his writings, his speeches, all project in a most vivid and convincing manner the joy of finding out new things about "life."

Prof was always a very hard worker with an unusual ability to concentrate on the formulation of questions in the simplest possible way. Frequently, one felt almost annoyed by the oversimplification or the naïvete of his propositions. But this feeling was immediately counteracted by the knowledge of how frequently he has been right and how frequently his most important contributions have been initially met with incredulity. Thus, one did not feel entirely at liberty to turn down his suggestions without first considering them very seriously. Even when he was wrong, he did it in a glorious way. The grace, wit and spontaneity were coupled with hard work and with an insistence that one interpret simple observations from a new angle and relate them to basic principles of nature. Sometimes the mistakes were great; but so were the rewards.

GENERAL CONSIDERATIONS

The two important facets of muscle physiology are the mechanism of contraction and how it is prevented. Most skeletal muscles are at rest most of the time during the life of the animal and can be vigorously activated at a moment's notice. At rest heat production is about 10^4–10^5 times lower than in contracting muscles. In resting muscle, actin and myosin interaction is prevented by a regulatory system. Contraction is triggered by the release of calcium ions from membranous compartments strategically positioned within the muscle fibers. Calcium reacts with control proteins and reverses their inhibitory function. At sarcoplasmic calcium concentrations of about 10^{-6} *M*, myosin readily combines with actin, and the cyclic reaction sequence at the cross-bridge leads to tension generation and to motion at the expense of ATP hydrolysis. At the cessation of the stimulus, calcium is reconcentrated within the membranous compartments and the free calcium concentration in the sarcoplasm falls below 10^{-8} *M*. Once the control proteins lose their bound calcium, they become inhibitory again, actomyosin is no longer formed, and the resting state is restored.

Actomyosin formation may be prevented either by blocking reactive sites on actin or by interfering with sites on myosin. Regulation is thus actin- or myosin-linked. The basic general features of muscle control have been clarified by the detailed studies on the classic regulatory proteins, tropomyosin and troponin. These regulatory proteins, discovered by Ebashi and colleagues (see Ebashi and Endo, 1968; Weber and Murray, 1973), react with actin and alter its affinity to myosin in the absence of calcium. The properties of the troponin–tropomyosin system is discussed by Professor Ebashi in this volume and will be referred to here only to the extent that is needed for clarity. In molluscan muscles however, troponin is not found. In these muscles, contraction is controlled by one of its light chains regulating the myosin molecule which has an "on" or "off" state, depending on the presence or absence of calcium ions. In this paper I will describe the main features of myosin control and will speculate on how myosin may be regulated by calcium. This paper is based entirely on experiments that my colleagues, J. Kendrick-Jones, W. Lehman, and E. M. Szentkiralyi, and I have conducted during the past years (Kendrick-Jones *et al.*, 1970, 1972; Lehman *et al.*, 1972; Szent-Györgyi *et al.*, 1973; Lehman and Szent-Györgyi, 1975; Szent-Györgyi, 1975).

COMPARISON OF THE TWO CONTROL SYSTEMS OF MUSCLES

Both actin-linked and myosin-linked regulatory systems function by interfering with cross-bridge formation between myosin and actin in the absence of calcium. Both regulatory systems require the presence of ATP for regulation. The species of myosin that participates in both types of regulation is the one that dominates the myosin population in the presence of ATP, i.e., the myosin intermediate, My**(ADP, Pi). Rigor links that are formed in the absence of ATP between actin and myosin are not prevented by either of the two regulatory systems. Although the final result of the two regulations is similar, i.e., no cross-bridges are formed, the components of regulations and the way they function are quite different. Actin control requires the presence of tropomyosin. It is the tropomyosin molecule that blocks sterically the myosin binding sites on the actin monomers in the absence of calcium and moves toward the center of the actin groove in the presence of calcium, thus making these sites available to react with myosin. The movement of tropomyosin is regulated by troponin found in an equimolar ratio with tropomyosin on the thin filaments associating with tropomyosin and actin every 385 Å. Troponin in turn is controlled by calcium that binds to troponin C, one of its three subunits. The conformational changes induced by calcium on troponin C are relayed via troponin T and troponin I to actin and tropomyosin resulting in the movement of tropomyosin.

Using a variety of combinations of proteins, the following has been found. Myosin control does not require the presence of tropomyosin. A particular light chain is the only regulatory component so far identified in myosin-linked regulation. The components of myosin control interact only with myosin, whereas the components of actin control interact with actin and tropomyosin. The regulatory light chains do not replace troponin components in actin-linked regulation; neither do any of the troponin components replace the regulatory light chains in myosin-linked regulation. In contrast to troponin C, the isolated regulatory light chains do not bind calcium, although the amino acid sequences indicate homologies between the regulatory and other light chains as well as troponin C (Collins, 1974; Weeds and McLachlan, 1974; Tufty and Kretsinger, 1975; Jakes *et al.*, 1976). The components of the two regulatory systems do not hybridize. They act independently of each other although their effects may be additive.

The different requirements of the two regulatory systems form the basis of the functional tests used to probe for the presence of a particular regulation in different muscles. The *in vitro* test for contractile activity is the actin-activated ATPase activity of myosin. This is a cyclic reaction sequence that involves the making of an actomyosin cross-link, a conformational change, a myosin intermediate, and the breaking of the cross-link. The calcium dependence of the actin-activated ATPase activity is the measure of regulation. The action of calcium is very specific; it is bound selectively from solutions containing 10^{-7} *M* $CaCl_2$ and 10^{-3}–10^{-2} *M* $MgCl_2$.

Pure rabbit actin and rabbit myosin lack control functions and interact readily even in the absence of calcium. These purified preparations can be used to probe for the presence of control in actins and myosins of different animals by making hybrid actomyosins. Troponin and tropomyosin act by reducing the affinity of actin to myosin. The presence of regulated actin does not interfere with the reactions of myosin with pure actin. Thus, one may probe for the presence of myosin control in an unknown actomyosin preparation by flooding it with pure rabbit actin in the absence of calcium. If the myosin is regulated, the ATPase activity remains unchanged by excess pure actin. If the myosin in question is unregulated, pure actin activates. One may probe for the presence of actin control by using purified rabbit myosin in the absence of calcium. A survey of the animal kingdom, using these functional tests on crude actomyosins and examining the properties of isolated myosins and thin filaments, indicates that both actin control and myosin control are found widely (Lehman and Szent-Györgyi, 1975).

Actin control was reported also in the slime mold, *Physarum* (Nachmias and Asch, 1974; Kato and Tonomura, 1975). Many animals have muscles with dual regulation, e.g., muscles of nematodes, priapulids, some sipunculids, annelids, most crustaceans, chelicerates and insects. A single myosin control was observed in the muscles of nemertine worms, brachiopods, echiuroids, echinoderms, and molluscs. In most of these muscles actin control is missing since they are deficient in troponin. In vertebrate striated muscle and in the fast muscle of the crustacean decapods, only actin control could be demonstrated (Lehman and Szent-Györgyi, 1975). Myosin control was found, however, in smooth muscle of vertebrates (Bremel, 1974; Mrwa and Rüegg, 1975) and in the slow muscle of decapods (Lehman and Szent-Györgyi, 1975).

PROPERTIES OF MYOSIN CONTROL

A mole of the regulatory light chain is detached from scallop myosin by reducing the concentration of divalent cations with EDTA. In the presence of magnesium ions the regulatory light chain readily recombines with the stripped scallop myosin (Szent-Györgyi *et al.*, 1973). The actin-activated ATPase activity of a stripped scallop myosin preparation is no longer calcium-dependent, being high both in the presence and in the absence of calcium. In such preparations the number of calcium binding sites is reduced from about two to one. When the light chain is recombined with the desensitized myosin, both calcium sensitivity and calcium binding are fully regained. Thus calcium control depends on the presence of the full complement of the regulatory light chains.

The subunit structure of scallop myosin is very similar to rabbit myosin (cf. Lowey *et al.*, 1969). The scallop myosin also consists of two heavy chains with a chain weight of about 190,000 and four light chains of two kinds with a chain weight of about 18,000 (Szent-Györgyi *et al.*, 1973; Kendrick-Jones *et al.*, 1976). The heavy chains have globular regions at one end of the molecule from which emerges a rod region about 1500 Å long, forming a fully coiled α-helix consisting of two chains. The molecule can be fragmented by trypsin into heavy and light meromyosins; the heavy meromyosin is a soluble portion of the myosin molecule representing the two globular regions connected by a short tail. Papain snips the molecules at the region where the globular ends join the helical rod region. The S-1 fragment obtained represents single globular regions of the myosin molecules. Both heavy meromyosin and the S-1 subfragment have ATPase activity and contain the light chains. The S-1 obtained from scallop also binds calcium provided the fragment retains the regulatory light chains.

There are two regulatory light chains in each scallop myosin molecule. However, only one of these is removed when the concentration of divalent cations is reduced by EDTA treatment; nevertheless, calcium sensitivity, as already mentioned, is completely lost. The presence of both regulatory light chains is needed to maintain the "off" state of myosin. The removal of one light chain does not yield a mixture of desensitized and regulated myosin molecules. The second regulatory light chain left behind after EDTA treatment can be selectively obtained by treatment with dithiodinitrobenzoic acid; such a treatment, however, irreversibly inactivates myosin. The two fractions of regulatory light chains yield identical tryptic peptide maps. Both of the fractions recombine with desensitized myosin preparations and restore regulation with the same affinity. Thus, there appears to be an obligatory cooperativity between the two myosin heads, and both regulatory light chains are needed to prevent the combination of the myosin with actin. The mechanism of this cooperativity is unclear. The findings indicate that the myosin heads interact, possibly through their regulatory light chains, and that this interaction is required to place the light chains in an inhibitory position at the actin binding sites of the heavy chains. Alternatively, the effect of the light chains may be more indirect; they may simply fix the relative position of the two heads so that they cannot combine with actin.

The requirement for a cooperation between the myosin heads during regulation gains further support from studies on the soluble myosin fragments. Heavy meromyosin preparations are calcium-sensitive, whereas the single-headed S-1 subfragments lack calcium sensitivity although they bind calcium and retain the regulatory light chains. S-1 preparations lack calcium sensitivity even when combined with regulatory light chains obtained from intact myosin molecules.

There appears to be a simple proportionality between calcium binding and light chain content. Desensitized myosin preparations that lack a mole of light chain contain half of their original calcium binding sites. Calcium binding is regained when the myosin is recombined with its light chain. Isolated regulatory light chains do not bind calcium. The lack of high affinity calcium binding sites on the light chains may be due to a change in the conformation of the light chains once detached from the heavy chains. It is not excluded, however, that the heavy chains have a more direct role in calcium binding.

Both magnesium and calcium ions play an important role in the interactions of light and heavy chains. The detachment and the reassociation of the light chains depend on free magnesium ion concentrations. It is likely that magnesium participates in one of the multiple attachment sites between light and heavy chains. One may assume that the magnesium-mediated link is invariant and ensures that the regulatory light chain remains attached to the heavy chain during activity. There must also be a calcium-mediated link that is a variable one and depends on the presence or absence of calcium. The addition of calcium results in the uncovering of myosin sites. It is this calcium-dependent interaction that would be responsible for the "on" and "off" state of myosin. In addition to these links, there are other hydrophobic and electrostatic interactions that may explain why regulatory light chains are not removed by EDTA treatment in a mole to mole ratio from myosins other than from scallop. Although the presence of regulatory light chains are the features of many myosins, desensitized preparations have only been obtained from scallop at present. The regulatory properties of the light chains of different myosins can be shown at present only by their ability to hybridize with desensitized scallop myofibrils and restore the calcium requirement of the ATPase activity.

Comparative studies of the various regulatory light chains yield some clues on how they function and help to elucidate physiological events. Different molluscan myosins all contain regulatory light chains that hybridize with desensitized scallop myosin and restore their calcium regulation. Surprisingly, light chains with regulatory functions can also be obtained from rabbit myosin (Kendrick-

Jones, 1974). These light chains hybridize with desensitized scallop myofibrils and restore their calcium sensitivity. This finding is unexpected since functional tests of rabbit skeletal actomyosins do not indicate the presence of myosin control. Subsequent studies have shown that rabbit myosin is not unique in this respect.

Regulatory light chains are present in the myosins of other striated and cardiac vertebrate muscles and of fast decapod muscles, although these muscles lack myosin control (Kendrick-Jones *et al.*, 1976).

These regulatory light chains obtained from different molluscan muscles, vertebrate skeletal muscles, chicken gizzard, beef cardiac and lobster tail muscles all combine with desensitized scallop myofibrils with great affinity in a ratio of one mole of light chain to a mole of myosin. These light chains differ in amino acid composition and in tryptic peptide maps. Nevertheless, the region of the regulatory light chain that combines with the heavy chains must be homologous in all the light chains tested. This is the site that is likely to involve magnesium ions. One surmises that the complementary binding site of the heavy chain may also retain a degree of homology in different myosins. Interestingly, this binding site is not identical with the actin binding site since the regulatory light chain is readily removed from myofibrils and from actomyosins. Also, light chains from various sources readily recombine with desensitized scallop myofibrils. The presence of actin does not significantly hinder the removal and re-uptake of the regulatory light chain.

The regulatory light chains obtained from various myosins can be assigned into two different classes on the basis of their effect on calcium binding and on their ability to restore calcium sensitivity when combined with scallop myosin. The molluscan types of regulatory light chains behave like the regulatory light chains of scallop; they resensitize equally well myosin and myofibril preparations and restore calcium binding fully. This class contains the regulatory light chains from molluscs, such as the surf clam, quahog, squid, and, interestingly, the regulatory light chains of chicken gizzard. The light chains obtained from vertebrate skeletal muscles, from beef cardiac muscles and from the fast tail muscles of lobster do not increase the calcium binding of desensitized myofibrils that remains at the lowered level corresponding to about one mole of calcium for a myosin molecule present in the muscle. There are also more stringent requirements for restoring calcium control with this class of light chains. Actin must be present during recombination for the preparation to become calcium-sensitive. Pure desensitized myosin also binds these light chains; however, calcium sensitivity is not restored. Evidently the presence of actin positions the heavy chains in a particularly favorable position so that recombination yields a calcium-sensitive product.

It is somewhat paradoxical that calcium sensitivity is restored by a preparation that does not bind calcium with high affinity on its own or when it is combined with the heavy chain. However, desensitized scallop myofibrils retain a mole of scallop regulatory light chain and the hybrids contain an equimolar amount of scallop regulatory light chain in addition to the one introduced. Interaction with calcium may be mediated by the residual scallop regulatory light chain that contributes to the remaining high affinity calcium binding site, while the light chain used for hybridization may play a "supporting" role indicated by the obligatory cooperativity and may not respond directly to calcium. This situation may be rather delicate and that is the reason for the requirement for actin during light chain–heavy chain recombination.

This explanation may be overly speculative. Nevertheless, the interesting fact remains that the regulatory light chains obtained from myosins of muscles that lack myosin control differ from the regulatory light chains of muscles having a clearly established myosin-linked regulation. The regulatory light chain from chicken gizzard is a rather striking example. According to this interpretation, the light chains found in vertebrate skeletal, cardiac and fast lobster myosins are not on their own capable of conferring calcium control. It is noteworthy that regulatory light chains that appear to be not fully "competent" are obtained from myosins of muscles that do not show myosin control in the different functional tests performed *in vitro.*

The possibility of the presence of an *in vivo* myosin control in vertebrate skeletal muscles is worth entertaining since several different studies indicate an effect of calcium on vertebrate myosins. Thus calcium in concentrations of about 10 times higher than required for triggering contraction can alter the viscosity and the sedimentation of the thick filaments (Morimoto and Harrington, 1974), bind to myosin inhibiting somewhat the actin-activated ATPase activity (Bremel and Weber, 1975), and reduce the affinity of myosin to actin provided the regulatory light chain remains intact (Margossian *et al.,* 1975). Importantly, x-ray evidence indicates structural changes in the thick filaments on activation in muscles that are stretched beyond overlap, and thus are unable to form cross-bridges (Haselgrove, 1975). Nevertheless, pure myosin and actin of these muscles do not form an actomyosin that has a calcium-dependent ATPase activity. It appears that one of the reasons for the lack of myosin control in these muscles is due to their light chains and that the properties of light chains may be directly related to physiological performance.

The structural aspects of myosin control are not known. Some important features of the control mechanism are slowly emerging. Although each of the two heads of myosin is associated with a regulatory light chain, both light chains and both myosin heads are necessary to maintain the "off" state of myosin. There are multiple interactions between the regulatory light chains and the heavy chains. Some of these require magnesium ions; some of these depend on calcium ions. The magnesium-dependent site involves a region on the light chain that is a recurrent feature of different regulatory light chains and a site on the heavy chain that is not at the actin binding site. Little is known about where the calcium-dependent interactions take place. Regulatory light chains may interfere with actin combination by directly interacting with the actin binding region of the heavy chains in the absence of calcium. However, the steric effect of the light chains may be more indirect and they may simply fix the two myosin heads relative to each other in a way that the binding regions are not available to actin combination.

It is interesting to recall Prof's proposal in 1946 that contraction and the ATPase activity of myosin are regulated by small molecular weight proteins that are heat and acid resistant. He named these "protins." This was at a time when protein structure was still a mystery; subunit organization, regulation by subunits, or allosteric mechanisms were not known. It may turn out to be that the story of regulation of myosin by light chains is an additional example of the remarkable flashes of insight that are so characteristic of Prof's thinking.

REFERENCES

Bremel, R. D. (1974). *Nature (London)* **252,** 405–407.

Bremel, R. D., and Weber, A. (1975). *Biochim. Biophys. Acta* **376,** 366–374.

Collins, J. H. (1974). *Biochem. Biophys. Res. Commun.* **58,** 301–308.

Ebashi, S., and Endo, M. (1968). *Prog. Biophys. Mol. Biol.* **18,** 123–183.

Haselgrove, J. C. (1975). *J. Mol. Biol.* **92,** 113–143.

Jakes, R., Northrop, F., and Kendrick-Jones, J. (1976). *FEBS Lett.* **70,** 229–234.

Kato, T., and Tonomura, Y. (1975). *J. Biochem. (Tokyo),* **78,** 583–588.

Kendrick-Jones, J. (1974). *Nature (London)* **249,** 631–623.

Kendrick-Jones, J., Lehman, W., and Szent-Györgyi, A. G. (1970). *J. Mol. Biol.* **54,** 313–326.

Kendrick-Jones, J., Szentkiralyi, E. M., and Szent-Györgyi, A. G. (1972). *Cold Spring Habor Symp. Quant. Biol.* **37,** 47–54.

Kendrick-Jones, J., Szentkiralyi, E. M., and Szent-Györgyi, A. G. (1976). *J. Mol. Biol.* **104,** 747–775.

Lehman, W., Kendrick-Jones, J., and Szent-Györgyi, A. G. (1972). *Cold Spring Harbor Symp. Quant. Biol.* **37,** 319–330.

Lehman, W., and Szent-Györgyi, A. G. (1975). *J. Gen. Physiol.* **66,** 1–30.

Lowey, S., Slayter, H. S., Weeds, A. G., and Baker, H. (1969). *J. Mol. Biol.* **42,** 1–29.

Margossian, S. S., Lowey, S., and Barshop, B. (1975). *Nature (London)* **258,** 163–166.

Morimoto, K., and Harrington, W. F. (1974). *J. Mol. Biol.* **83,** 83–97.

Mrwa, V., and Rüegg, J. C. (1975). *FEBS Lett.* **60,** 81–84.

Nachmias, V., and Asch, A. (1974). *Biochem. Biophys. Res. Commun.* **60,** 656–663.

Szent-Györgyi, A. G. (1975). *Biophys. J.* **15,** 707–723.

Szent-Györgyi, A. G., Szentkiralyi, E. M., and Kendrick-Jones, J. (1973). *J. Mol. Biol.* **74,** 179–203.

Tufty, R. M., and Kretsinger, R. H. (1975). *Science* **187,** 167–169.

Weber, A., and Murray, J. M. (1973). *Physiol. Rev.* **53**, 612–673.
Weeds, A. G., and McLachlan, A. D. (1974). *Nature (London)* **252**, 646–649.

ACKNOWLEDGMENT

The experiments on which this article is based are supported by a grant from the United States Public Health Service AM 15963.

11

From Uterine Actomyosin to Parturition

ARPAD I. CSAPO

Today's question is not
Whether there is life after death
But whether there is life after birth.
Albert Szent-Györgyi

He warned us, long before it had been broadly recognized, that uncontrolled population growth accompanied by prematurity and perinatal complications severely threaten our civilization by undermining the quality of life and man. He foresaw that the key to these problems lies buried in the complex regulatory mechanism of pregnancy accessible only through basic experiments in animal models. Thus in 1946, when I asked his permission to undertake the isolation and quantitation of uterine actomyosin, he not only approved my request but shared with me his vast experience and the sanctuary of his laboratory with its precious tools in war-torn Hungary: a porcelain mortar, a viscosimeter, quartz sand and ATP. Had he, a doctor at heart, foreseen that the achievements of basic muscle research would first be utilized in postconceptional therapy? He never claimed that he did, yet this article is a testament to his impact on reproductive medicine (for references, see Csapo, 1948, 1954, 1956, 1959, 1970a, b, c, 1971, 1973, 1975, 1976).

A STEP TOWARD QUANTITATIVE UTERINE PHYSIOLOGY

> When after persistent prayers Heaven did open for Him,
> Allowing a glance inside, what the Holy Man saw was another Heaven.
> *Albert Szent-Györgyi*

The profound human experience of attending 2000 deliveries provided me with the skill and a license to practice obstetrics, but offered no insight into a central problem of reproduction: the mystery of how pregnancy is maintained and terminated. Quantitative examination of uterine function first exposed and defined this persistent problem, beginning with the isolation of uterine actomyosin and the *in vitro* reconstruction of myometrial function as seen in the ATP–induced contraction of the actomyosin thread. The fibrous protein complex, actomyosin, turned out to be the generator of the force of parturition.

The isolation and partial characterization of uterine actomyosin prompted questions demanding further exploration of the contractile system of the uterus. It became obligatory to examine estradiol- and stretch-induced actomyosin synthesis, actomyosin-determined working capacity, bioenergetics and fine structure. Further experiments led to the demonstration in the myometrium of the classic relationships between (a) the strength and duration of electric stimulation (in triggering a threshold response); (b) length and tension; and (c) load and shortening velocity. These studies prepared the way to *quantitative uterine physiology* and to the functional definitions of pregnancy maintenance and parturition, the subjects of this review. Inevitably the questions prompted answers and the answers in turn raised new questions. Indeed, the prayers which opened the gates of Heaven were rewarded by leading to another Heaven.

The reconstruction of the molecular sequence of the contraction-relaxation process was made possible by elucidation of the role of calcium and the regulatory proteins, tropomyosin and troponin. For these studies, the structurally more precise cross-striated muscle offered a far better experimental medium than did the uterus. This exciting phase of muscle biochemistry will be described in this volume by its eminent leaders.

In contrast, the uterus promised to be an excellent choice for studying the regulation of muscle function by various physiological factors including hormones.

THE PROBLEM

> When the child is grown big and the mother cannot continue
> To provide him with enough nourishment; he becomes agitated,
> Breaks through the membranes and incontinently passes out
> Into the external world free from any bonds.
> *Hippocrates*

The problem of how pregnancy is maintained and terminated is as old as man's ability to ask the question. No final solution has been found as yet, but steps of simplification resolved the complexity and exposed parts of the problem. It had been recognized early that while the maintenance of pregnancy and parturition are strikingly different biological processes, they have the same controlling mechanism, for the termination of pregnancy is a failure of its maintenance and vice versa. A point of strategic significance has been the further recognition that the functional state of the myometrium, and not the circulating hormones per se, maintains and terminates pregnancy.

Impressed by the dramatic outburst of uterine pains, signalling the onset of clinical labor, investigators in the 1950's were searching for a circulating hormone which abruptly makes its appearance at term and triggers parturition in an "inert uterus." In spite of repeated failures at attempts to obtain experimental support for this "oxytocin theory" (Caldeyro-Barcia, 1964), it has survived to the present day, except that oxytocin has been conceptually replaced by prostaglandins. However, the concept of an "inert" pregnant uterus stands unsupported, since not even massive oxytocic stimulation can promptly evoke high level cyclic intrauterine pressure and guarantee the safe induction of labor, until the reactivity of the myometrium to stimulants is increased. Evidently, a change in the reactivity of the uterus, rather than the changing concentration of a stimulant, is the critical regulatory step in the preparation for parturition. This concept remained undocumented, however, until it was demonstrated that the increase in uterine "sensitivity" results from a decrease in myometrial threshold, a consequence of effective progesterone withdrawal (Pw). It also became apparent that high level uterine wall tension (wT), rather than circulating hormones, terminates gestation and conversely tension suppression maintains pregnancy. Nevertheless both maternal and fetal hormones were carefully considered in terms of their link (direct or indirect) to changes in uterine function (Csapo and Wood, 1968).

In this review, discussions will be limited to those regulatory factors which directly affect uterine activity. This limitation stresses the point that unless a proposed controlling factor of the pregnant uterus is tied definitively to myometrial function, the extreme complexity of the feto-maternal endocrine system will continue to obstruct the resolution of its regulatory significance.

A "MODEL" OF UTERINE FUNCTION

> If one key opens several locks
> Those locks cannot be very different.
> *Albert Szent-Györgyi*

The "key" he offered was the interreaction between actomyosin, ATP and ions. The "locks" were the unknown mechanisms through which the various cross-striated and smooth muscles perform their contractile functions. The more we learned about them, the more we appreciated his penetrating foresight, predicting what is evident today, that the basic principles of contractile function are obeyed by all muscles.

The Single Myometrial Cell

The myometrium is a community of billions of spindle-shaped, smooth muscle cells, only ~5–10 μm wide and ~200 μm long. They are held together in an intricate network of fiber bundles, embedded in a connective tissue framework. Each myometrial cell is a contractile system surrounded by an excitable membrane.

The intracellular microelectrode technique (Kuriyama and Csapo, 1961) (Fig.

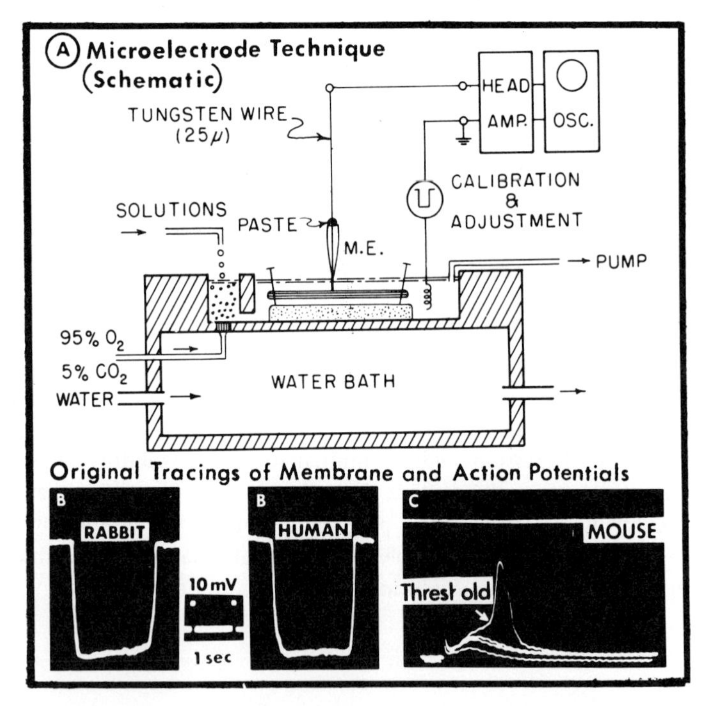

Fig. 1. The membrane potential and threshold of the myometrial cell. (A) Schematic illustration of the "flexible" intracellular microelectrode technique (as used by Csapo, Goto, Kuriyama and Abe). (B) Note the ~50 mV resting potential, recorded in the rabbit and human myometrium when the microelectrode penetrates the cell. (C) Note also that the myometrial cell generates no action potential until depolarization (with electric current) reaches a critical "threshold" value (for references, see Csapo, 1971). This microelectrode technique was used in experiments in Figs. 2–5.

1A) complemented by the "sucrose-gap" method (Marshall and Csapo, 1961) demonstrated that the membrane potential of the resting myometrial cell is sustained, as in other muscles, in a polarized state by the electrochemical gradients (Fig. 1B). However, in comparison with cross-striated muscle, the membrane potential was found to be low, close to the critical threshold value at which rapid depolarization triggers an action potential (Fig. 1C) which is propagated to neighboring regions. This limited membrane potential "excess" (over the threshold value) is a characteristic feature of involuntary smooth muscles, capable of spontaneous rhythmic activity (for references, see Marshall, 1973).

One action potential only triggers a twitch (Fig. 2A); a series of spikes of low repetition frequency induces an incomplete tetanus (Fig. 2B); and only long trains of high discharge frequency provoke a complete tetanus (Fig. 2C). The short-lived "active state" of the uterine contractile system has to be sustained for several seconds to achieve maximum activity.

Evidently, a control over the discharge frequency and the duration of the train permits the adjustment of uterine function to any desired level. Fig. 3 illustrates varying discharge frequencies and duration in response to stretch and oxytocin.

The pacemaker cell can be identified by the rhythmic generation of slow depolarization waves, which precede the spike discharge (Fig. 4A). The spike discharge either propagates rapidly to distant uterine regions (Fig. 4B) or its propagation remains restricted to its site of origin. In both cases, the activated region generates wT. But of course, the physiological consequences of "local" (nonpropagating) and "propagating" electric activity are different. As will be discussed presently under Progesterone (see below), regulating mechanisms determine whether the electric activity is "local" or "propagated" in the pregnant uterus, depending on the physiological demand. A significant achievement has

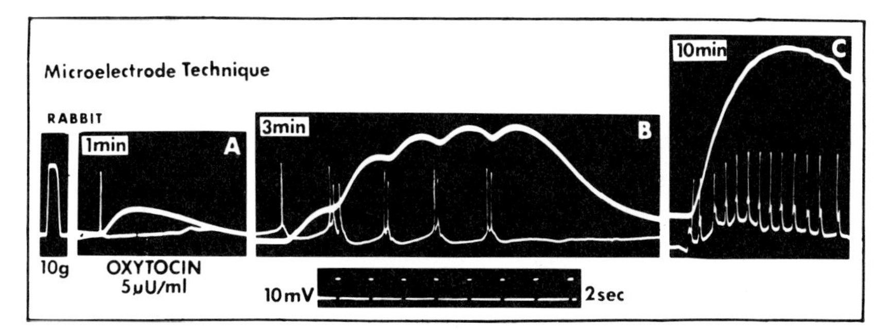

Fig. 2. The relationship between electric and mechanical activity. Parturient rabbit uterus *in vitro*. The intimate relationship (as in cross-striated muscle) is illustrated between the frequency of the action potentials and wall tension (wT). Note the twitch (A), incomplete (B), and complete tetanus (C) (Kuriyama and Csapo, 1961).

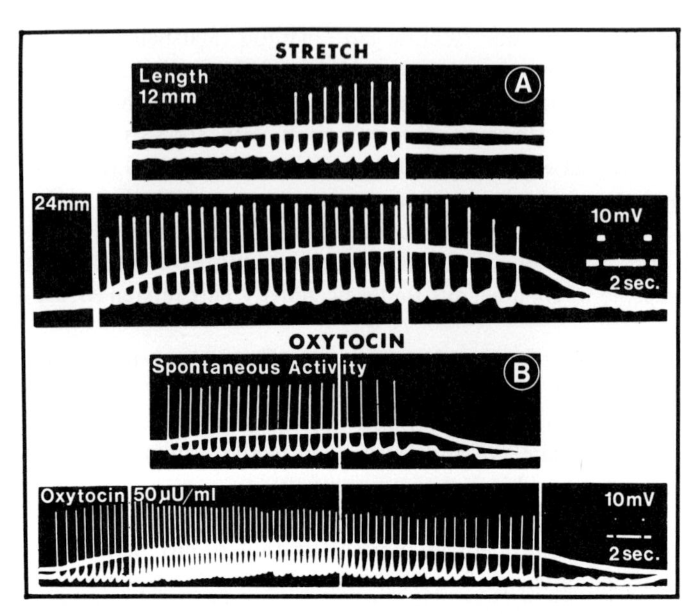

Fig. 3. The effects of stretch and oxytocin on the train discharge and tension. Parturient rat uterus *in vitro.* Note (A) that stretching the excised strip from 12 to 24 mm, or (B) exposing it to 50 μU/ml oxytocin, increases the frequency of spikes, the duration of the train discharge and isometric tension (Kuriyama and Csapo, 1961).

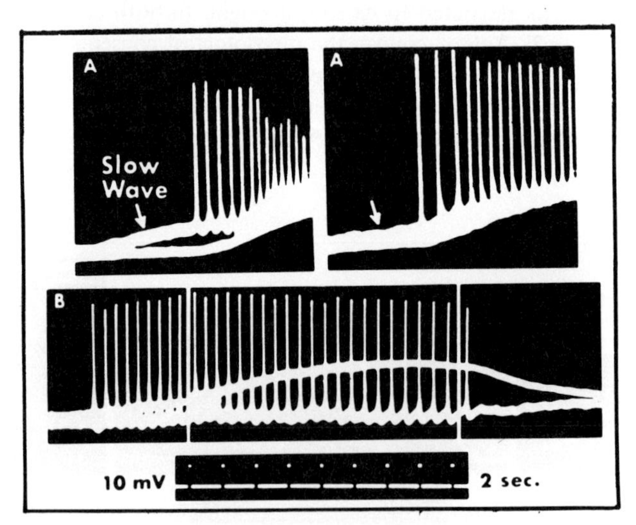

Fig. 4. "Pacemaker" and "propagated" action potentials. Parturient rat uterus. Note the "slow waves" (arrows) of depolarization preceding the train discharges of the pacemaker cells (A) and the propagated potentials (without slow waves) of the nonpacemaker cells (B) (Kuriyama and Csapo, 1961).

been the recent demonstration that these controlling actions can be therapeutically modified in a predictable manner. However, clinical implementation is yet to be accomplished.

The Activation of the Myometrial Cell

Excitation-contraction coupling will be discussed by other authors in this volume. It is generally accepted that membrane excitation liberates calcium ions which bind to troponin-tropomyosin, or myosin light chains, resulting in myosin cross-bridge interaction with actin. This calcium-activated "switch" between excitation and contraction is a central problem of uterine physiology and regulatory biology.

The influence of calcium on contraction of the isolated uterus was demonstrated (Csapo, 1959) in the following way:

(a) A Ca-free Krebs' solution reversibly suspends (Fig. 5A) normal excitability of the uterus elicited by a 1.5 V/cm, ac stimulus or by 100 mU/ml oxytocin (Fig. 5B).

(b) Partial replacement of calcium results in the regaining of normal electric excitability (Fig. 5A) and pharmacological reactivity (Fig. 5B) as a function of the concentration of calcium.

(c) While a Ca-deficient uterus is inexcitable by a 1.5 V/cm stimulus and is refractory to 100 mU/ml oxytocin, it retains near-maximum contractility when activated by a ~10 V/cm electric field, which presumably liberates calcium from intracellular sites (Fig. 5B).

(d) This step, of triggering the actomyosin-ATP interreaction by calcium, can be effectively induced by direct stimulation of the myoplasm with ~10 V/cm field, even when the membrane activity of the uterus is abolished through depolarization with excess potassium (rather than through Ca-deficiency).

(e) The state of "Ca-deficiency" and "inexcitability" is reached slowly in the pregnant (Fig. 5A) and rapidly in the postpartum uterus (Fig. 5B), while excitability is restored (in normal Krebs' solution) rapidly in the pregnant and slowly in the postpartum uterus.

These findings led to the conclusion that "Muscle can be 'uncoupled,' that is rendered inexcitable temporarily and reversibly at a membrane potential close to its normal value, by the displacement of a fraction of its labile calcium, suggesting that depolarization and myoplasmic activity are not linked directly, but are connected by at least one intermediate step in which calcium is involved" (Csapo, 1959).

The additional demonstration (Fig. 5C) that Ca deficiency abolishes the train discharge (Kuriyama and Csapo, 1961) substantiated the premise (Csapo, 1956) that (a) under progesterone dominance, the "activator" calcium is strongly

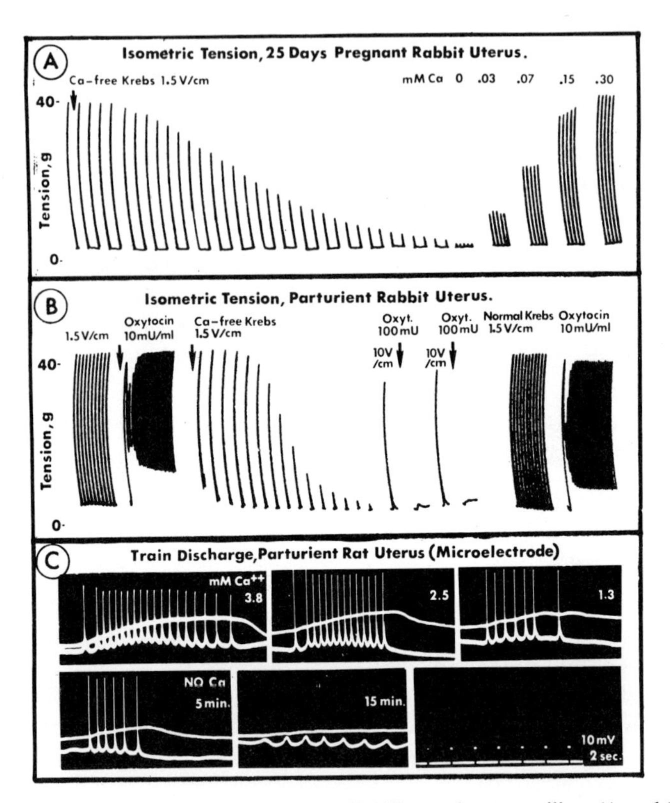

Fig. 5. The effect of Ca-deficiency on excitability and contractility. (A and B) Rabbit uterus; (C) rat uterus. In (A) note the gradual decline in wT of the 25-day pregnant uterus when exposed to Ca-free Krebs' solution, a failure of the 1.5 V/cm stimulus in fully activating the muscle and the recovery of excitability as a function of the replacement of calcium. In (B) note the more rapid loss of excitability in Ca-free Krebs' of the parturient uterus, the refractoriness to oxytocin in Ca-deficiency, but the ability to develop near-maximal wT when exposed to the strong 10 V/cm electric field. Note also that this state of Ca-deficiency is fully reversible and that the uterus regains its excitability and pharmacological reactivity in normal Krebs'. In (C) note the suspension of the train discharge when the rat uterus is rendered Ca-deficient. In comparing the properties of the 25-day pregnant and parturient uterus, note that the respective progesterone concentrations of these muscles are 14±1 and 4±1 ng/gm tissue (Csapo, 1971).

bound, creating a "bottleneck" in excitation; and (b) by curtailing the movement of membrane-bound activator calcium, progesterone increases threshold and thus suppresses excitability, pharmacological reactivity and propagation.

Recently (Csapo 1976; Torok and Csapo, 1976) these conclusions were supported by more direct evidence. Measurements of calcium in the inexcitable Ca-deficient uterus showed only a small loss, corresponding to extracellular calcium. Efflux studies with ^{45}Ca revealed that progesterone drastically decreases efflux, substantiating the premise of a strong binding of the activator calcium. Apparently, increased threshold and suppressed excitability during progesterone dominance are due to a decrease in the mobility of the activator calcium. These quantitative studies of the myometrium, when complemented with reliable prostaglandin and progesterone assays, exposed significant features of the molecular mechanism of uterine function and regulation, as will be discussed presently.

Uterine Function *in Situ*

The microelectrode method for studying single cells had to be complemented by other techniques in order to study the regulation of the whole uterus. Hence "suction" (Csapo and Takeda, 1965; Talo and Csapo, 1970) and "implanted" macroelectrode techniques (de Paiva and Csapo, 1973) were applied successfully to the excised uterus *in vitro* as well as to the intact organ *in situ,* and also in combination with an extraovular microballoon. The combined techniques described the topography of the electric activity of the whole myometrium together with the intrauterine pressure (IUP) (Fig. 6). Macroelectrodes have even been used to describe the electric activity of the human uterus *in situ,* providing reassurance that the relationship observed between train discharges and IUP in animal "models" is operative in patients (Csapo and Takeda, 1963). Apparently, the same "key" does open several "locks," as Szent-Györgyi predicted.

The contractile activity of billions of cells in the pregnant uterus is determined by the degree of electric "communication" between cells during the contraction time. To facilitate communication, the action potentials, generated by the pacemaker region, must rapidly sweep over to distant regions in order to trigger physiologically significant wall tension (wT). During normal pregnancy (Figs. 6 and 7, upper panels), irregular spikes are generated in different uterine regions, but these electric discharges remain confined to their point of origin. They do not propagate to neighboring regions and, therefore, uterine activity is asynchronic and IUP is reduced to low level oscillations (Fig. 6B and C).

This suppression of excitation is suspended at term. Trains of action potentials (generated at the pacemaker regions) readily propagate to distant regions, despite the shortness of myometrial cells which impedes rapid propagation by conduction delays between cells and by the dissipation of spikes at a certain distance from their origin. During parturition, propagation improves sufficiently to provoke synchronic activity and high level cyclic IUP (Fig. 6E).

Under physiological conditions, the incompressibility of the uterine contents prevents the shortening of the uterus. Only internal shortening and lengthening (between nearby regions) occurs, rendering myometrial function isometric. However, this mechanical "silence" does not imply that the uterus is inactive or relaxed. A certain number of cells irregularly discharge electric activity locally and asynchronously, not only during normal gestation but even during meno-

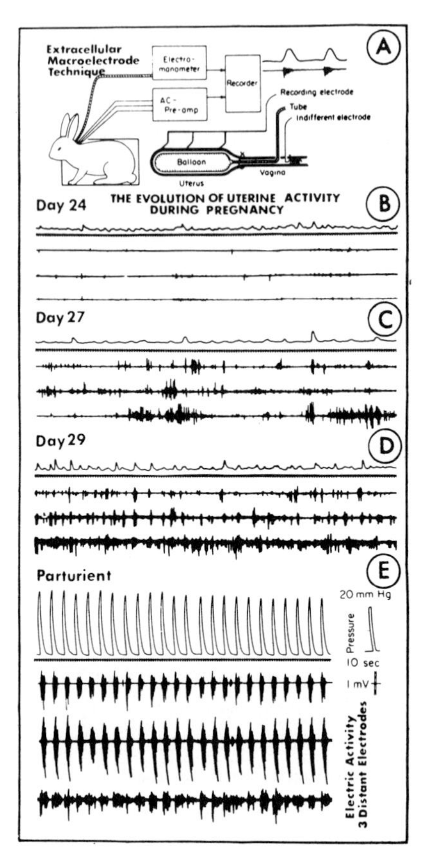

Fig. 6. The evolution of electric activity and intrauterine pressure *in situ.* Nonanesthetized rabbit. The combined implanted macroelectrode and microballoon techniques (as used by Csapo and Takeda, 1965). (A) Schematic illustration of the technique. (B) Note the occasional, asynchronic, nonpropagating, electric discharges recorded at the ovarian, middle and cervical uterine regions with the macroelectrode assembly and the low level oscillation in IUP recorded with the microballoon on day 24 of pregnancy. (C and D) Note that as the uterine P levels decrease from 14±1 to 4±1 ng/gm with the approach of term, electric activity increases (days 27 and 29), but since it remains asynchronic and nonpropagating, it does not increase IUP. (E) Note also that only the parturient uterus has synchronic, propagating electric activity and high level cyclic IUP (Csapo and Takeda, 1965).

pause. The magnitude of contractile activity generated is controlled by the degree of propagation of the electric train discharges, because the passive lengthening of inactive regions reduces wT, generated by shortening of the active myometrial regions. Evidently, only propagating synchronic activity ensures the evolution of high level wT, and only advanced wT can overcome normal cervical resistance. This contention that cervical dilatation is caused by high level cyclic wT (rather than by a mysterious mechanism, directly acting upon the cervix) has been validated by hypertonic saline-induced abortions. The hard, closed, long

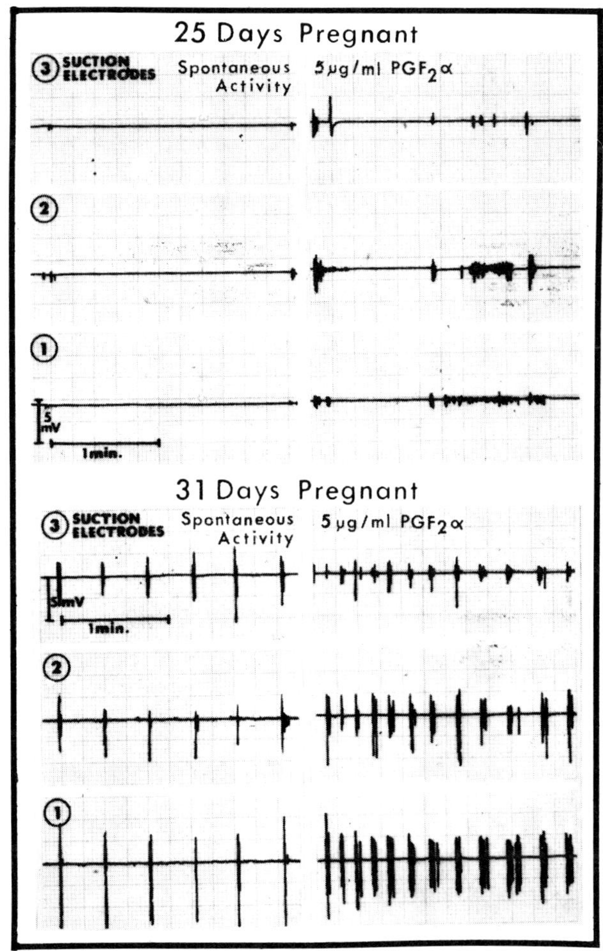

Fig. 7. The nonpropagating and propagating electric activity of the uterus *in vitro*. Rabbit uterus, "suction electrode" technique (as used by Talo and Csapo, 1970). Compare the occasional asynchronic discharges of the 25-day pregnant uterus (P = 14±1 ng/gm) with the synchronic propagating electric activity of the 31-day pregnant uterus (P = 5±1 ng/gm). Note that while $PGF_{2\alpha}$ increases the electric activity of both uteri, it does not provoke regular and synchronic train discharges in the 25-day pregnant uterus. Note also that the three suction electrodes are placed along the length of the muscle 15 mm apart. (Talo and Csapo, 1970).

and conic cervixes of mid-trimester patients readily dilated in 5 hr, once the IUP provoked by the saline-induced regulatory imbalance reached over 60 mm Hg (Csapo, 1970b).

In both animal models and human patients, investigators measure IUP rather than wT, although it is well established that the latter determines the fate of pregnancy. It is far easier to reliably measure IUP than wT, using the noninvasive and hydrodynamically sound extraovular microballoon technique *in situ* (Csapo, 1970a). However, wall tension can be derived from accurate pressure measurements according to the LaPlace equation (wT = IUP $\times$ $R/2$). The effect of R (uterine radius) on wT is marked since R increases significantly during pregnancy. For example, the wT of the parturient uterus at ~70 mm Hg IUP is comparable with that of the first trimester pregnant uterus when its IUP is over 200 mm Hg.

The "resting pressure" (RP), "active pressure" (AP), "rate of rise in pressure" (AP/Tr) and "frequency" (F), meaningfully define uterine activity as long as they are quantitated and treated separately (Csapo, 1970a). Expressions such as AP $\times$ F ("Montevideo Unit", MU) are misleading, since they do not distinguish 10 contraction cycles of 50 mm Hg/unit time (500 MU) from 50 contraction cycles of 10 mm Hg/unit time (500 MU).

The Intrinsic Myometrial Stimulant

Prostaglandin (PG) has recently been identified as a stimulant involved in the generation of rhythmic spontaneous contractions of the isolated uterus. Isometric recordings revealed (Fig. 8) that an inhibitor of PG synthesis (Naproxen-sodium) suspends spontaneous cyclic wT which is restored by PG but not by oxytocin (Csapo and Csapo, 1974).

The distention of the postpartum uterus *in situ* simultaneously increased PG synthesis and IUP while release from stretch suspended these effects (Fig. 9). Our unpublished observations show that *in situ* stretch rapidly increases the

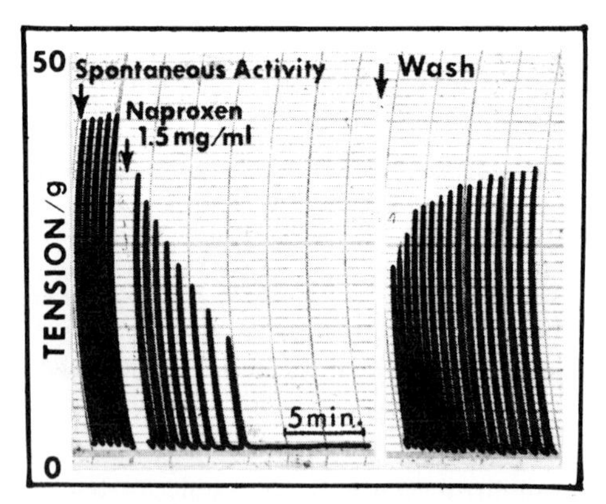

Fig. 8. The suppression of spontaneous activity by inhibition of PG synthesis. Postpartum rabbit uterus *in vitro,* isometric recording. Note the near-maximal wT, cyclically induced by the intrinsic uterine stimulant. Note also the suspension of wT by the PG synthesis inhibitor, Naproxen, and the reversibility of this effect when Naproxen is removed by washing the excised uterus (Csapo and Csapo, 1974).

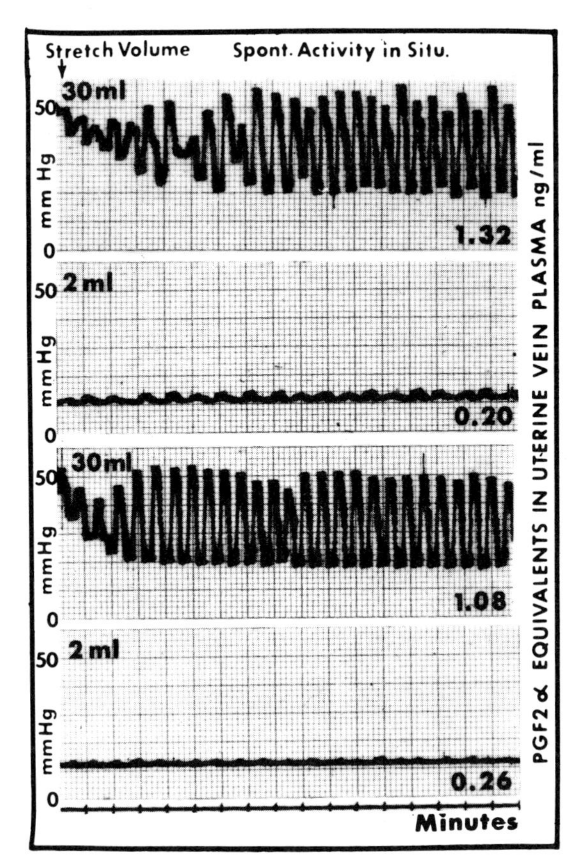

Fig. 9. The effect of *in situ* stretch on PG synthesis and IUP. Rabbit uterus postpartum. Note the simultaneous increase in uterine vein PG's and IUP, when the uterus is stretched by distending it with a rubber balloon containing 30 ml fluid. Note also that release from stretch by decreasing uterine volume to 2 ml, reduces both PG and IUP (Csapo, 1973).

concentrations of PGE (from 11±1 to 14±2 ng/gm) and PGF (from 23±3 to 61±13 ng/gm) in the postpartum uterus. Stretch might also be related to the significantly higher PG levels in the 25-day-pregnant (PGE = 19±2 ng/gm; PGF = 43±6 ng/gm) as compared to postpartum uteri (PGE = 10±2 ng/gm; PGF = 24±4 ng/gm).

Apparently, throughout pregnancy the PG concentration of the uterus is higher than the level required for the activation of the myometrium, and therefore the salient question regarding the mechanism of the initiation of labor is not what triggers uterine activity at term, but rather what protects the undeveloped fetus against abortion or premature delivery. The remainder of this article is focused on this question.

REGULATION

Economic Nature did not invent new mechanisms
For the resolution of the same problem.
Albert Szent-Györgyi

Progesterone

Corner and Allen (1930) discovered, using the rabbit as an experimental model, that pregnancy can be terminated by a mere progesterone withdrawal (Pw). Recognizing the fundamental nature of this discovery, George W. Corner asked me in 1949: "How does progesterone do it?" To answer him meaning-

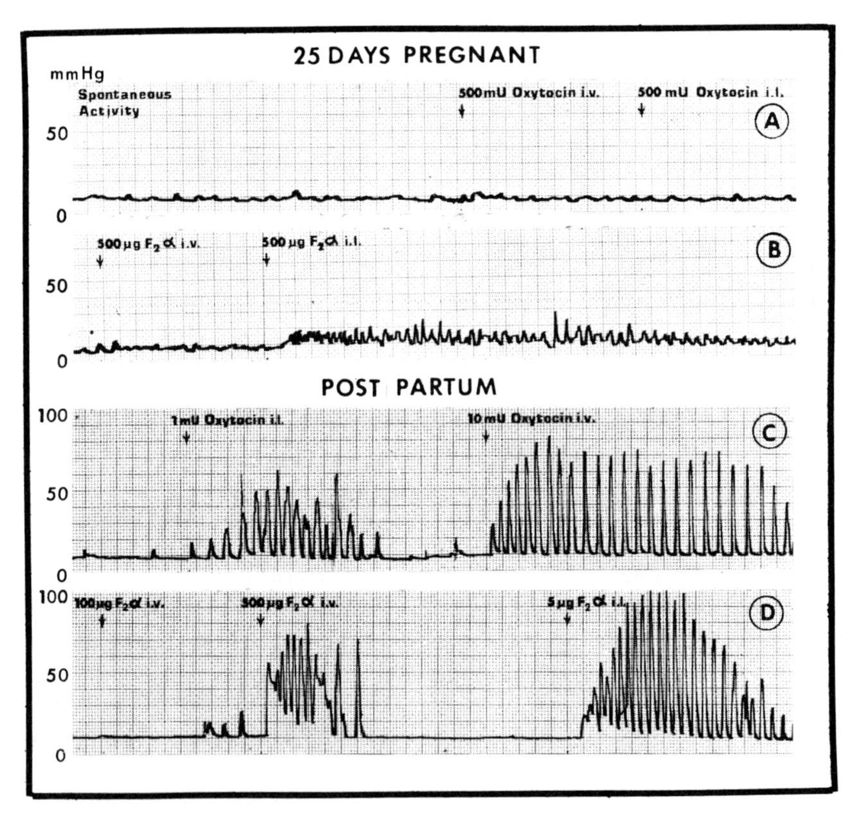

Fig. 10. The effects of oxytocin and $PGF_{2\alpha}$ on IUP at different stages of pregnancy. Note that at day 25 of pregnancy (when the uterine P level is 14±1 ng/ml) neither 500 mU oxytocin (A) nor 500 μg $PGF_{2\alpha}$ (B) provokes high level cyclic IUP, regardless of intravenous (iv) or intraluminal (il) administration. In contrast, postpartum (when the uterine P level is 3±1 ng/ml) as little as 1 mU oxytocin il or 10 mU iv (C) and 5 μg $PGF_{2\alpha}$ il or 500 μg iv (D) provoke high level cyclic IUP (Csapo, 1973).

fully, the mechanism of the P effect had to be examined quantitatively at different levels of organization of the myometrial cell, a study which is still in progress.

On day 25 of pregnancy when the concentration of P is 14±1 ng/gm, neither 500 mU oxytocin (Fig. 10A) nor 500 μg $PGF_{2\alpha}$ (Fig. 10B) provoked distinct cyclic IUP responses, regardless of whether administration was intravenous or intraluminal. The massive intraluminal dose of 500 μg $PGF_{2\alpha}$ only provoked an unphysiological and sustained contracture. In sharp contrast, the postpartum uterus (P = 3±1 ng/gm) responded with high level cyclic IUP to both oxytocin and PG. More specifically, distinct responses were provoked by only 10 mU intravenous (iv) and 1 mU intraluminal (il) administered oxytocin (Fig. 10C) and 500 μg iv and 5 μg il administered $PGF_{2\alpha}$ (Fig. 10D). This marked increase in the reactivity of the postpartum uterus is thus not primarily determined by endogenous and exogenous PG's but by the uterine concentration of the "suppressor", P.

PG promotes spontaneous uterine activity through the lowering of threshold (Csapo, 1976). However, while this effect (Fig. 11A2 and A3) was readily induced *in vitro* with $PGF_{2\alpha}$ at pg/ml concentrations in the postpartum rabbit uterus (P = 3±1 ng/gm), a quantitatively comparable effect could not be provoked with thousand-fold greater PG concentrations (ng/ml) in the 25-day-pregnant uterus (P = 14±1 ng/gm). This action of PG on threshold was reversible, since a single wash of the uterine strip restored pretreatment threshold and suspended spontaneous activity (Fig. 11A4).

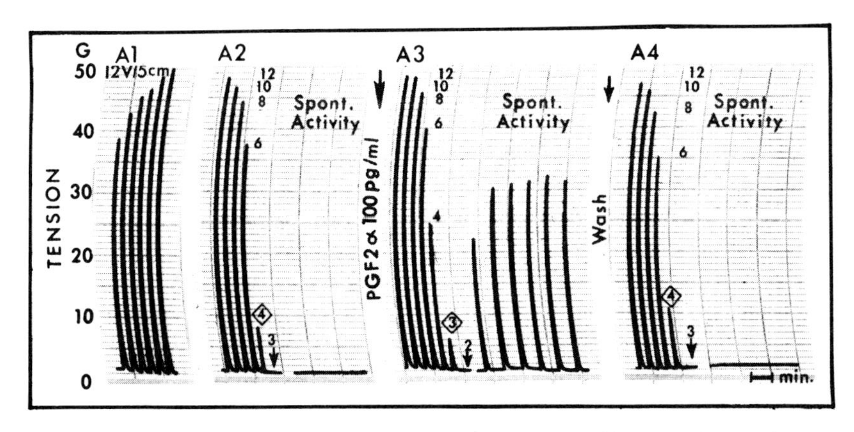

Fig. 11. The effect of $PGF_{2\alpha}$ on threshold. Postpartum rabbit uterus, isometric recording, *in vitro.* Note that when wT reaches steady state after repeated stimulation with 12 V/5 cm electric field (A1), the threshold of the uterus is 4 V/5 cm and there is no spontaneous activity immediately after the discontinuation of electric stimulation (A2). Treatment with 100 pg/ml $PGF_{2\alpha}$ lowers threshold to 3 V/5cm and triggers spontaneous wT (A3). A single wash restores threshold (4 V/5cm) and suspends spontaneous activity (A4) (Csapo, 1976).

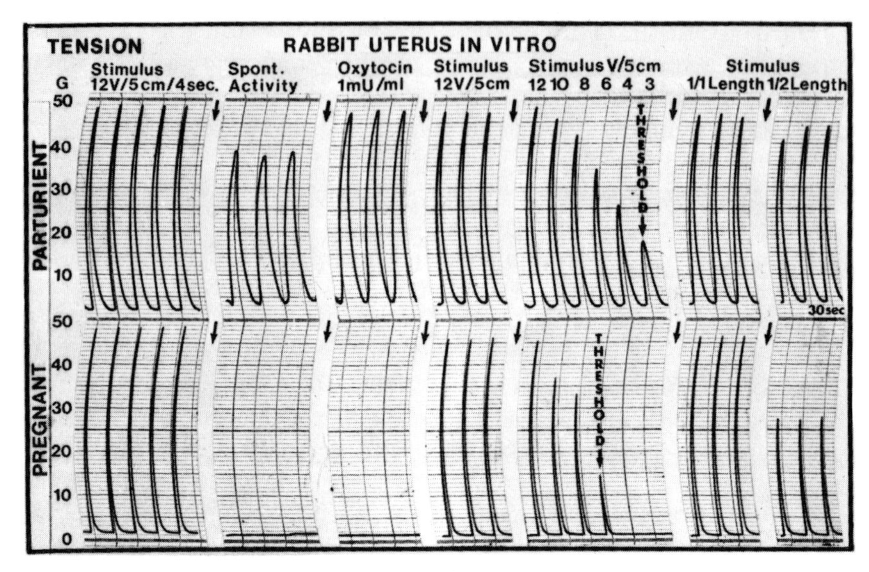

Fig. 12. The excitability and tension of the pregnant and parturient rabbit uteri. Isometric recording *in vitro.* Note that both (~0.5 gm) uterine strips develop maximal (~50 gm) wT when tetanized at all points simultaneously in an electric field of 12 V/5 cm. Yet only the parturient uterus develops (at some time after the discontinuation of electric stimulation) high level and regular cyclic wT and only this muscle responds to 1 mU/ml oxytocin. Note also that threshold of the pregnant uterus (P = 14±1 ng/gm) is high (6 V/5 cm), while that of the parturient uterus (P = 4±1 ng/gm) is low (3 V/5 cm) and that half-length stimulation only provokes half of the maximum wT in the pregnant uterus (for references, see Csapo, 1971).

On the other hand, the working capacities of the 25-day-pregnant and the postpartum uteri were similar when tetanized in an electric field of 12 V/5 cm (Fig. 12), in spite of the high P levels of the pregnant uterus and its refractoriness to endogenous and pharmacological stimulants. Evidently, P does not suppress contractility, but some regulatory process. When electric stimulation was discontinued, the postpartum, but not the 25-day-pregnant uterus, developed high and regular wT, irrespective of the presence or absence of oxytocin (Fig. 12).

More importantly, during pregnancy, spontaneous activity and reactivity were also suppressed *in situ* (Fig. 13A). This suppression, however, was readily suspended by ovariectomy (Fig. 13B) which reduced the plasma P levels from 9 to 2 ng/ml (Fig. 14). Since only the near-term uterus, both *in vitro* (Fig. 12) and *in situ* (Fig. 13) responded to physiological oxytocin stimulation, it had been concluded that high P levels sustain the pregnant uterus in a refractory state. Studies in the rat and the human extended these observations and were complemented by the current finding that not only are the *in situ* uterine PG levels higher during late pregnancy than postpartum, but they also remain higher after *in vitro* exposure. Evidently, suppressed spontaneous activity and reactivity of the pregnant uterus to stimulants (Figs. 12 and 13) cannot be explained by insufficient PG levels, but rather by high P levels.

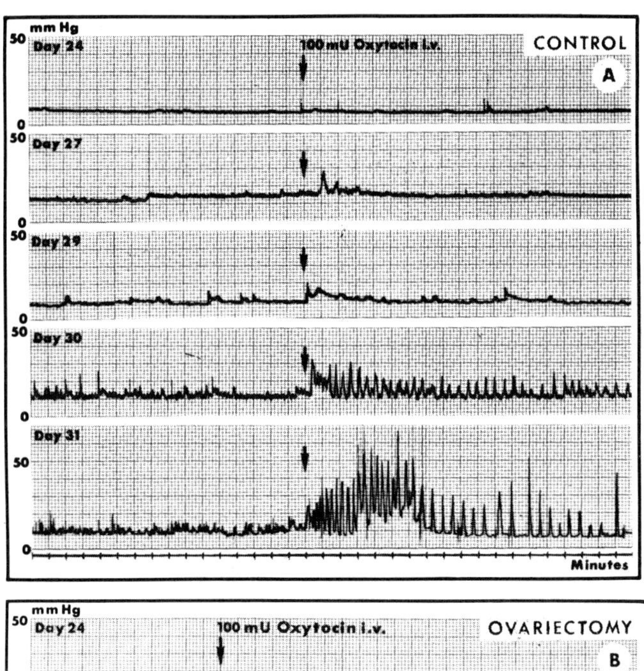

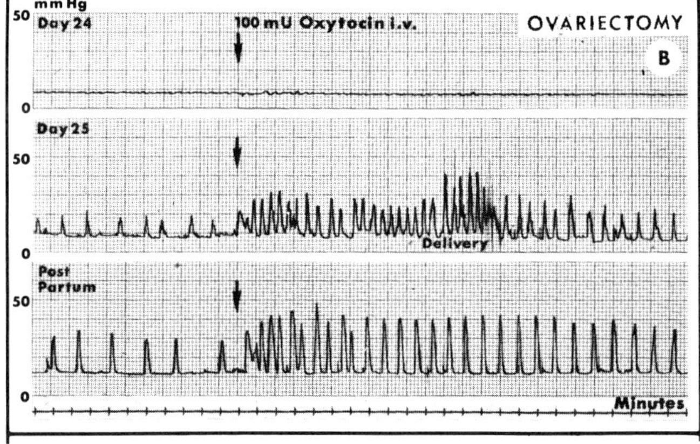

Fig. 13. Spontaneous activity and reactivity of the pregnant rabbit uterus *in situ* in the presence and absence of functional corpora lutea. Extraovular microballoon technique. Note the low level oscillation in IUP and the refractoriness (to 100 mU oxytocin) of the normal pregnant uterus until the P levels decreased from 14±1 ng/gm to 5±1 ng/gm at day 31 of gestation. Note also that this state of suppression is readily suspended and the uterus is converted to a spontaneously active and reactive organ by the removal of the corpora lutea in this instance on day 25 of gestation (for references, see Csapo, 1971).

By rendering the myometrium refractory to physiological stimulants and by increasing threshold (Fig. 11), P curtails the response of the pregnant uterus to oxytocic agents, including its own intrinsic stimulant PG (Fig. 10B). Only massive pharmacological doses of oxytocin or PG can break through this "defense mechanism" of pregnancy. The protective action of P was verified experimentally by P withdrawal (Pw) in 25-day pregnant rabbits through ovariectomy with and without replacement therapy (Fig. 14). Conversion of the refractory pregnant uterus to a reactive organ occurred only when plasma P decreased to a critical (~2 ng/ml) value. P substitution therapy prevented this conversion and sustained the refractory state. The sensitivity, specificity and precision of the

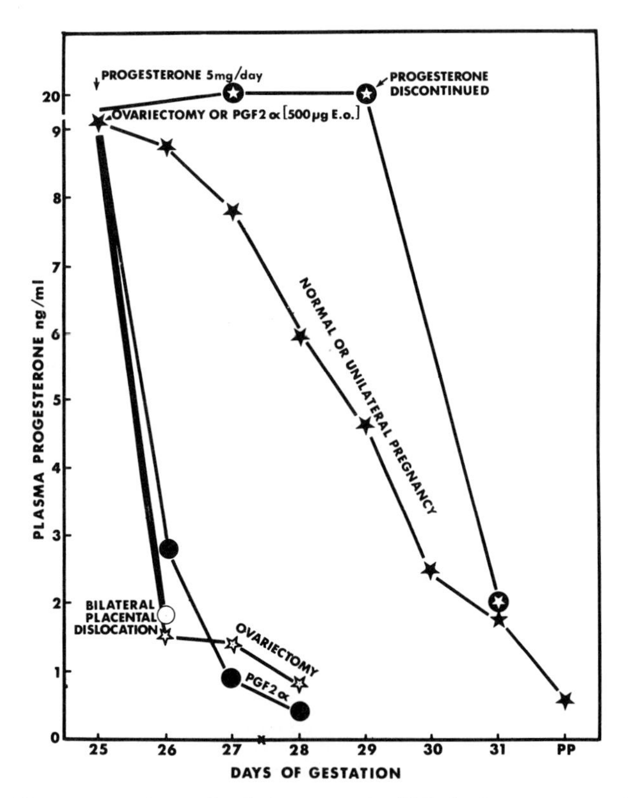

Fig. 14. Plasma progesterone levels in pregnant rabbits in the presence and absence of functional corpora lutea. Radioimmunoassay. Note the gradual decrease in P levels with the advance of term; the precipitous Pw when P synthesis is suspended directly by ovariectomy, or indirectly through placental dislocation, or $PGF_{2\alpha}$ treatment (both of which discontinue the luteotrophic support by the feto-placental unit); the sustenance of P levels by P substitution therapy; and the rapid Pw provoked by the discontinuation of P treatment. Note also the similarity (in the rabbit) between plasma and uterine P levels, by comparing the values described in this and in earlier figures) (Csapo and Wiest, 1972).

physiological techniques permitted the demonstration of the Pw-induced functional conversion of the uterus 16 years before it could be confirmed by radioimmunoassays (Csapo, 1956; Csapo and Wiest, 1972).

The "P Block"

The effect of P could be mediated by regulatory actions on the generation or propagation of the train discharges. Studies showed that P blocks propagation, but its effect on the generation of electric activity has not been definitively established.

The propagation "block" was first recognized by comparing the response of the pregnant and postpartum uteri to full– and half-length electric stimulation. In the postpartum uterus (P concentration = 3±1 ng/gm), half-length stimulation of the excised strip only reduced wT slightly (Fig. 12) indicating that electric activity did propagate over to the unstimulated half in this P-deficient uterus. In contrast, half-length stimulation of the P-dominated pregnant uterus (P concentration = 14±1 ng/gm) only provoked one-half of the maximal wT (Fig. 12) indicating that high P levels render the uterus nonpropagating and prohibit the conduction of electrical activity from the stimulated to the unstimulated region.

Similarly the action potentials, generated at one end of the paturient uterus, rapidly propagated to distant regions (Fig. 15A) in contrast to the pregnant uterus in which the impulse did not spread, not even to the nearest recording electrode (Fig. 15B).

The Pregnant Uterus

Evidently the factors required for the initiation of labor are present at all times, but their activating action during normal pregnancy is blocked by P. This block can be released throughout gestation resulting in spontaneous or induced miscarriage or premature labor.

In spite of the P block, the pregnant uterus is neither inert nor inactive. It generates electrical activity (usually by several pacemakers) which, however, remains local, nonpropagating and asynchronic, resulting in slight increase in IUP of high frequency. Near term, Pw promotes propagation, a gradual improvement in the synchrony of the discharges and an increase in the magnitude and regularity of IUP, resulting finally in parturition.

In addition to Pw (usually measured in the circulating blood) other variables are introduced, by the "luteo-placental shift" in P genesis, during the transition from pregnancy maintenance to the initiation of labor. During early pregnancy, P originates from the corpus luteum in all species and luteal P is transported to the myometrium systemically through the circulating blood. Therefore, the concentration of P is uniform throughout the uterus.

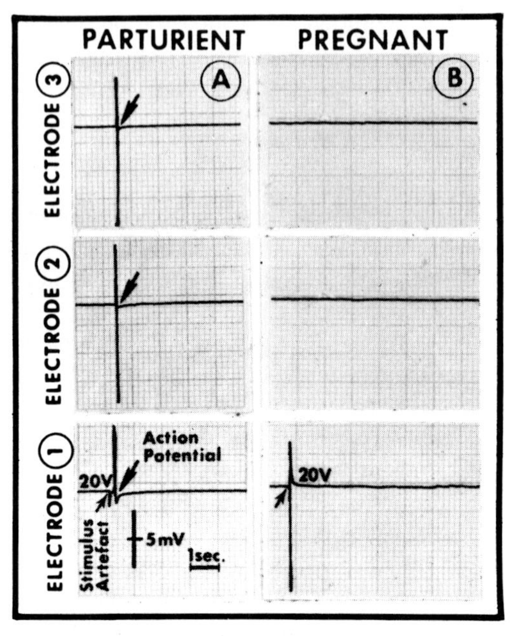

Fig. 15. The progesterone block. Parturient (P = 4±1 ng/gm) and 25-day pregnant (P = 14±1 ng/gm) rabbit uteri. The uterine strips are stimulated *in vitro* with a pair of stimulating electrodes at one end of the 60 mm long strips. The action potential generated is recorded with three suction electrodes (placed 15 mm apart) as it propagates along the length of the muscles. Note that in the parturient uterus, the application of the stimulus (indicated by a "stimulus artifact") triggers an action potential which rapidly appears at the first, second and third electrodes after only short delays. In contrast, note that in the 25-day pregnant uterus only the "stimulus artifact" is recorded and no action potential appears, not even at the first electrode (Talo and Csapo, 1970).

However, in several species (including the human and, to a lesser degree and at a later gestational time, the rat) the "luteo-placental shift" in P genesis permits variability in the concentration of P which is now transported locally through lipid-rich fetal membranes rather than through the systemic circulation. A P gradient in the myometrium, thus created, was first demonstrated electrophysiologically (Csapo, 1969) and later by direct P assays. Near and distant regions of the myometrium (Zander *et al.,* 1969) and fetal membranes (Pulkkinen and Enkola, 1972) from the placenta showed a P gradient resulting in a functional asymmetry of the uterus. Since the pacemaker region is generally located most distant from the placenta, a functional "polarity" can develop in the myometrium, topographically dependent on the placental implantation site. Hence the site of pacemaker activity and the direction in the spread of uterine activity are influenced by the location of the placenta in the uterine cavity. Thus, variations in placental location influence the onset and progress of labor (Csapo, 1971).

THE SEESAW THEORY

> There is but one safe way to avoid mistakes:
> To do nothing or at least,
> To avoid doing something new.
> *Albert Szent-Györgyi*

In its austere form, the "seesaw theory" stated (Csapo, 1975) that as long as the prenate has to be protected from the harmful consequences of advanced IUP (Csapo, 1970c), the myometrial stimulant PG and the suppressor P must be in regulatory balance (Fig. 16, first panel). Furthermore, spontaneous labor at or before term is initiated by a regulatory imbalance, resulting from P deficiency (Fig. 16, third panel).

Theoretically, both an increase in PG or a decrease in P could upset the seesaw and provoke a regulatory imbalance. However, systematic studies showed that during normal pregnancy (when P levels are far above the critical value at which advanced IUP evolves) PG can only "bend" but not "shift" the seesaw (Fig. 16, second panel). Under this unphysiological condition, the uterine response to PG is an unphysiological contracture, rather than high level cyclical activity. Of course if pregnant women are given massive PG treatment for hours to terminate pregnancy, the regulatory mechanism eventually becomes defective and through the resultant Pw, the seesaw shifts (Fig. 16, third panel).

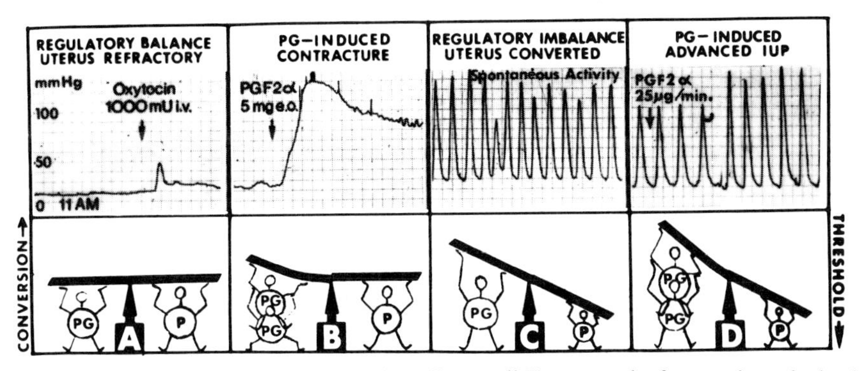

Fig. 16. The four basic positions of the "seesaw." Upper panels: four tracings obtained during PG-induced abortion of a mid-trimester pregnant patient. Lower panels: four schematic drawings illustrating the seesaw theory. Panel (A) illustrates the regulatory balance of PG and P during normal pregnancy, as reflected by the refractoriness of the uterus to massive stimulation. Note the only slight and transient contracture (rather than advanced cyclic IUP) response of the uterus to 1000 mU of oxytocin. Panel (B) illustrates that in this condition of regulatory balance the response of the normal pregnant uterus is a contracture to PG. The seesaw is only "bent" but not "shifted." Panel (C) illustrates that high level cyclic IUP is only superimposed on this contracture response when several hours after the "PG impact" Pw had already shifted the seesaw. Panel (D) illustrates that after Pw the uterus is spontaneously active and reactive to exogenous PG (Csapo, 1975).

However, of critical theoretical interest is the demonstration that under endogenous PG stimulation high level cyclic IUP evolves only when Pw reaches a critical value (characteristic of the species and of the gestational period). This relationship between Pw and spontaneous uterine activity is illustrated (Fig. 17). Not only does Pw control spontaneous activity but also the reactivity of the early pregnant uterus to exogenous oxytocin and $PGF_{2\alpha}$ (Fig. 17). Apparently, during early pregnancy, high level cyclic IUP only evolves when P decreases

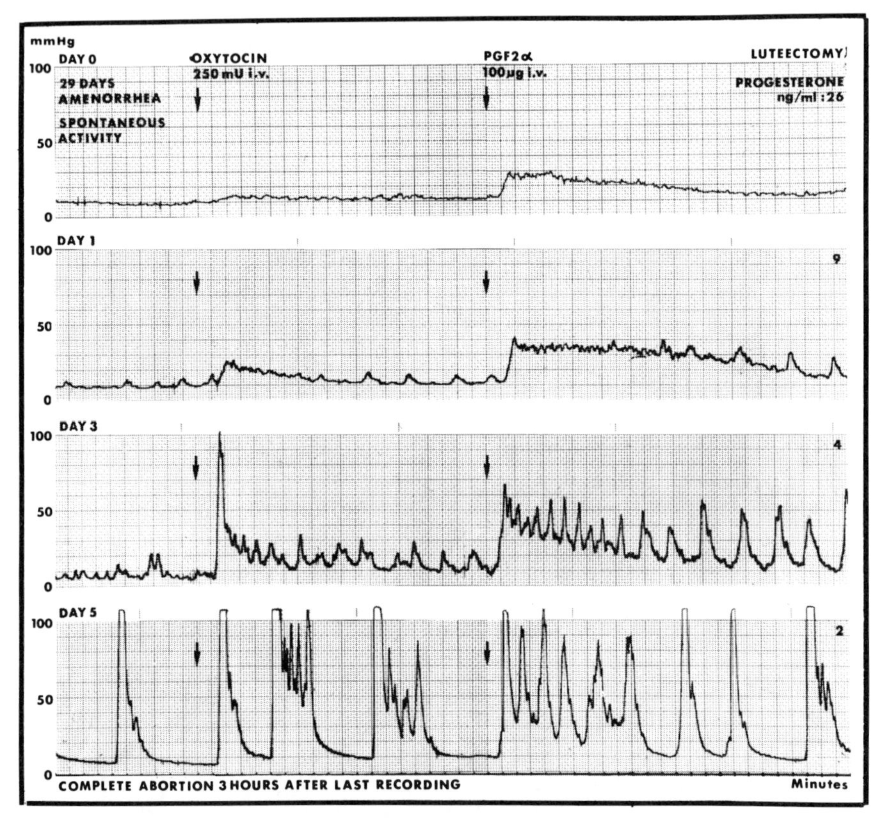

Fig. 17. The effect of luteectomy-induced progesterone withdrawal on the spontaneous activity and reactivity of the early pregnant human uterus. Luteectomy is performed at 6 weeks after conception. IUP is recorded sequentially with the extraovular microballoon technique and P is measured with radioimmunoassay. Note that until the plasma P concentration is reduced below 5 ng/ml, spontaneous activity is suppressed to low level oscillations and the uterine response to the massive iv doses of 250 mU oxytocin and 100 µg $PGF_{2\alpha}$ is only an unphysiological contracture, rather than high level cyclic IUP. Note also that moderate cyclic IUP (spontaneous or induced) is only recorded at ~4 ng/ml P concentration; over 100 mm Hg cyclic IUP (extending the scale) at ~2 ng/ml P concentration and that at this low P level the patient aborted completely. Evidently, the "sensitivity" of the uterus to stimulants is controlled by P (Csapo *et al.*, 1973d).

below ~4 ng/ml. This P-dependent response of the pregnant uterus to PG's (Fig. 17) complemented the already massive evidence that P is the key regulator of both processes: the maintenance and termination of pregnancy. Recognizing the merits of its premises, the seesaw theory was extensively tested to determine the validity and clinical benefits of its predictions.

1. The prediction that a mere Pw to a critical value terminates pregnancy (in the presence of endogenous uterine PG sufficient to activate the myometrium throughout gestation) was recently validated by systematic studies in 57 early pregnant patients.

Rigorously controlled clinical trials (Fig. 18) showed that during the initial 5 wk after conception, the corpus luteum is the main source of P. Luteectomy alone (without exogenous stimulation) readily provoked the evolution of IUP and terminated pregnancy (Fig. 18B) as it did in animal models, whenever Pw

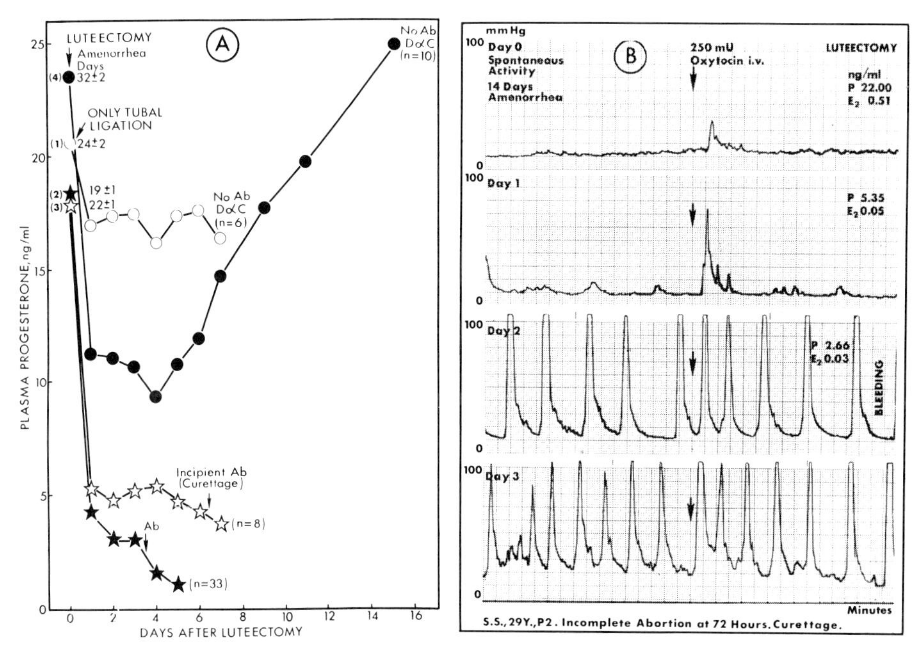

Fig. 18. A summary of the mean changes in plasma progesterone (A) and a typical example of IUP (B) induced by luteectomy during early human pregnancy. A total of 57 patients. In (A) note that rapid Pw in ~5-week pregnant patients to ~3 ng/ml provokes abortion on the third day after luteectomy (2); Pw to ~5 ng/ml leads to incipient abortion on the sixth day (3); partial Pw in ~7-week pregnant patients to ~10 ng/ml has no effect on pregnancy maintenance (4); and tubal sterilization without luteectomy does not affect P levels and pregnancy (1).Note also in (B) that a massive dose of 250 mU oxytocin only triggered a single response when the P was ~5 ng/ml. At lower levels of P (~3 ng/ml) many spontaneous cyclical contractions reached levels of IUP of over 100 mm Hg; with oxytocin these levels were more consistent.

reached the critical ~3 ng/ml value (Fig. 18A). This decisive evidence was complemented by two additional findings. In 5-week-pregnant women, P replacement therapy after luteectomy prevented the evolution of IUP and abortion as it did in animal models. Furthermore, during the sixth to seventh week of pregnancy, when the "luteo-placental shift" in P genesis reached such a degree that luteectomy could no longer reduce plasma P levels below ~10 ng/ml (Fig. 18A) uterine activity and reactivity remained suppressed and pregnancy continued undisturbed in spite of luteectomy (Csapo *et al*, 1972, 1973a).

These studies demonstrated conclusively that in the human, P plays the same key role in pregnancy maintenance as it does in laboratory animals.

2. Normal pregnancy maintenance demands the sustenance of the regulatory balance between PG and P not only before but also after the completion of the luteo-placental shift in P genesis. For example, intraamniotic injection of hypertonic saline suppressed placental P genesis leading to about ~50% Pw (Fig. 19A). This excessive Pw provoked the evolution of high level cyclic IUP (Fig. 19B) and terminated pregnancy (Csapo *et al.*, 1969; Tyack *et al.*, 1973). Thus therapeutically-induced Pw became the rational therapy for the management of legal abortions (Kerenyi *et al.*, 1973).

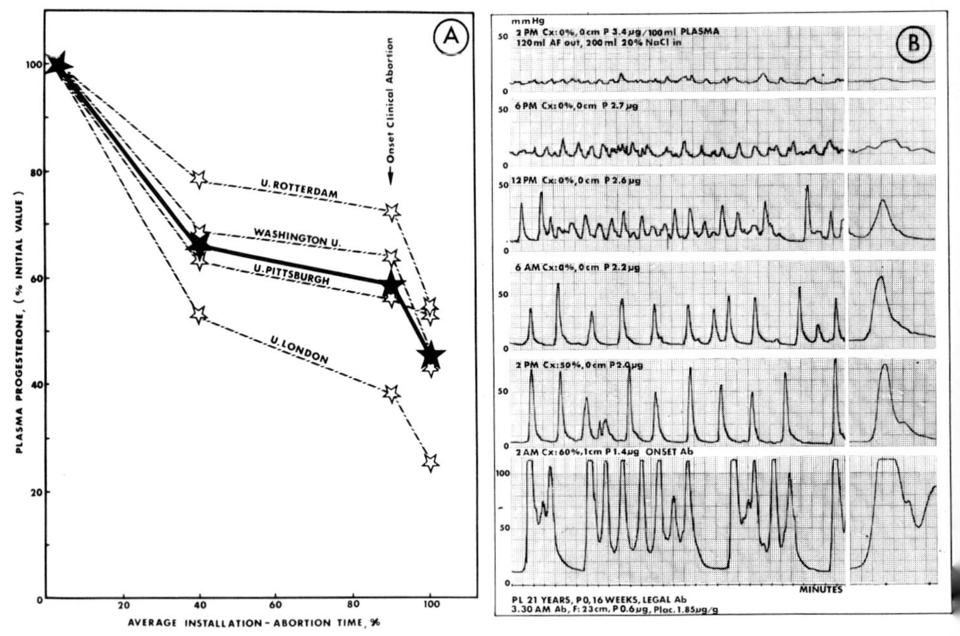

Fig. 19. The changes induced by hypertonic saline treatment in plasma progesterone and intrauterine pressure during mid-trimester human pregnancy. Note in (A) that the exchange of amniotic fluid with 200 ml 20% NaCl induces during the instillation-abortion time ~50% decrease in plasma P, measured in four different laboratories. Note also (B) that as Pw advances, high level cyclic IUP evolves and pregnancy terminates (Csapo *et al.*, 1969).

It was resonable to assume that the suppression of feto-placental endocrine function predictably terminates pregnancy, even if the source of P is the corpus luteum, for luteal P genesis is critically dependent on feto-placental luteotrophic support. This premise was first validated in rabbits by "placental dislocation" (Csapo and Lloyd-Jacob, 1962) and by intraamniotic hypertonic saline treatment (Porter *et al.*, 1968). The experiments showed that if one feto-placental unit is left intact, the luteotrophic action of this single endocrine organ can sustain luteal P genesis and pregnancy in spite of the exposure of both uterine horns to placental dislocation or hypertonic saline. Dislocation of all placentas terminates pregnancy as a result of the complete suspension of P synthesis.

3. Regardless of the mechanism through which Pw is provoked, pregnancy terminates whenever Pw reaches the critical abortion value. Unfortunately, the supportive evidence of this prediction was not considered during the conduct of the early clinical trials with PG's. Nor was it recognized that the pregnant uterus is a suppressed rather than an inert muscle. Thus, investigators infused $PGF_{2\alpha}$ intravenously with the intent of stimulating the uterus. In spite of excessive (over 100 mg) doses, which provoked clinically unacceptable side effects, this regimen had a low success rate in terminating pregnancy (Transcript, 1971).

The intravenous approach was corrected by the observation (Fig. 20) that the effective end point of PG therapy is an over 50% Pw. This demonstration led to the development of the therapeutic principle "PG impact" (PGI) and the intraamniotic, extraovular and intrauterine applications of PG's. It is now firmly established that PGI markedly reduces the effective therapeutic dose of $PGF_{2\alpha}$ and that the reduction in dose also reduces side effects (for references, see Csapo *et al.*, 1976). This clinical success in terminating pregnancy by the therapeutic application of an endogenous compound is an achievement of basic research which clarified the regulatory significance of endogenous PG's and the relationship between effective PG treatment (Fig. 20B) and Pw (Fig. 20A).

Of considerable academic interest is the repeated demonstration that PG administered during normal gestation does not provoke high level cyclic IUP and does not terminate pregnancy, not even if the doses are excessive until P is significantly reduced (Csapo, 1973; Csapo *et al.*, 1976; Saldana *et al.*, 1973; Enkola, 1974; Pitkanen and Rauramo, 1974; Tyack *et al.*, 1974; Zoltan *et al.*, 1974; Craft *et al.*, 1974; Abstracts, 1975). This effect of PG on P levels is best explained by a suppression of utero-placental blood flow and feto-placental endocrine function (Csapo, 1973; Pulkkinen *et al.*, 1975) and the induced contracture (Fig. 20B) which further suppresses utero-placental blood flow and placental P genesis (Fig. 20A). Once Pw decreases threshold, endogenous PG effectively activates the uterus and cyclic IUP increases in spite of the rapid elimination of exogenous $PGF_{2\alpha}$. If the PG-impact only provokes insufficient Pw, supplementary PG or oxytocin treatment is needed to accentuate it in the promotion of uterine activity in abortion (Fig. 20B). Apparently, the trials in human patients with PG's exposed a new therapeutic principle of fertility

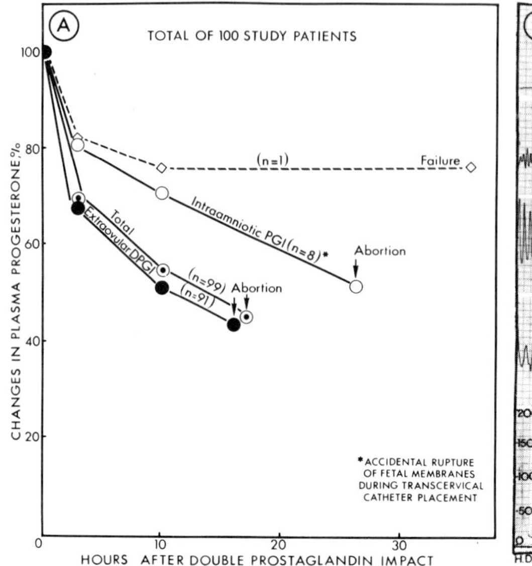

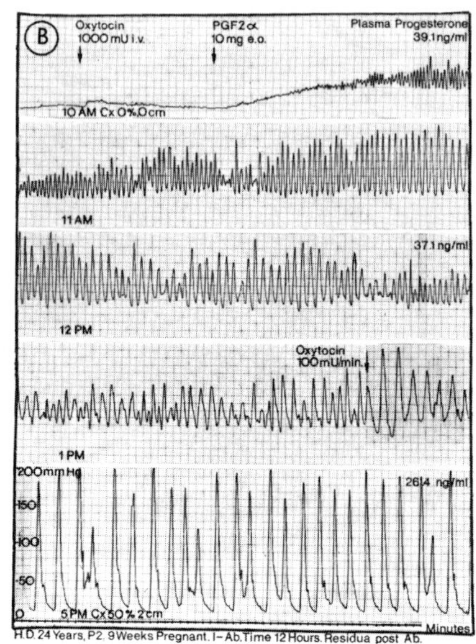

Fig. 20. The changes induced by the "PG-impact" in plasma progesterone, intrauterine pressure and pregnancy. One hundred first and second trimester patients. Progesterone is measured sequentially by radioimmunoassay and intrauterine pressure by the extraovular microballoon technique. Note (A) that 99 of the 100 patients who received extraovularly 19.0±1.5 mg $PGF_{2\alpha}$, showed over 50% Pw and aborted in 16.9±0.9 hr. Note also (B) the initial contracture response of the uterus to 10 mg $PGF_{2\alpha}$; the gradual evolution of cyclic IUP following the PG-induced Pw; and the conversion of the refractory uterus to a reactive organ, as indicated by the oxytocin tests (Csapo *et al.*, 1976).

control, an acute intervention with the regulatory balance through pharmacological doses of an endogenous compound.

4. A crucial prediction was that Pw is a prerequisite for the initiation of normal spontaneous labor. This prediction, long since validated in a variety of animal models (Csapo, 1971), has been documented recently in humans by sequential radioimmunoassays of plasma P levels in the same group of patients at different stages of gestation (Csapo *et al.*, 1971; Turnbull *et al.*, 1974) (Fig. 21).

The critical demonstration that myometrial P levels decrease prior to parturition also has been achieved by sequential radioimmunoassays in the rat (Csapo and Wiest, 1972) and in the rabbit (Csapo, 1966). Only single myometrial samples can be obtained in patients during Cesarean section. In the rat uterine P decreased during the last 8 days of gestation from 164±14 ng/gm to 16±3 ng/gm, and in the rabbit during the last 5 days of gestation from 14±1 ng/gm to 4±1 ng/gm. However, uterine vein P levels showed a marked decrease in monkeys (*Macaca mulatta*) when measured (Lanman *et al.*, 1975) before the 100th day of gestation (range 11–71 ng/ml) and after the 130th day of gestation

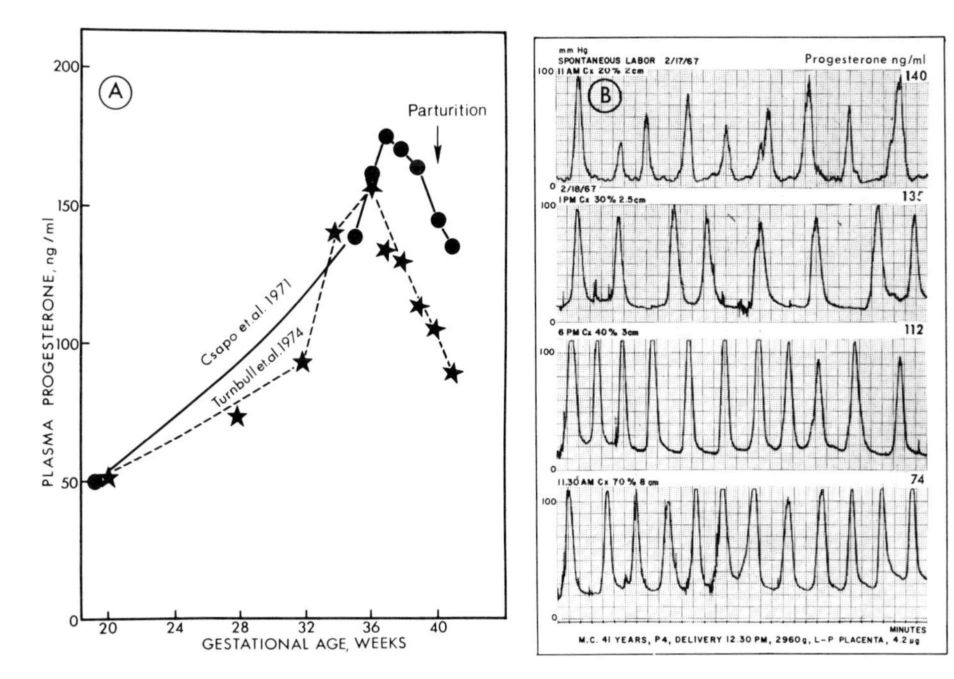

Fig. 21. Progesterone withdrawal preceding high level cyclic IUP and spontaneous labor at term in human patients. (A) Plasma P is measured by radioimmunoassay in two groups of third trimester pregnant patients. Note the good agreement between the two independent studies, showing that Pw precedes clinical labor. (B) The evolution of IUP during clinical labor is measured by the extraovular microballoon method. Note the relationship between Pw, the increase in the magnitude, frequency and regularity in IUP and the progress of clinical labor (Csapo *et al.*, 1971; Turnbull *et al.*, 1974).

(range 2–3 ng/ml). Also, uterine vein P levels showed a significant decrease (from 478 to 250 ng/ml) in term pregnant patients prior to and during clinical labor (A. I. Csapo and M. Pulkkinen unpublished, 1976). This change in the P levels of the uterine vein (draining the placenta) is comparable in magnitude to that observed in the peripheral plasma and uterus of the rabbit and rat models. Apparently spontaneous labor in term patients is preceded by a marked Pw, as predicted by the seesaw theory.

However, since P is synthesized by the placenta, the peripheral or even the uterine vein P level may not fully describe the magnitude of Pw. Due to the P gradient in the myometrium, there is a relative P deficiency of the nonplacental uterine regions in comparison with the placental regions. This explains why the human uterus responds partially to stimulants several weeks before term. Since myometrial P values can only be obtained in human patients under exceptional circumstances, the elegant demonstration (Aleem and Schulman, 1975) that the termination of pregnancy is preceded by a far greater Pw than indicated by the

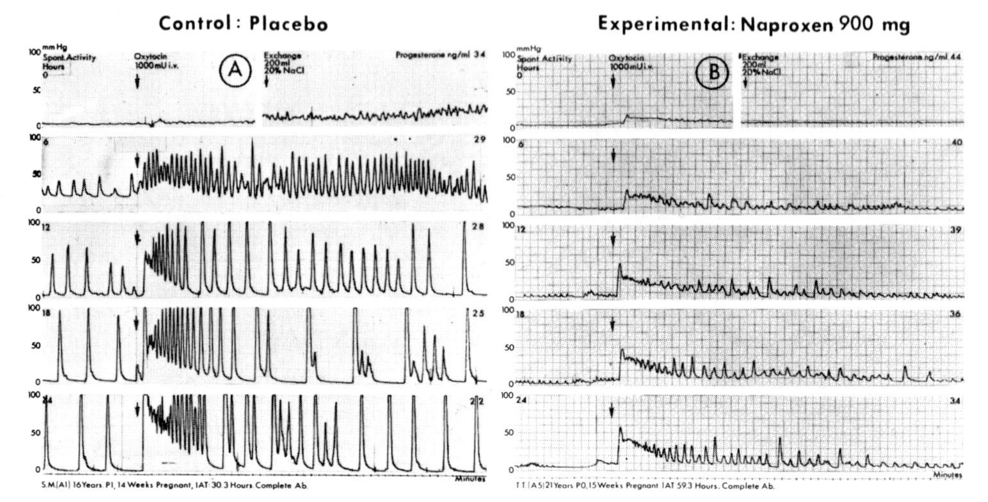

Fig. 22. The effect of the inhibition of PG synthesis on spontaneous uterine activity and reactivity. Mid-trimester pregnant patients, treated either with placebo (A) or 900 mg Naproxen, an inhibitor of PG synthesis (B). The evolution of uterine activity and reactivity is provoked in both the control and experimental groups by hypertonic saline-induced Pw. The IUP is measured sequentially with the extraovular microballoon and plasma P by radioimmunoassay. Note in (A) the rapid evolution and in (B) the delayed evolution of IUP, oxytocin response and abortion, in spite of only 12 hr of Naproxen treatment. Note also that Pw occurred in both patients (Csapo *et al.*, 1974).

peripheral plasma P levels is of considerable significance. In mid-trimester patients, the P concentration of the surgically removed normal placentae was 5.2±0.3 μg/gm, while in those delivered during hypertonic saline-induced abortion it was 2.0±0.1 μg/gm and during PG-induced abortion 1.5±0.1 μg/gm (62 and 71% withdrawals, respectively).

While Pw does not initiate labor directly, it nevertheless provides the key regulatory step by allowing the refractory pregnant uterus to react to endogenous PG's with high level cyclic IUP.

5. Abortion provoked by Pw can be prevented by rebalancing the seesaw at a lower level through an inhibition of PG synthesis (Fig. 22). This rebalancing, first examined in the rat model (Csapo *et al.*, 1973b, c), has been validated in human patients. By using a PG synthesis inhibitor the evolution of uterine activity was suppressed and pregnancy maintained in spite of drastic Pw (Csapo *et al.*, 1974). Premature labor was also prevented by a synthetic P (Johnson *et al.*, 1975).

The studies described illustrate that the predictions of the seesaw theory have been broadly validated in animal models as well as in patients under a variety of regulatory and clinical conditions. Whether regulatory factors other than PG and P affect myometrial function directly is an outstanding question.

THE MAINTENANCE AND TERMINATION OF PREGNANCY

> Cures for disease flow out of progress
> As the natural fruits of knowledge.
> *Albert Szent-Györgyi*

If Pw is the critical regulatory step for the initiation of spontaneous labor, what is the role of the fetus in the control of this process? While there is no evidence that the human fetus is directly involved in myometrial regulation (except through stretch-induced PG synthesis) careful clinical observations established that it affects the timing of parturition through a hypothalamic–pituitary–adrenal action (for references, see Csapo and Wood, 1968). Since the seesaw theory is primarily concerned with direct regulatory actions on the myometrium itself and since a direct myometrial effect of the fetus has not been demonstrated as yet, in Fig. 23 the fetus (F) is positioned with a question mark at a distance from the seesaw. The same question mark is placed under estradiol (E_2) and oxytocin (Oxy) since it is not established with certainty that these regulators can directly promote uterine activity, for example, when the myometrium is depleted of PG. As mentioned earlier, myometrial PG's increase gradually as pregnancy advances (Willman and Collins, 1975) probably due to increasing uterine volume (V) (Csapo, 1973) and estradiol stimulation (Liggins *et al.*, 1973). These effects are indicated in Fig. 23.

Recent experiments in sheep (Liggins *et al.*, 1972) and goats (Thorburn *et al.*, 1972) clarified that the fetal contribution to the timing of parturition is mediated by cortisol release and a subsequent Pw. In the goat, a species in which P is of luteal origin and in which PGF provokes luteolysis, the link between fetal cortisol and Pw seems to be PGF (Currie, 1974).

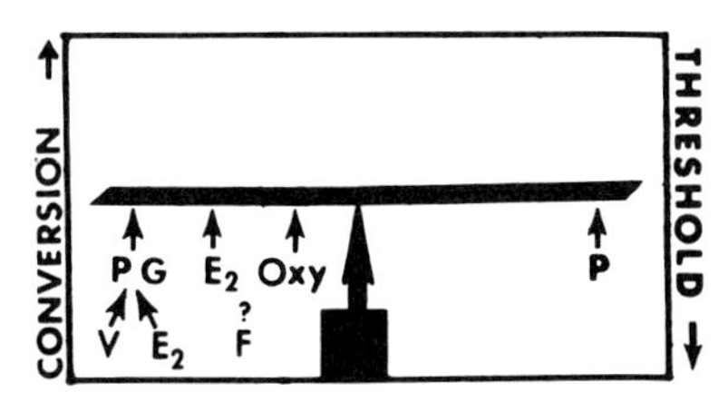

Fig. 23. The complexity of the regulatory mechanisms of the pregnant uterus. The schematic drawing illustrates the balance between stimulants and suppressor and the relationship between direct and indirect regulators. While endogenous and exogenous compounds only promote or suppress activity, the mechanism of control is complex, because a large number of regulatory factors act upon the uterus directly or indirectly. The action of PG and P on the myometrium is direct, while that of estradiol (E_2), oxytocin (Oxy), uterine volume (V), and the fetus (F) is not clarified as yet. The effect of these regulators may well be indirect, mediated through actions on PG or P synthesis. (Csapo, 1975).

The sustenance of P levels at a high value during the major part of gestation might well be a prerequisite of normal prenatal life, for experiments in rats showed that prolonged P deficiency (insufficient for terminating pregnancy) provokes a variety of fetal defects mediated by the premature evolution of partial uterine activity (Csapo, 1970c). Thus fetal control over Pw is a desirable mechanism for the appropriate timing of the initiation of labor. A better understanding of this mechanism might offer rational and predictable therapy for the synchronization of labor with fetal maturity and for the shortening of prelabor and clinical labor.

Estrogen (E_2) provokes (Csapo and Wiest, 1972) and promotes (Csapo *et al.*, 1973c) premature labor in the rat, the goat (Currie *et al.*, 1976) and sheep (Hindson *et al.*, 1967). However, these findings do not define the action of E_2 in the initiation of labor, since E_2 precipitously decreases after fetal death *in utero* and yet parturition is accomplished (Csapo and Wood, 1968). Thus the regulatory role of E_2 in the control of uterine function remains to be clarified before it can be definitively established whether E_2 directly affects the myometrium when labor begins, or only indirectly through PG synthesis (Fig. 23). It is, of course, well established that both E_2 and stretch have indispensable actions on the contractile system as well as on excitability (Csapo, 1971).

Several investigators proposed without evidence that oxytocin triggers the initiation of labor (Caldeyro-Barcia, 1964). A recent study stimulated considerable interest in oxytocin by demonstrating that it increases PG synthesis (Mitchell *et al.*, 1975). How significant the contribution of oxytocin is to normal parturition still remains to be determined. *In vitro* studies showed that oxytocin is ineffective in stimulating the uterus when the muscle is depleted of PG (Csapo and Csapo, 1974).

The effects of the fetus (F), E_2 and oxytocin (Oxy) are recognized but their direct action on the myometrium (the left side of the seesaw in Fig. 23) remains to be demonstrated. It appears that nature designed the backbone of the basic regulatory mechanism of the pregnant uterus in such a way that in case of emergency (for example, fetal death *in utero*) the myometrium is equipped for the delivery of its contents without fetal contributions or a coincidental rise in estradiol or oxytocin levels. In these instances the uterus has to operate with the essential regulatory step, the imbalance between PG and P, the timing of which is not critically important when the fetus is dead. If so, the fetus provides no players but only the umpire to the regulatory tournament to ensure that the PG/P imbalance occurs at a time most compatible with pre- and post-natal life and development.

At a time when serious questions are raised about the desirability and clinical merits of the routine practice of fetal monitoring (Haverkamp, 1975) it is of interest that the much simpler method of monitored induced labor yielded clinical results unmatched by other forms of labor management. A study of

1000 control and 1000 experimental patients demonstrated that if uterine activity is set at an optimum by carefully controlled oxytocin stimulation in term patients, nature's mechanism for the control of spontaneous labor can be improved to such an extent that all major complications of parturition are significantly reduced (Csapo, 1970b).

No definitive therapy has been developed as yet for the management of premature labor. However as predicted, the inhibition of PG synthesis in P-deficient patients, "rebalanced the seesaw at a lower level" (Fig. 24) suppressed the evolution of IUP, oxytocin response and prolonged pregnancy (Fig. 21). When it can be demonstrated that treatment with inhibitors of PG synthesis for several days or weeks has no harmful effects on mother and fetus, a rational therapy will be available for the first time for the prevention of premature labor. Using the same rationale, progesterone replacement therapy as a method for rebalancing the seesaw, and thus preventing premature labor, gained new momentum through a recent study with a synthetic gestagen (Johnson *et al.*, 1975).

The timing of the luteo-placental shift in P genesis (Csapo *et al.*, 1972, 1973a) has set the stage for the rational therapeutic use of luteolytic agents in the management of fertility and for progesterone treatment in cases of luteal insufficiency and habitual abortion.

Menstrual induction has been achieved with clinical symptoms of menstrua-

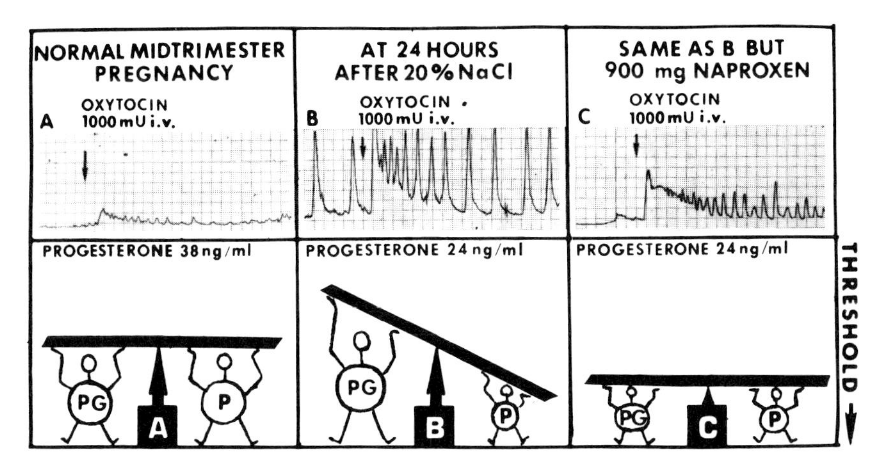

Fig. 24. The rebalancing of the seesaw. Upper panels: Three tracings obtained during hypertonic saline-induced abortions, with and without anti-PG treatment. Lower panels: Three schematic drawings illustrating the theory. Panel A illustrates the short-lived oxytocin-induced contracture of the normal pregnant uterus, reflecting regulatory balance. Note the P level. Panel B illustrates the advanced oxytocin response of the same uterus after the seesaw has been shifted by saline-induced Pw. Panel C illustrates that, despite Pw, the oxytocin response is suppressed by "rebalancing" the seesaw (at a lower level) by anti-PG (Naproxen) treatment. (Csapo, 1975).

tion rather than abortion in over 500 early pregnant women by provoking Pw with the PG impact (Mocsary and Csapo, 1975).

The PG impact also terminated first and second trimester pregnancies successfully whenever it provoked a Pw of critical degree. Leading medical centers adopted the PG method as the technique of choice for the nonsurgical termination of advanced pregnancy (for references, see Csapo *et al.*, 1976).

Regardless of whether the therapeutic end point is the termination of unwanted pregnancy or the protection of the wanted child, quantitative uterine physiology continues to serve as a reliable guide during the basic experiments in animal models and the pilot clinical trials. Hopefully a less hypocritical new generation will apply it more effectively to protect the quality of life, an obligation hazardously neglected by their elders.

A PERSONAL NOTE

Albert Szent-Györgyi had been appointed Professor of Medical Chemistry in my hometown of Szeged, Hungary, when as a high school student I had to decide about values and priorities in preparation for a life assignment. The rare privilege of winning from him a game of tennis gave me the courage to ask his advice. He answered unhesitantly: "Whatever you do, be sure to be constructive even on the toilet." Through this profound advice he became my hero, for he could not hide the depth of his message behind a Villonian twist. During the subsequent 40 years, he rigorously adhered to his own dictum, the magic of his unmatched productivity and leadership.

ACKNOWLEDGMENT

This review, summarizing 30 years of work of my laboratory, could not have been written without the dedicated work of my pupils, who provided the factual data. I gratefully acknowledge their respective contributions, which far exceed the quotations listed in legends and references. Elise F. Csapo not only helped me in every phase of the preparation of the article, but shared with me over 10 years my respect, admiration, and affection for Albert Szent-Györgyi. Susan Hayden labored patiently with typing during the long evolution of the final form of the manuscript and Bruce Currie and Frederick Sweet advised me kindly but critically on presentation.

REFERENCES

Abstracts. (1975). *Int. Conf. Prostaglandins, 1975.*
Aleem, F. A., and Schulman, H. (1975). *Prostaglandins* **9**, 495.

Caldeyro-Barcia, R. (1964). *In* "Muscle, Proceedings of the Symposium held at the Faculty of Medicine, University of Alberta" (W. M. Paul *et al.*, eds.), pp. 317–347. Pergamon, Oxford.
Corner, G. W., and Allen, W. M. (1930). *Proc. Soc. Exp. Biol. Med.* **27**, 403–405.
Craft, I., Carriere, E., and Youssefnejadian, E. (1974). *Obstet. Gynecol.* **44**, 135–141.
Csapo, A. I. (1948). *Nature (London)* **162**, 218–219.
Csapo, A. I. (1954). *Nature (London)* **173**, 1019–1026.
Csapo, A. I. (1956). *Recent Prog. Horm. Res.* **12**, 405–431.
Csapo, A. I. (1959). *Ann. N.Y. Acad. Sci.* **81**, 453–467.
Csapo, A. I. (1969). *Postgrad. Med. J.* **45**, 57–64.
Csapo, A. I. (1970a). *Obstet. & Gynecol. Surv.* **25**, 403–435.
Csapo, A. I. (1970b). *Obstet. & Gynecol. Surv.* **25**, 515–543.
Csapo, A. I. (1970c). *In* "Prenatal Life" (H. C. Mack, ed.), pp. 155–182. Wayne State Univ. Press, Detroit, Michigan.
Csapo, A. I. (1971). *In* "Contractile Proteins and Muscles" (K. Laki, ed.), pp. 413–482. Dekker, New York.
Csapo, A. I. (1973). *Prostaglandins* **3**, 245–289.
Csapo, A. I. (1975). *Am. J. Obstet. Gynecol.* **121**, 578–581.
Csapo, A. I. (1976a). *Am. J. Obstet. Gynecol.* **124**, 367–379.
Csapo, A. I. (1976b). *Prostaglandins* **12**, 149.
Csapo, A. I., and Csapo, E. E. (1974). *Life Sci.* **14**, 719–724.
Csapo, A. I., and Lloyd-Jacob, M. A. (1962). *Am. J. Obstet. Gynecol.* **83**, 1073–1082.
Csapo, A. I., and Takeda, H. (1963). *Nature (London)* **200**, 680–682.
Caspo, A. I., and Takeda, H. (1965). *Am. J. Obstet. Gynecol.* **91**, 221–231.
Csapo, A. I., and Wiest, W. G. (1972). *Prostaglandins* **1**, 157–165.
Csapo, A. I., and Wood, C. (1968). *In* "Recent Advances in Endocrinology" (V. H. T. James, ed.), Chapter 7, pp. 207–239. Churchill, London.
Csapo, A. I., Knobil, E., Pulkkinen, M. O., van der Molen, H. J., Sommerville, I. F., and Wiest, W. G. (1969). *Am. J. Obstet. Gynecol.* **105**, 1132–1134.
Csapo, A. I., Knobil, E., van der Molen, H. J., and Wiest, W. G. (1971). *Am. J. Obstet. Gynecol.* **110**, 630–632.
Csapo, A. I., Pulkkinen, M. O., Ruttner, B., Sauvage, J. P., and Wiest, W. G. (1972). *Am. J. Obstet. Gynecol.* **112**, 1061–1067.
Csapo, A. I., Pulkkinen, M. O., and Wiest, W. G. (1973a). *Am. J. Obstet. Gynecol.* **115**, 759–765.
Csapo, A. I., Csapo, E. F., Fay, E., Henzl, M. R., and Salau, G. (1973b). *Prostaglandins* **3**, 827–837.
Csapo, A. I., Csapo, E. F., Fay, E., Henzel, M. R., and Salau, G. (1973c). *Prostaglandins* **3**, 839–846.
Csapo, A. I., Pulkkinen, M. O., and Kaihola, H. L. (1973d). *Prostaglandins* **4**, 421–429.
Csapo, A. I., Henzl, M. R., Kaihola, H. L., Kivikoski, A., and Pulkkinen, M. O. (1974). *Prostaglandins* **7**, 38–47.
Csapo, A. I., Herczeg, J., Pulkkinen, M. O., Kaihola, H. L., Zoltan, I., Csillag, M., and Mocsary, P. (1976). *Am. J. Obstet. Gynecol.* **124**, 1–13.
Currie, W. B. (1974). *J. Reprod. Fertil.* **36**, 481–482.
Currie, W. B., Cox, R. I., and Thorburn, G. D. (1976). *Prostaglandins* **12**, 1093.
de Paiva, C. E. N., and Csapo, A. I. (1973). *Prostaglandins* **4**, 173–188.
Endkola, K. (1974). *Prostaglandins* **5**, 115–121.
Haverkamp, A. D. (1975). *Obstet. Gynecol. News* **10**, No. 24, 1.
Hindson, J. C., Schofield, B. M., and Turner, C. B. (1967). *Res. Vet. Sci.* **8**, 353–360.

Johnson, J. W. C., Austin, K. L., Jones, G. S., Davis, G. H., and King, T. M. (1975). *N. Engl. J. Med.* **293**, 675–680.

Kerenyi, T. D., Mendelman, N., and Sherman, D. (1973). *Am. J. Obstet. Gynecol.* **116**, 593–600.

Kuriyama, H., and Csapo, A. I. (1961). *Endocrinology* **68**, 1010 and 1025.

Lanman, J. T., Thau, R., and Brinson, A. (1975). *8th Ann. Meet. Soc. Study Reprod.* Abstract No. 113.

Liggins, G. C., Grieves, S. A., Kendall, J. Z., and Knox, B. S. (1972). *J. Reprod. Fertil., Suppl.* **16**, 85–103.

Liggins, G. C., Fairclough, R. J., Grieves, S. A., Kendall, J. Z., and Knox, B. S. (1973). *Recent Prog. Horm. Res.* **29**, 111–159.

Marshall, J. M. (1973). *Ann. Rev. Pharmacol.* **13**, 19.

Marshall, J. M., and Csapo, A. I. (1961). *Endocrinology* **68**, 1026–1035.

Mitchell, M. D., Flint, A. P. F., and Turnbull, A. C. (1975). *Prostaglandins* **9**, 47–55.

Mocsary, P., and Csapo, A. I. (1975). *Prostaglandins* **10**, 545–547.

Pitkanen, Y., and Rauramo, L. (1974). *Prostaglandins* **5**, 269–274.

Porter, D. G., Becker, R., and Csapo, A. I. (1968). *J. Reprod. Fertil.* **17**, 433–442.

Pulkkinen, M. O., and Enkola, K. (1972). *Int.J. Gynecol. Obstet.* **10**, 93–94.

Pulkkinen, M. O., Pitkanen, Y., Ojala, A., and Hannelin, H. (1975). *Prostaglandins* **9**, 61–66.

Saldana, L., Schulman, H., and Yang, W. H. (1973). *Prostaglandins* **3**, 847–858.

Talo, A., and Csapo, A. I. (1970). *Physiol. Chem. Phys.* **2**, 489–494.

Thorburn, G. D., Nicol, D. H., Bassett, J. M., Shutt, D. A., and Cox, R. I. (1972). *J. Reprod. Fertil., Suppl.* **16**, 61–84.

Torok, I., and Csapo, A. I. (1976). *Prostaglandins* **12**, 253.

Transcript. (1971). "Symposium on Prostaglandins in Fertility Control." World Health Organization Research and Training Center, Karolinska Institute.

Turnbull, A. C., Flint, A. P. F., Jeremy, J. Y., Patten, P. T., Keirse, J. J. N. C., and Anderson, A. B. M. (1974). *Lancet* **1**, 101–104.

Tyack, A. J., Parsons, R. J., Millar, D. R., and Pennington, G. (1973). *J. Obstet. Gynaecol. Br. Commonw.* **80**, 548–552.

Tyack, A. J., Lambadarios, C., Parson, R. J., Stewart, C. R., and Cooke, I. D. (1974). *J. Obstet. Gynaecol. Br. Commonw.* **81**, 52–56.

Willman, E. A., and Collins, W. P. (1975). *Br. J. Obstet. Gynecol.* **82**, 843–846.

Zander, J., Holzmann, K., von Munstermann, A. M., Runnebaum, B., and Sieber, W. (1969). *In* "The Feto-Placental Unit" (A. Pecile and C. Finzi, eds.), Int. Congr. Ser. No. 183, pp. 162–175. Excerpta Med. Found., Amsterdam.

Zoltan, I., Csillag, M., Zsolnai, B., Zubek, L., Moksony, I., and Matanyi, S. (1974). *Prostaglandins* **6**, 211–216.

12

From Uterine Actomyosin to Uterine Steroid Hormone "Receptors"

TAMAS ERDOS

INTRODUCTION

Actomyosin of striated muscle was studied in the early 1940's by Szent-Györgyi and his collaborators in Szeged (Annau *et al.,* 1941–1943). Uterine muscle actomyosin was isolated and quantitated in 1948 (Csapo, 1948). The physicochemical properties of actomyosin from both types of muscle were characterized in Svedberg's laboratory at Uppsala (Erdos and Snellman, 1948; Snellmann and Erdos, 1948a,b; Csapo *et al.,* 1950).

While uterine actomyosin and the mechanism of myometrial contraction were studied solely by Csapo and his associates in the 1940's and early 1950's, the actomyosin and contraction of striated muscle were widely examined, since the highly ordered structure of striated muscle provides the best model for the study of the mechanism of muscular contraction. The myometrium, however, offers the best model for the study of the regulation of contractile function. Accordingly, Csapo (1950) took advantage of this model and examined the differentiation of the myometrial cells, the estradiol-induced synthesis of uterine actomyosin and the hormonal control of the myometrial contractile process.

The mechanisms of action of estradiol and progesterone on the myometrium have been extensively studied (for recent reviews, see Csapo, 1971; Finn and Porter, 1975). However, the action of estradiol at the molecular level has been studied only recently (Dupont-Mairesse and Galand, 1975), although its regulatory effect on uterine actomyosin synthesis was known in 1948 (Csapo, 1948). While the regulatory effect of progesterone on the uterus *in situ* and *in vitro* has been examined and described (for references, see Csapo, 1971), the precise molecular mechanism of action remains obscure.

An important step toward the understanding of the mode of action of steroid hormones was the discovery of the uterine estradiol "receptor," a cytoplasmic protein which binds estradiol specifically (for a review, see Jensen and Jacobson, 1962; Liao, 1975). This was subsequently followed by the description of the uterine progesterone receptor (for a review, see Jänne *et al.*, 1975). Most of these experiments were conducted on the intact uterus, although estradiol and progesterone receptors are present in both the myometrium and the endometrium.

It is now generally accepted that steroid hormones, bound to their specific receptor proteins, affect the regulatory processes by acting at the level of genetic expression of their target cells. But hormonal regulation alone does not account for maximal myometrial growth. Under physiological conditions of pregnancy, myometrial growth and function are regulated by both hormone stimulation and a response to uterine stretch, the latter being imposed on the myometrium by the growing uterine contents. Indeed, under experimental conditions, distension alone (without hormonal support) promotes myometrial growth and function (for reviews, see Csapo, 1971; Finn and Porter, 1975).

During the past 10 years our laboratory, in collaboration with A. Csapo (Washington University, St. Louis, Missouri) directed our studies toward an understanding of (a) the effect of distension on myometrial growth; (b) the properties of uterine estradiol and progesterone receptors; and (c) the relationship between plasma progesterone levels and pregnancy maintenance and termination.

THE EFFECT OF STRETCH ON MYOMETRIAL GROWTH AND FUNCTION

Ten years after my last "contact" with muscle proteins (Erdos, 1954), I returned to the field of my earlier interest. In collaboration with A. Csapo, we reinvestigated the challenging problem of stretch-induced myometrial hypertrophy, described as early as 1909 (Blair-Bell and Hick, 1909; Reynolds, 1949).

Our experiments (Csapo *et al.*, 1965; Mattos *et al.*, 1967) showed that when one of the uterine horns of an ovariectomized rabbit is distended with an intrauterine balloon, the uterus maintains its parturient weight if the animal is studied postpartum or it increases in weight if the animal is not pregnant. These studies demonstrated that the pregnancy-induced myometrial hypertrophy is maintained in the former, while it is induced in the latter case. The undistended (control) horn of the same animal undergoes involution or castrate atrophy. The quantitative data showed that myometrial growth is a function of the extent and duration of imposed stretch. Further experiments have shown that stretch provokes the same weight increase regardless of whether it is applied at 0, 5, 10 or 15 days after ovariectomy. It would appear to have been reasonable to continue these experiments and examine the problem at the molecular level, but progress in uterine estradiol receptors dominated our interest. Thus, the puzzle of how a *physical* factor, chronic stretch, can maintain the uterine contractile proteins in a functional state and even promote their synthesis (or inhibit the dramatic protein breakdown during involution) remains to be solved.

THE UTERINE ESTRADIOL RECEPTOR

The estradiol receptor is a cytoplasmic protein that binds estrogenic substances with high affinity and specificity. It is present in both the endometrium and the myometrium (for a review, see Jensen and De Sombre, 1972). Though native receptor has not yet been isolated in pure form, the remarkable stability of the receptor–hormone complex (half-life of about 20 days at 0°C; Erdos *et al.*, 1968) and the availability of tritiated estradiol with a specific activity of up to 100 Ci/mM permits its study in a impure form (Erdos, 1968; Erdos *et al.*, 1970; Best-Belpomme *et al.*, 1970). Data on the receptor's molecular parameters are still debated due to the fact that it exhibits multiple molecular forms which depend on the experimental conditions used (Erdos, 1971a,b; Réti and Erdos, 1971). Nevertheless, whatever its apparent native molecular weight and form, in denaturing solvents this protein exhibits a molecular weight of about 60,000 (Erdos and Fries, 1974). This value is similar to that of the smallest form of undenatured estradiol receptor, which retains its properties *vis-a-vis* the hormone (Rat *et al.*, 1974).

When the cytoplasmic receptor binds estradiol, the protein undergoes a "transformation" or "activation," the complex penetrates the cell nucleus, and it binds to chromatin. The mechanism of this complex reaction, the exact nature of the acceptor-site(s) on the chromatin and binding site(s) on the receptor, is still unknown. Work in our laboratory has shown that (a) the stability of the receptor-hormone complex is increased by a factor of 30 through binding to the chromatin (i.e., its half-life is 30 times higher than when unbound (Sala-Trepat and Réti, 1974); (b) chromatin-binding sites of the cytoplasmic receptor can be destroyed by mild proteolysis, leaving the estradiol binding sites intact (Sala-

Trepat and Vallet-Strouve, 1974); (c) the receptor–estradiol complex is bound preferentially to a small, heavy fraction of the chromatin, corresponding to 5% of total DNA (Sala-Trepat *et al.,* 1977).

The mechanism by which the chromatin-bound receptor affects the regulatory processes at the level of genetic expression is not known at present. Yet this question is one of the most exciting problems of molecular endocrinology, and so is the nature of the cytoplasmic receptor, its reaction with the hormone and the identity or nonidentity of the endometrial and myometrial estradiol receptor. Probably the purification of the receptor would promote the solution of these problems. However, this is a formidable task, since a 100,000X purification would be required of the protein which has a strong tendency to form irreversible aggregates. Appreciating that native receptor is not available yet in a pure state, we have recently developed in collaboration with F. Sweet (Washington University, St. Louis, Missouri) a series of affinity chromatography columns for the purification of steroid-binding proteins. Among these the estrone 17-carboxymethyl oximinobisaminoethyldisulfide monoamide-agarose holds promise. This column is stable and the specifically adsorbed receptor is easily eluted by splitting disulfide linkage with 2-mercaptoethanol (Sweet and Szabados, 1977).

PROGESTERONE AND PREGNANCY MAINTENANCE

Pioneering experiments in 1930 predicted that the indispensability of progesterone for pregnancy maintenance is a biological law of broad validity (Corner and Allen, 1930). The "blocking action" of progesterone on myometrial excitability was described in 1956 (Csapo, 1956). Extensive studies in animal "models" and human patients, complemented by reliable progesterone radioimmunoassays, led to the concept of the "critical progesterone value," i.e., the progesterone concentration at which normal pregnancy could not be maintained. The critical progesterone value was found, of course, to be different at different stages of gestation and in different species as predicted by the "seesaw theory" (Csapo, 1971, 1975), which made the universal application of this concept difficult. A typical example of this follows: rats treated with antibodies to progesterone aborted predictably only when antibody was given at day 11 of gestation, a finding which was interpreted as an indication that the regulatory mechanism of pregnancy is only sensitive to antibody-induced withdrawal of progesterone at this critical day of gestation (Raziano *et al.,* 1972). Therefore, in collaboration with A. Csapo, we gave the universality of the concept of a critical progesterone value a new trial.

Groups of day 6 and day 10 pregnant rats were injected with antiprogesterone antibodies. The concentrations of antibody, total progesterone, progesterone bound to antibody, and unbound progesterone were measured in the plasma samples collected sequentially after antibody treatment, using the strictly quantitative methods developed for the purpose of this study (Dessen and Erdos, 1976a). The results showed that antibody-induced withdrawal of progesterone provokes abortion over a broad gestation range provided that unbound progesterone is reduced to a "critical abortion value." However, at day 6 this value is lower than at day 10, and the rate of progesterone synthesis is higher. Therefore, 4 times more antibody was needed for reducing unbound progesterone to the critical value at day 6 than at day 10 (Csapo *et al.*, 1974, 1975; Csapo and Erdos, 1976). Effective antibody treatment reduced unbound progesterone to the critical value within 3–6 hr and kept it at this low value until the animals started to bleed and then aborted 24 hr later. Thus the universality of the concept of the critical progesterone value received additional support. We believe that the strictly quantitative information obtained in these studies represents a solid commencement of the inquiry into the problem of how progesterone acts at the molecular level.

Evidently, the continued binding of progesterone by antibody reduces unbound progesterone to a critically low level, which is incompatible with pregnancy maintenance. Yet a computer simulation of the biological experiments indicated that the situation is somewhat more complicated. Computer simulation was based on the assumptions that the rates of progesterone synthesis and clearance remain constant during the experiments. The results indicated that the antibody doses used in the biological experiments only provoked drastic decreases in unbound progesterone during the initial 20–40 min after injection of antibody. Subsequently unbound progesterone increased gradually as antibody became saturated with the constantly synthesized progesterone. Thus, unbound progesterone inevitably returned to its original value within 3–6 hr (Dessen and Erdos, 1976b).

Consideration of these results led to the suggestion that the decisive step, induced by progesterone withdrawal which ultimately provokes abortion (observed 24–48 hr later) must have occurred during this short-lived initial decrease in unbound progesterone. Most probably, the initial decrease of unbound progesterone compromises the feto-placental unit and consequently the maternal organism "switches over" to a low rate of progesterone synthesis corresponding to that of a nonpregnant animal. This interpretation explains why unbound progesterone is low when antibody is already saturated. Further experiments substantiated this interpretation by revealing that the initial effect of progesterone withdrawal is reversible (A. I. Csapo and T. Erdos, unpublished). A dose of 5 mg progesterone injected 3 hr after antibody treatment protected pregnancy in spite of the reduced unbound progesterone for the preceding 3 hr, while progesterone injected 6 hr after antibody treatment was ineffective and the animals aborted. Thus, the possibility cannot be ignored that the target of the initial effect of progesterone withdrawal is the selective reversible inhibition of progesterone synthesis. We may logically assume that the initial progesterone withdrawal suppresses the synthesis of placental luteotropin.

WHAT IS IN THE BLACK BOX?

Myometrial growth, which includes the massive synthesis of contractile proteins, is under the double control of estradiol and stretch, provoked by increasing estradiol levels and the growing uterine contents during pregnancy. Experimentally, myometrial growth can be induced by either estradiol or by uterine distension in pregnant as well as nonpregnant animals. Indirect evidence suggests that the action of estradiol (indeed all steroid hormones) is mediated through a specific "receptor" protein. However, between the steroid receptor and its observed effect, there remains a "black box," which is truly unilluminated in the case of the myometrium. The way in which stretch promotes myometrial growth is a further puzzle. Yet to ask, "What is in this black box?" is no less than asking how protein synthesis is regulated in eukaryotic cells, a problem which is certain to keep molecular biologists busily occupied in the coming decade.

Are steroid hormone target organs, in general, and the myometrium, in particular, good systems for the study of regulation of protein synthesis? The answer seems to be self-evident, since knowing the specific signal which turns on protein synthesis is a considerable advantage. The myometrium would seem to be a superior organ for the study, since it is simpler to examine a system that responds to a signal with massive synthesis of well known proteins such as actin and myosin than one with cell proliferation. The study of a similar response (the massive ovalbumin synthesis of the chick oviduct after estradiol treatment) so far furnished the best information about estradiol-induced protein synthesis (O'Malley and Means, 1974; Schimke, 1975).

The initial events of estradiol action in the myometrium and in the endometrium (and for that matter, of all steroid hormones in their target cells) seem to be identical: the hormone bound to the receptor protein penetrates into the nucleus and "induces" the synthesis of a protein, the function of which is as yet unknown. Yet, the ultimate response of the two tissues to estradiol is profoundly different: massive protein synthesis in the myometrium and cell proliferation in the endometrium. Evidently the two tissues are best studied separately.

The stimulatory effect of estradiol on phospholipid, RNA and protein synthesis in the total uterus has been studied extensively since 1958 (Mueller *et al.*, 1958). It would be desirable to conduct similar studies on the effects of estradiol and of stretch on the myometrium, since (regardless of the early events which follow these two different treatments) the final result, massive protein synthesis, is the same. Methods to identify and prepare myosin mRNA from striated muscle have been described (Heywood and Nwaguru, 1969). Uteri of immature, adult and even pregnant cow and sheep (available from slaughterhouses) could provide ample material for preparing uterine myosin mRNA and open a new field of investigation with a reconstructed acellular system.

Another attractive study is the myometrial cell in tissue culture. Myoblasts, isolated from striated muscle, cultured as an established cell line (Yaffé, 1968), retain their ability to differentiate into myotubes. In primary cell cultures, myosin mRNA is synthesized in the myoblasts prior to their differentiation into myotubes (Buckingham *et al.,* 1974). It is quite likely that by similar studies with myometrial cells, the effect of estradiol on differentiation and on myosin synthesis can be clarified.

The properties of the myometrial progesterone receptor are similar to those described for other steroid hormone receptors (Jänne *et al.,* 1975). Since the molecular mechanism of the myometrial progesterone block is not yet known, it is uncertain whether it is mediated through a receptor. Similarly unknown is whether the synthesis of a specific protein is required for the effect. It would probably be simpler to devise an experiment for answering these questions (using inhibitors of RNA and/or protein synthesis) than to enumerate arguments for or against the various possibilities. Should these experiments indicate that RNA and/or protein synthesis is needed for the progesterone block, one would at least know what stands on both sides of the "black box." However, a profoundly different experimental strategy (designed to penetrate into this black box) may be needed than the one used for the study of estradiol action. To ask how myosin synthesis is "induced" by estradiol is practically the same as asking how protein synthesis is regulated in eukaryotic cells. In the case of the progesterone block, different questions must be asked. For instance: What are the kinetics of the development of the progesterone block? Can they be compared with the kinetics of protein synthesis? Does the extent of cytoplasmic and nuclear binding of progesterone correspond with the changes induced in the excitability of the myometrium?

If the progesterone block is independent of protein synthesis, the question arises: What is the biological role of the myometrial progesterone receptor? and, even more important: Through what mechanism does progesterone act? Does progesterone interfere directly with the activation of the uterus? Is there a membrane site for its action? If so, what is the relationship between this site and the binding sites of prostaglandins (Wakeling and Wyngarten, 1974) or oxytocin (Soloff, 1975)? Experimental approaches to these problems are within the realm of possibility even before more information becomes available on the fundamental problem: the regulation of protein synthesis in eukaryotic cells.

Szent-Györgyi wrote in his magnificent little book, "The Living State" (1972): "Science is built on the premise that nature answers intelligent questions intelligently; so if no answer exists, there must be something wrong with the question." We can only hope that some of our questions will be occasionally answered intelligently.

REFERENCES

Annau, E., Banga, I., Erdos, T., Laki, K., Mommaerts, W., Straub, F., and Szent-Györgyi, A. (1941–1943). *Stud. Inst. Med. Chem. Univ. Szeged.*

Best-Belpomme, M., Fries, J., and Erdos, T. (1970). *Eur. J. Biochem.* **17,** 425.

Blair-Bell, W., and Hick, P. (1909). Cited by Finn and Porter (1975).

Buckingham, M. E., Caput, D., Cohen, A., Whalen, R. G., and Gros, F. (1974). *Proc. Natl. Acad. Sci. U.S.A.* **71,** 1466.

Corner, G. W., and Allen, W. M. (1930). *Proc. Soc. Exp. Biol. Med.* **27,** 403.

Csapo, A. I. (1948). *Nature (London)* **162,** 218.

Csapo, A. I. (1950). *Am. J. Physiol.* **162,** 406.

Csapo, A. I. (1956) *Am. J. Anat.* **98,** 273.

Csapo, A. I. (1971). *In* "Contractile Proteins and Muscle" (K. Laki, ed.), p. 413. Dekker, New York.

Csapo, A. I. (1975). *Am. J. Obstet. Gynecol.* **121,** 578.

Csapo, A. I., and Erdos, T. (1976). *Am. J. Obstet. Gynecol.* **126,** 598.

Csapo, A. I., Erdos, T., Naeslund, I., and Snellmann, O. (1950). *Biochim. Biophys. Acta* **5,** 53.

Csapo, A., Erdos, T., Mattos, C. R., Gramss, E., and Moscowitz, C. (1965). *Nature (London)* **207,** 1378.

Csapo, A. I., Dray, F., and Erdos, T. (1974). *Am. J. Obstet. Gynecol.* **120,** 572.

Csapo, A. I., Dray, F., and Erdos, T. (1975). *Endocrinology* **97,** 603.

Dessen, P., and Erdos, T. (1976a). *Biochimie* **58,** 455.

Dessen, P., and Erdos, T. (1976b). *Biochimie* **58,** 1227.

Dupont-Mairesse, N., and Galand, P. (1975). *Endocrinology* **96,** 1587.

Erdos, T. (1954). *Acta Physiol. Acad. Sci. Hung.* **7,** 1.

Erdos, T. (1968). *Biochem. Biophys. Res. Commun.* **32,** 338.

Erdos, T. (1971a). *Horm. Steroids, Proc. Int. Congr., 3rd, 1970* Int. Congr. Ser. No. 219, Vol. 3, p. 364.

Erdos, T. (1971b). *Adv. Biosci.* **7,** 120–135.

Erdos, T., and Friès, J. (1974). *Biochem. Biophys. Res. Commun.* **58,** 932.

Erdos, T., and Snellmann, O. (1948). *Biochim. Biophys. Acta* **2,** 642.

Erdos, T., Gospodarowicz, D., Bessada, R., and Friès, J. (1968). *C. R. Hebd. Seances Acad. Sci.* **266,** 2164.

Erdos, T., Best-Belpomme, M., and Bessada, R. (1970). *Anal. Biochem.* **37,** 244.

Finn, C. A., and Porter, D. G. (1975). *In* "The Uterus" (C. A. Finn, ed.). Elek Science, London.

Heywood, S. M., and Nwaguru, M. (1969). *Biochemistry* **8,** 3839.

Jänne, O., Kontula, K., Luukainen, T., and Vihko, R. (1975). *J. Steroid Biochem.* **6,** 501.

Jensen, E. V., and De Sombre, E. R. (1972). *Annu. Rev. Biochem.* **41,** 203.

Jensen, E. V., and Jacobson, H. I. (1962). *Recent Prog. Horm. Res.* **18,** 387.

Liao, S. (1975). *Int. Rev. Cytol.* **41,** 87.

Mattos, de C. E. R., Kempson, R. L., Erdos, T., and Csapo, A. (1967). *Fert. Steril.* **18,** 545.

Mueller, G. C., Herranen, A. M., and Jervell, K. F. (1958). *Recent Prog. Horm. Res.* **14,** 95.

O'Malley, B. W., and Means, A. R. (1974). *Science* **183**, 610.
Rat, R. L., Vallet-Strouve, C., and Erdos, T. (1974). *Biochimie* **56**, 1387.
Raziano, J., Ferin, M., and Van de Wiele, R. L. (1972). *Endocrinology* **90**, 1133.
Réti, E., and Erdos, T. (1971). *Biochimie* **53**, 435.
Reynolds, S. R. M. (1949). "Physiology of the Uterus." Harper (Hoeber), New York.
Sala-Trepat, J. M., and Réti, E. (1974). *Biochim. Biophys. Acta* **338**, 92.
Sala-Trepat, J. M., and Vallet-Strouve, C. (1974). *Biochim. Biophys. Acta* **371**, 186.
Sala-Trepat, J. M., Hibner, U., and Vallet-Strouve, C. (1977). *Nucleic Acids Res.* **4**, 649.
Schimke, R. T. (1975). *Recent Prog. Horm. Res.* **31**, 175.
Snellmann, O., and Erdos, T. (1948a). *Biochim. Biophys. Acta* **2**, 650 and 660.
Snellmann, O., and Erdos, T. (1948b). *Biochim. Biophys. Acta* **3**, 523.
Soloff, M. S. (1975). *Biochem. Biophys. Res. Commun.* **65**, 205.
Sweet, F., and Szabados, L. (1977). *Steroids* **29**, 127.
Szent-Györgyi, A. (1972). "The Living State with Observations on Cancer," p. 1. Academic Press, New York.
Wakeling, A. E., and Wyngarten, L. J. (1974). *Endocrinology* **95**, 55.
Yaffé, D. (1968). *Proc. Natl. Acad. Sci. U.S.A.* **61**, 447.

13

Effects of Heavy Water on Muscle

BENJAMIN KAMINER

The impact of Albert Szent-Györgyi's writings is immeasurable, particularly in his little books (Szent-Györgyi, 1947, 1951, 1953, 1957, 1960, 1968, 1972) in which he weaves his discoveries, concepts, and beliefs into a colorful fabric. His excitement of discovery as he sails unchartered seas, and his intuitive, imaginative insights have stimulated and fascinated students and young investigators far and wide. His penetrating questions and insights into the biological role of water, the "mater of life," particularly in terms of its possible role in muscle contraction, stimulated my interest and led to the simple question: would heavy water, with properties different from normal water, influence muscle contraction and be a useful tool in elucidating aspects of the mechanisms of contraction?

Following the discovery of deuterium by Urey (1933) and his colleagues the prediction was immediately made by these investigators and others on possible biochemical effects of deuterium substitution in view of the lower zero point energy of deuterium as compared to hydrogen; being lower by 1.2 to 1.5 kcal per mole, the deuterium bond is more stable than the hydrogen bond, requiring a greater energy of activation. Besides this primary isotope effect, substitution of heavy water (D_2O) for normal water (H_2O) may lead to effects associated with the changed solvent properties. D_2O for example is about 10% more dense than H_2O, has a higher melting point (3.79°C), is more viscous (about 18% higher than H_2O at 38°C), is a poorer solvent, its dissociation constant is 1.95×10^{-15}, i.e., about 20% lower than that of H_2O, and salts dissolved in D_2O have a lower electrical conductivity (Kirshenbaum, 1951).

Following up these predictions, Lewis and Goody (1933) demonstrated retardation of germination of tobacco seeds grown in D_2O in the first biological experiment. The production of D_2O in sufficient quantities led to the testing of its effect on a variety of biochemical reactions and biological preparations from viruses to mammals. The early work was well reviewed by Morowitz and Brown (1953) and later studies were covered in a symposium (Kritchevsky, 1960) and in a monograph by Thomson (1963).

Our question led to studies on the effect of heavy water on heart, smooth, and striated muscle (Kaminer, 1960, 1966; Kaminer and Kimura, 1972).

HEART MUSCLE

Tension and surface electroatriograms were recorded from spontaneously beating isolated atria of *Xenopus laevis* (African toad). After a control period in a temperature controlled chamber, the normal frog Ringer's solution was replaced with ones containing D_2O in concentrations of 25%, 50%, 75%, and 99.8%. The rate and force of contraction were markedly affected as was observed previously in frog hearts (Barnes and Warren, 1935; Brandt, 1935; Verzar and Haffter, 1935). Figure 1 illustrates a progressive decrease in frequency of contraction with increasing concentrations of D_2O. A noticeable reduction of frequency was observed 50 sec after immersion in the D_2O and the frequency was reduced by 75% within 15 min in 75% or 99.8% D_2O. Thereafter the atria in 99.8% D_2O either stopped beating, or slowed down to 1 beat every 5 or 7 min.

Similarly a reduction in tension occurred within 50 sec in D_2O. In 99.8% D_2O, the tension decreased by more than 90% of its basal value after 2.5 min (Fig. 2). The maximum reduction in tension occurred in 5 min in all concentrations of D_2O. In 25% D_2O there was a subsequent rise in tension after 10–20 min but it remained 23% below the basal value (Fig. 2). This suggests some degree of adaptation in this lower D_2O concentration. The velocities of contraction and relaxation were also reduced.

Simultaneous recordings of the surface electroatriogram (Fig. 3) revealed that 99.8% D_2O increased the duration of depolarization by 40% and repolarization by 60% within 5 min. The interval between the onset of the electrical impulse and contraction (E-C interval) increased by about 90% in 99.8% D_2O (Fig. 4). The duration of the E-C interval increased progressively with time, and after 20 min in 25% and 50% D_2O, increased by 22% and 55%, respectively (Fig. 4). All the effects of D_2O referred to above are reversible on resubstitution with H_2O-Ringer's solution (Fig. 3).

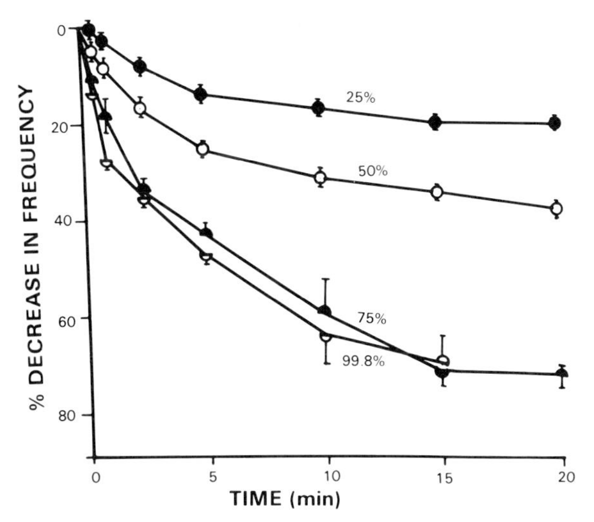

Fig. 1. The percentage decrease of frequency of the spontaneously beating isolated atria of *Xenopus laevis* in varying concentrations of heavy water (D_2O) in Ringer's solution for 20 min. The means and standard deviations are depicted for 10 preparations in each concentration of D_2O. Closed circles–25% D_2O; open circles–50%; closed semicircles–75%; open semicircles–99.8%. D_2O was obtained from Norsk Co., Norway. Frog atria were subsequently re-examined in D_2O obtained from BioRad, California. Atria in Ringer's made with glass-distilled D_2O also ceased to contract. Similarly glass-distilled D_2O produced inhibition of the rabbit heart and Purkinje fibers described in the text. The findings on frog sartorius muscle were also confirmed using glass-distilled D_2O (Sandow *et al.*, 1976). Hence the suggestion by Huxtable and Bressler (1974) that the D_2O inhibition might be due to a contaminant is extremely unlikely.

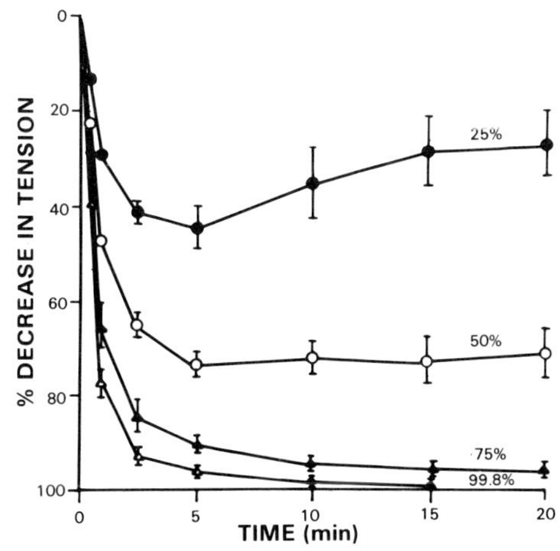

Fig. 2. The percentage decrease of tension of the spontaneously beating atria as in Fig. 1. Methods of recording are described under Fig. 3.

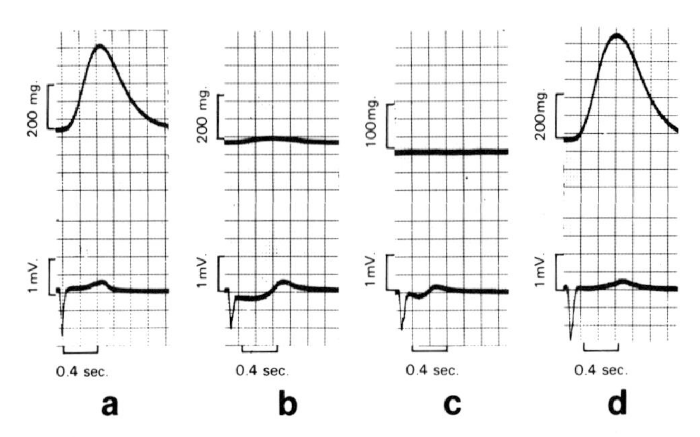

Fig. 3. Simultaneous recordings of tension (upper tracing) obtained with an RCA 5734 transducer and unipolar surface electroatriogram (lower tracing). (a) Basal recording in H_2O-Ringer's solution; (b) after 5 min in 99.8% D_2O-Ringer's solution; (c) after 17 min in 99.8% D_2O-Ringer's solution (note, no recordable contraction with increase in gain of amplifier); (d) recovery after 15 min in H_2O-Ringer's solution (note the force of contraction is greater than in the basal recording). (From Kaminer, 1960.)

A striking feature in about 40% of the atria immersed in 99.8% D_2O was the complete lack of a contractile response to the infrequent spontaneous electrical impulse (Fig. 3). Atria, completely inhibited, were stimulated electrically, resulting in a propagated electrical impulse, but importantly, there was no contractile response. Weak, slowly contracting atria responded to stimulation with a weak contraction never greater than the preceding spontaneous contractions; increas-

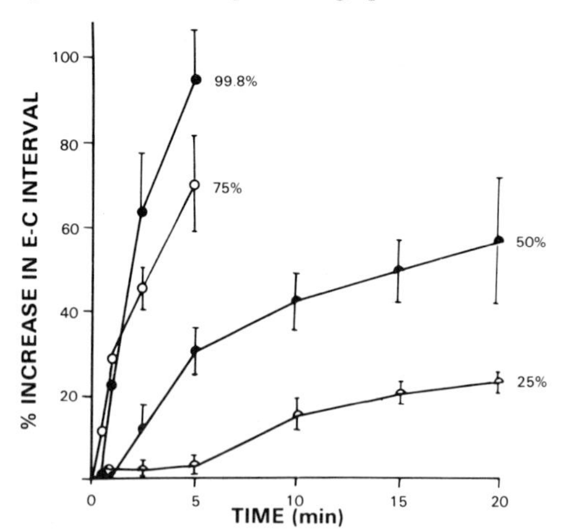

Fig. 4. The percentage increase of the excitation-contraction interval (E-C interval) in the varying concentrations of D_2O as in Fig. 1. The interval is measured from the onset of the depolarization wave to the onset of the contraction as seen in Fig. 3. After 5 min, this interval was difficult to measure accurately from recordings obtained from atria in 75% and 99.8% D_2O because of the very slow rise of the tension graph.

ing the strength of the electrical stimulation did not augment the force of contraction. Hence, excitation was not significantly inhibited and the reduced tension in D_2O appeared to be due to inhibition of either the excitation-contraction coupling process or the contractile mechanism. Evidence will be presented later in support of the former proposition.

The slowing of the heart rate and complete cessation in some instances suggest an inhibition of pacemaker activity. Pacemaker cells of the frog sinus venosus and calf Purkinje fibers are therefore being currently investigated by Rebecca Biegon, a graduate student, and William L. Hardy, in our department. Recording intracellularly (Fig. 5), they have shown that 90% D_2O will decrease the frequency of the spontaneous action potentials. In Purkinje fibers, for example, activity will cease completely within several minutes. There is an associated progressive decline in the slope of the prepotential and an increase in duration of the plateau and repolarization phases. The D_2O may therefore in some way modify changes in outward current. Electrical stimulation of Purkinje fibers, completely inhibited in D_2O, results in action potentials with normal amplitudes (Fig. 6) and the threshold is not increased, suggesting no significant effect on the inward conductance and current. These interpretations correspond with those of Jenerick (1964) in his findings on the effect of D_2O on the action potential of frog striated muscle. Electrically induced action potentials of Purkinje fibers in 90% D_2O showed a complete flattening of the prepotential even in the presence of epinephrine (Fig. 6b); resubstitution with H_2O–Ringer's

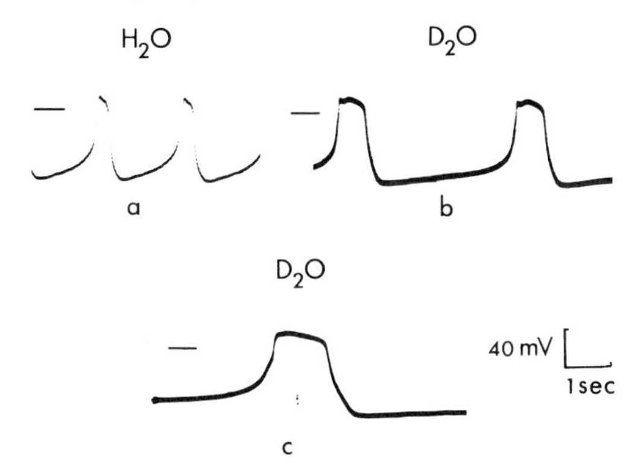

Fig. 5. Effects of D_2O on spontaneous action potentials from isolated Purkinje fibers (R. Biegon and W. L. Hardy, personal communication). (a) Intracellular recordings of calf Purkinje fibers in H_2O-Tyrode's solution containing 5m*M* epinephrine (temperature 35.5–36°C). (b) Action potentials within 1 min after replacement of H_2O-Tyrode's solution with 90% D_2O-Tyrode's solution under the same conditions as in (a). (note the decreased frequency, the decline in prepotential slope and the increased duration of the plateau.) (c) 1.5 min after replacement with D_2O-Tyrode's solution. The prepotential slope is markedly diminished and the plateau and the repolarization phases of the action potential are markedly increased in duration. Immediately after this recording the potentials ceased. Resubstitution with H_2O-Tyrode's solution led to recovery. Bar represents 0 mV.

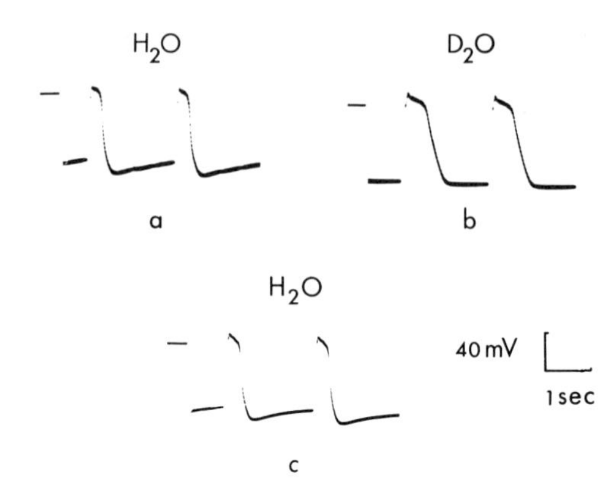

Fig. 6. Effects of D_2O on electrically induced action potentials from isolated Purkinje fibers (R. Biegon and W. L. Hardy, personal communication). (a) Action potentials induced by field electrical stimuli at rate of 0.5 per sec (0.5 msec duration). Conditions are as in Fig. 5a except for the concentration of epinephrine being 0.1 m*M*. (b) Action potentials 5 min after replacement of H_2O-Tyrode's solution with D_2O-Tyrode's solution containing 0.1 m*M* epinephrine. Note the complete flattening of the prepotential and increased duration of the plateau and repolarization phases of the action potentials. (c) Action potentials 60 min after replacement with H_2O-Tyrode's solution containing 0.24 m*M* epinephrine. Note recovery of the prepotential; durations of the other phases of the action potential are now comparable with those in (a). Bar represents 0 mV.

solution resulted in recovery of the prepotential (Fig. 6c). The prepotential in Purkinje fibers is related to "deactivation" of potassium conductance (Noble, 1975) and how D_2O inhibits this particular event will require further investigation. These findings suggest that the decreased frequency and complete cessation of the electrical impulse in the *Xenopus* atria in D_2O is due mainly to the inhibition of the pacemaker potential.

A series of experiments was done on atria of *Xenopus laevis* to see if agents, known to augment the tension of heart muscle, would do so in the presence of D_2O, thus overcoming its inhibitory influence; epinephrine, norepinephrine, veratrine and calcium were selected.

Epinephrine ($1/2 \times 10^6$ dilution) which had the expected chronotropic effect on atria in H_2O-Ringer's solution had no such influence in the presence of 75% D_2O, thus it could not overcome the depressing influence of D_2O on the pacemaker which is in keeping with the above results on Purkinje fibers obtained by Biegon and Hardy. In control atria, epinephrine had, within 1–5 min, a marked inotropic action which then subsided, but lasted for about 15 min. In 75% D_2O, epinephrine reduced the D_2O inhibition by about 30%. This counter-effect against D_2O persisted for about 25 min as compared with the inotropic effect of 10–15 min in H_2O suggesting a greater stability of epinephrine in D_2O.

Norepinephrine ($1/2 \times 10^6$ dilution) did not have as marked on inotropic action as did epinephrine on the control atria, but the effect lasted for about 15 min. In 75% D_2O, norepinephrine had no significant inotropic action. Norepinephrine increased the atrial rate in H_2O-Ringer's solution but had no influence in D_2O.

Veratrine ($1/5 \times 10^5$ dilution) slowed down the atria and caused arrhythmias in H_2O-Ringer's solution. In 75% D_2O, the slowly contracting atria were slowed even further by veratrine and it also caused arrhythmias. In H_2O-Ringer's solution, veratrine increased the force of contraction in a period of 5 min, and thereafter had a negative inotropic effect. In 75% D_2O, veratrine had an inotropic effect which persisted for 25 min.

A high concentration of calcium (4.5 m*M*), sufficient to approximately double the force of contraction in control atria, had no inotropic influence on atria immersed in 75% D_2O. Andrew Wexler, a graduate student in our department, studying the effects of D_2O on the isolated perfused rabbit heart has, however, shown that high concentrations of calcium (11 m*M*) increases the force of the weakly contracting heart in 90% D_2O; similar findings have been reported by Vargas and Johnson (1975). This species difference remains to be explained.

SMOOTH MUSCLE

Seventy-five percent D_2O slowed down and decreased the force of spontaneously contracting rabbit duodenum (Figs. 7–9) and guinea pig uterus (Fig.

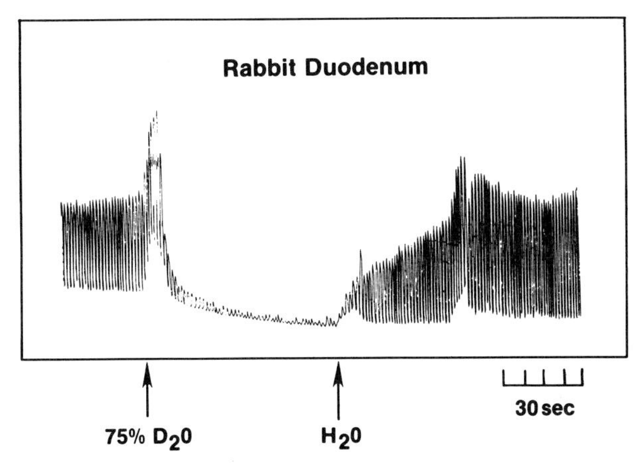

Fig. 7. Isotonic recordings of spontaneously contracting isolated rabbit duodenum. Replacement of H_2O-Tyrode's solution with 75% D_2O-Tyrode's solution leads to reduction in contraction, tone and frequency. The initial apparent rise in tone is an artifact associated with emptying and filling of chamber. Resubstitution with H_2O-Tyrode's solution leads to gradual recovery. The apparent increase in contraction in the middle of recovery tracing is due to a further change in the solution. Note the increased amplitude of contraction during this recovery phase.

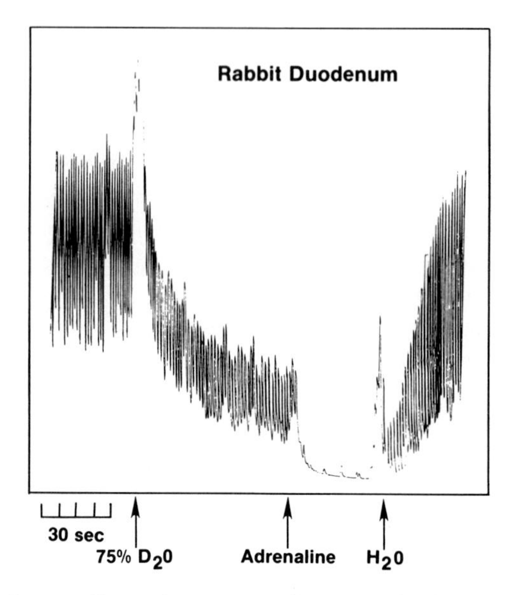

Fig. 8. Isotonic recordings of spontaneously contracting isolated rabbit duodenum. Response to 75% D_2O-Tyrode's solution is essentially as in Fig. 7 but the effect is not as marked. Addition of epinephrine ($1/1 \times 10^5$ dilution) in 75% D_2O leads to a further drop in tone and inhibition of contraction. Replacement with H_2O-Tyrode's solution leads to recovery. Artifacts are as described in Fig. 7.

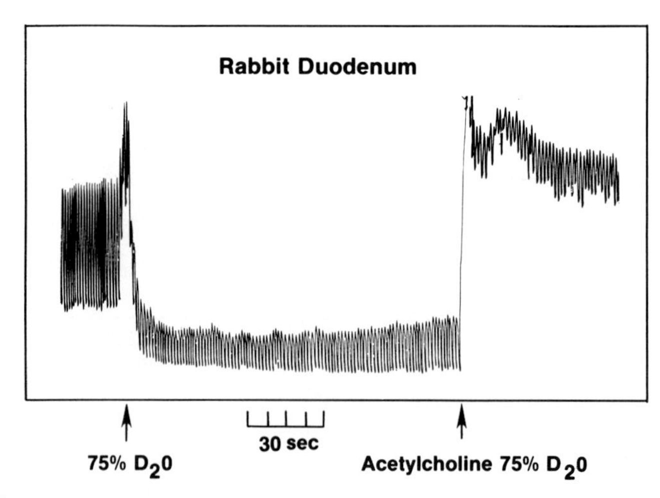

Fig. 9. Isotonic recordings of spontaneously contracting isolated rabbit duodenum. Response to 75% D_2O-Tyrode's solution as in Figs. 7 and 8. Addition of acetylcholine ($1/1 \times 10^6$ dilution) leads to a marked contraction.

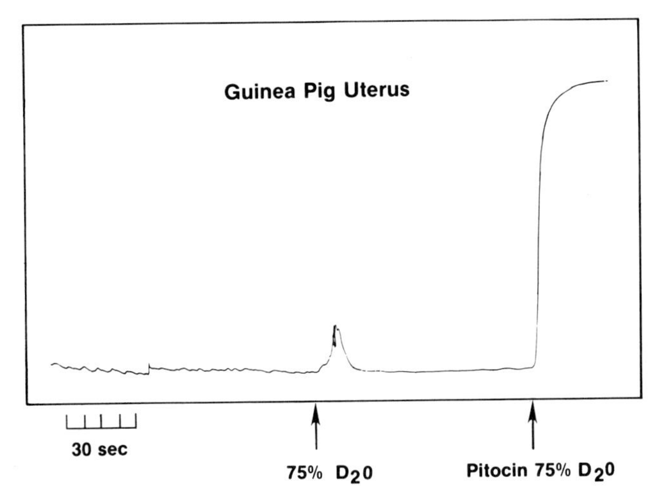

Fig. 10. Isotonic contraction of spontaneously contracting isolated guinea pig uterus. The contraction in H_2O-Tyrode's solution is weak. Replacement with 75% D_2O-Tyrode's solution inhibits contraction completely. Note the artifact due to emptying and filling of chamber. Addition of pitocin (0.04 IU) leads to a marked sustained contraction.

10). In the presence of D_2O, epinephrine decreased the tone and contraction of the duodenum (Fig. 8) and acetylcholine caused a marked contraction (Fig. 9). Pitocin induced a marked sustained contraction of the uterus in the presence of D_2O (Fig. 10).

In view of our above findings on heart and smooth muscle, consideration should be given to the possible influence of D_2O on the transport and release of Ca^{2+} and metabolism of cyclic nucleotides, the levels of which influence the contraction and relaxation cycle in both smooth and heart muscle (Rasmussen, 1970; Berridge, 1975). Is it possible that the negative inotropic and chronotropic effects of D_2O are due to direct or indirect retardation of transport or release of Ca^{2+} and of metabolism of cyclic nucleotide? The coupled Ca^{2+}–cyclic nucleotide interactions should also be investigated in relation to neurotransmitters, hormones and glycosides (Berridge, 1975) in the presence of D_2O, in view of the described effects of these agents in this investigation.

STRIATED MUSCLE

An obvious difficulty is to sort out the main sites of action of D_2O. One approach was to study its effect on glycerol-extracted psoas fibers (Szent-Györgyi, 1953) to see if the contractile mechanism was directly inhibited by D_2O. We first established that D_2O decreased markedly the force of contraction

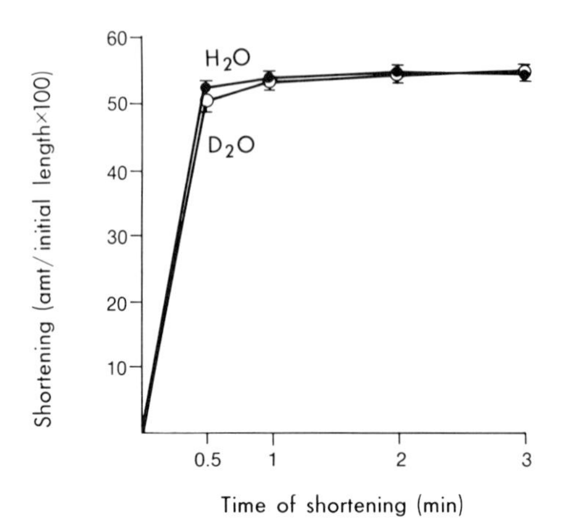

Fig. 11. Effect of D_2O on the degree of shortening of glycerol-extracted psoas fibers in response to ATP. Average values (with standard errors) are depicted of 10 experiments in H_2O-salt solution (full circles) and D_2O-salt solution (clear circles). Procedure and salt solution used are as described by Szent-Györgyi (1951). D_2O did not influence the response to ATP. (From Kaminer, 1960.)

of the living psoas bundles stimulated electrically; complete inhibition, however, did not occur. Figure 11 shows no influence of D_2O on the ATP-induced shortening of glycerol-extracted fibers. In subsequent experiments the degree of tension was likewise not decreased in 99.8% D_2O. Considering these results together with the lack of a significant effect of D_2O on the action potential, we concluded that D_2O affected predominantly the E-C coupling sites. Goodall (1958) as well as Gunther and Pfeiffer (1974) concluded that D_2O inhibited the contractile mechanism, whereas Svensmark (1961) remained uncertain as to the site of action.

To test our hypothesis further we studied the effect of D_2O on the giant muscle fibers of the giant barnacle (Kaminer and Kimura, 1972). Using this muscle Ashley and Ridgway (1970) had elegantly demonstrated the release of calcium following membrane excitation by having injected into the muscle, aequorin, a protein which luminesces in the presence of calcium (Shimomura *et al.*, 1962). We predicted that D_2O would, in this preparation, not significantly inhibit the excitation process but would retard or completely inhibit the release of calcium and contraction. If, however, calcium release would not be affected while contraction was inhibited, then the main D_2O inhibitory site would be closer to the actomyosin interaction, namely at the regulatory protein system. Our findings indeed turned out as predicted. In 99.8% D_2O the aequorin luminescence or "flash", the indicator of calcium release, disappeared and contraction was completely inhibited, but the electrotonic potential remained (Fig. 12). The flash and tension were not completely eliminated by D_2O in some

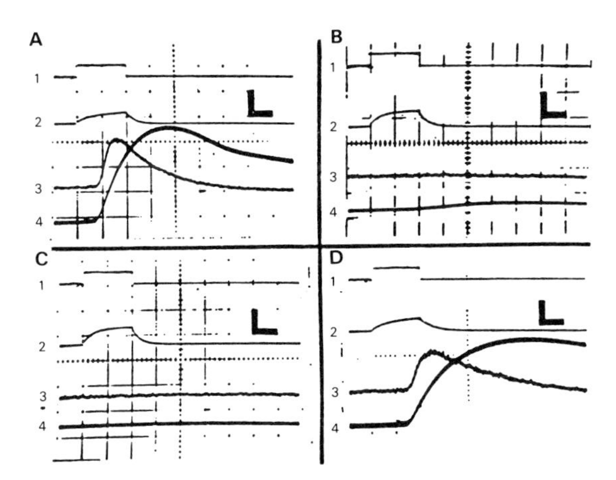

Fig. 12. Effects of 99.8% D_2O on a muscle fiber of *Balanus nubilis.* Recordings are depicted of a fiber in the following solutions: (A) H_2O-Ringer's. The solution was then replaced with 99.8% D_2O Ringer's and the fiber was incubated for 5 min (B) and 10 min (C). Thereafter the D_2O solution was replaced with one containing H_2O and the fiber was incubated for 10 min (D). Traces are: 1, the stimulus (45 μa, 200 msec); 2, membrane potential (20 mV/div.); 3, light emission of aequorin (1 μl of 0.1–0.5m*M* aequorin was injected, and emission was measured with an RCA 6342A photomultiplier tube); and 4, isometric tension (0.5 g/div.) measured with a Grass FT03 transducer. The horizontal calibration bar equals 100 msec. After the preparation was in D_2O for 5 min (B) the flash disappeared; the contraction was minimal at this time and disappeared after 10 min in D_2O (C). There were variations in the amplitude and shape of the membrane potential. In this example, recovery is not complete; in others, recovery was complete. (From Kaminer and Kimura, 1972. *Science* **176**, 406. Copyright 1972 by the American Association for the Advancement of Sciences.)

experiments. Once again the effects were reversible on replacement with H_2O-Ringer's solution (Fig. 12). Increasing the strength of stimulation tenfold did not induce a flash or contraction. We found that caffeine (10 m*M*) overcame the inhibition of contraction; in this experiment we only measured the tension. To exclude the possibility of D_2O inhibiting the luminescent reaction of aequorin and Ca^{2+}, we determined, by visual means, no difference in the flash of aequorin in H_2O and D_2O on exposure to Ca^{2+}. From all the evidence presented it appears that the *main* inhibitory effect of D_2O is on the E-C coupling process, but we are uncertain as to the precise point or points of action of D_2O.

Following these observations a number of other investigators pursued the problem. Using the frog sartorius muscle, Sandow *et al.* (1976) found that D_2O slowed the development of the latency relaxation besides reducing markedly the peak tension. They interpret the increase in time between stimulation and the start of the latency relaxation as a slowing of conduction along the T tubules and they consider the slowing down of both the development of latency relaxation and the subsequent rise in tension as a depressive effect of D_2O on Ca-release by the sarcoplasmic reticulum.

In studies on isolated rat muscle sarcoplasmic reticulum in D_2O, Huxtable and Bressler (1974) found a shift in the pD dependency curves for calcium binding and transport and for Ca/Mg-ATPase toward higher values relative to the corresponding pH dependency curves of preparations in H_2O. The maxima reached at the higher pD values were, however, similar to those reached in H_2O. The maximum for Ca-binding was at pH 6.5 and for Ca/Mg-ATPase (under their transport conditions) was at pH 7.2; at these corresponding pD values, D_2O reduced the Ca-binding by about 56% and the ATPase by about 65% as derived from the graphs presented. The authors state that the inhibitory effects of D_2O on the sarcoplasmic reticulum are a function of the pD but they also state that the sarcoplasmic reticulum is probably not the site of D_2O inhibition of excitation-contraction coupling.

Bezanilla and Horowicz (1975) found, on stimulating frog muscle stained with Nile blue, fluorescent intensity changes associated with excitation-contraction coupling and suggest that these changes could possibly be associated with potential changes in the sarcoplasmic reticulum. D_2O decreased the induced fluorescent intensity and thus in some way may have interfered with the potential changes in the sarcoplasmic reticulum.

Eastwood *et al.* (1975) have confirmed, in studies on skinned crayfish muscle fibers, that D_2O does not inhibit the contractile process directly. In intact fibers, increased electrical stimulation and depolarization could produce tension in D_2O equivalent to that in H_2O; the threshold, in this preparation, is thus increased by D_2O. From indirect evidence on both intact and skinned fibers they concluded that the sarcoplasmic reticulum in D_2O can accumulate and release calcium in sufficient quantities for production of maximum tension. Eastwood *et al.* claimed that 90% D_2O increases the affinity of the muscle for Ca^{2+} by about 1–2 orders of magnitude; a rather marked variation. On the other hand, Yagi and Endo (1976) in experiments on skinned muscle fibers found no such effect of D_2O on the affinity for Ca^{2+}. Yagi and Endo found some inhibition by D_2O of both Ca- and caffeine-induced calcium release; depolarization-induced Ca release was also slightly inhibited.

Although, as already mentioned, D_2O does not appear to influence the contractile mechanism directly, the following effects on Mg-ATPase activity must be noted. D_2O inhibits the ATPase of homogenates of glycerol-extracted muscle (at pH 7.0) by about 45% (Svensmark, 1961), whereas at pH 7.0–7.2 it does not influence the ATPase of natural actomyosin (myosin B) (Hotta and Morales, 1960); particularly at higher pH values beyond 8, however, the myosin B ATPase is depressed by D_2O. D_2O decreases the rate of heavy meromyosin ATPase in the steady state by 60% (Inoue *et al.*, 1975) and, in passing, it should be mentioned that D_2O inhibits the Na^+,K^+-ATPase of rat brain preparations (Ahmed and Foster, 1974).

Besides the inhibitory effects on muscle contraction, D_2O inhibits other forms of motility such as, for example, cell division (Katz *et al.*, 1957; Bennett *et al.*, 1958; Gross and Spindel, 1960; Marsland and Zimmerman, 1963). How D_2O affects the complicated series of movements during cell division is not clear. Gross and Spindel (1960) considered a number of possible sites and mechanisms of action and favored a "rigidification" of the cytoplasm and mitotic apparatus by substitution of deuterium for hydrogen. Marsland and Zimmerman (1963) described the mitotic apparatus as being "frozen" in D_2O and found the gel structure of the cytoplasm to be firmer. Inoue and Sato (1967) reported a twofold increase in the spindle birefringence produced by 45% D_2O and suggested a shift of the monomer-polymer equilibrium of microtubules towards the polymerized state. The initial rate of polymerization of brain tubulin is also increased by D_2O (Olmsted and Borisy, 1973). Hence heavy water may be a useful tool for further studies on mechanisms of cell motility.

While the numerous biological effects of D_2O may be related to its secondary or solvent properties, the substitution of deuterium for hydrogen, as mentioned at the outset, stabilizes hydrogen bonds. Since conformational changes and transformation of macromolecules occurs in numerous biological processes, the rate and degree of such changes will be influenced by deuterium substitution. Indeed in studies on temperature effects on helix-random coil transitions of macromolecules (ribonuclease and poly-γ-benzyl-L-glutamate) the helical coil was favored by heavy water (Scheraga, 1960).

The profound questions of Albert Szent-Györgyi on the biological role of water, particularly with respect to muscle contraction, which led to my investigations, remain unanswered. The findings have, as usual, raised more questions.

ACKNOWLEDGMENT

Portions of the work were supported by grants from NIH and NSF.

REFERENCES

Ahmed, K., and Foster, D. (1974). *Ann. N.Y. Acad. Sci.* **242,** 280.

Ashley, C. C., and Ridgway, E. B. (1970). *J. Physiol. (London)* **209,** 105.

Barnes, T. C., and Warren, J. (1935). *Science* **81,** 346.

Bennett, E. L., Holm-Hansen, O., Hughes, A. M., Lonberg-Holm, K., Moses, V., and Tolbert, B. M. (1958). *Science* **128,** 1142.

Berridge, M. J. (1975). *Adv. Cyclic Nucleotide Res.* **6,** 1.

Bezanilla, F., and Horowicz, P. (1975). *J. Physiol. (London)* **246**, 709.
Brandt, W. (1935). *Klin. Wochenschr.* **14**, 1597.
Eastwood, A. B., Grundfest, H., Brandt, P. W., and Reuben, J. P. (1975). *J. Memb. Biol.* **24**, 249.
Goodall, M. C. (1958). *Nature (London)* **182**, 677.
Gross, P. R., and Spindel, W. (1960). *Ann. N.Y. Acad. Sci.* **84**, 745.
Gunther, J., and Pfeiffer, C. (1974). *Acta Biol. Med. Ger.* **32**, 279.
Hotta, K., and Morales, M. F. (1960). *J. Biol. Chem.* **235**, PC61.
Huxtable, R., and Bressler, R. (1974). *J. Memb. Biol.* **17**, 189.
Inoue, A., Fukushima, Y., and Tonomura, Y. (1975). *J. Biochem. (Tokyo)* **78**, 1113.
Inoue, S., and Sato, H. (1967). *J. Gen. Physiol.* **50**, Suppl. 259.
Jenerick, H. (1964). *Am. J. Physiol.* **207**, 944.
Kaminer, B. (1960). *Nature (London)* **185**, 172.
Kaminer, B. (1966). *Nature (London)* **209**, 809.
Kaminer, B., and Kimura, J. (1972). *Science* **176**, 406.
Katz, J. J., Crespi, H. L., Hasterlik, R. J., Thomson, J. F., and Finkel, A. J. (1957). *J. Natl. Cancer Inst.* **18**, 641.
Kirshenbaum, I. (1951). "Physical Properties and Analysis of Heavy Water." McGraw-Hill, New York.
Kritchevsky, D. (1960). *Ann. N.Y. Acad. Sci.* **84**, Art. 16, 573.
Lewis, G. N., and Goody, T. C. (1933). *J. Am. Chem. Soc.* **55**, 3503.
Marsland, D., and Zimmerman, A. M. (1963). *Exp. Cell Res.* **30**, 23.
Morowitz, H. J., and Brown, L. M. (1953). "Biological Effects of Deuterium Compounds," Rep. No. 2179. Nat. Bur. Stand., Washington, D.C.
Noble, D. (1975). "The Initiation of the Heartbeat." Oxford Univ. Press (Clarendon), London and New York.
Olmsted, J. B., and Borisy, J. B. (1973). *Biochemistry* **12**, 4282.
Rasmussen, H. (1970). *Science* **170**, 404.
Sandow, A., Pagala, M. K. D., and Sphica, E. C. (1976). *Biochim. Biophys. Acta* **440**, 733.
Scheraga, H. A. (1960). *Ann. N.Y. Acad. Sci.* **84**, 608.
Shimomura, O., Johnson, F. H., and Saiga, Y. (1962). *J. Cell Comp. Physiol.* **59**, 223.
Svensmark, O. (1961). *Acta Physiol. Scand.* **53**, 75.
Szent-Györgyi, A. (1947). "Chemistry of Muscular Contraction," 1st ed. Academic Press, New York.
Szent-Györgyi, A. (1951). "Chemistry of Muscular Contraction," 2nd ed. Academic Press, New York.
Szent-Györgyi, A. (1953). "Chemical Physiology of Contraction in Body and Heart Muscle," Academic Press, New York.
Szent-Györgyi, A. (1957), "Bioenergetics." Academic Press, New York.
Szent-Györgyi, A. (1960), "Introduction to a Submolecular Biology." Academic Press, New York.
Szent-Györgyi, A. (1968). "Bioelectronics." Academic Press, New York.
Szent-Györgyi, A. (1972). "The Living State." Academic Press, New York.
Thomson, J. F. (1963). "Biological Effects of Deuterium." Pergamon, Oxford.
Urey, H. C. (1933). *Science* **78**, 566.
Vargas, F. F., and Johnson, J. A. (1975). *Physiologist* **18**, 429.
Verzar, F., and Haffter, C. (1935). *Pfluegers Arch. Gesamte Physiol. Menschen Tiere* **236**, 714.
Yagi, S., and Endo, M. (1976). *J. Physiol. Soc. Jpn.* **8**, 298.

CROSS-LINKING AND COMPLEXES

14

Biological Functions of Transamidating Enzymes Which Cross-Link Proteins

LASZLO LORAND

ENZYME-TRIGGERED PROTEIN ASSOCIATIONS

If the term "self-assembly" signifies the well-documented fact that a large number of proteins are capable of forming quaternary structures without any further enzymatic intervention beyond the translational events (see, e.g., Klotz *et al.*, 1975), there still remains a biologically important area of higher protein association which merits special attention. It could be described as the "enzymatically controlled post-translational assembly of proteins" which, like association by "self-assembly", must also proceed with an overall decrease in free energy. Hence, the distinction between the two processes is not a thermodynamic one.

In many ways, Linderstrøm-Lang anticipated the existence of "remodeling enzymes" when he proposed (1952) that they "may contribute by embroidering upon the canvas woven by hard working metabolic systems of the cell." In fact, he came rather close to the mark by suggesting that proteolytic enzymes could fulfill such a function. The work of Kunitz on the trypsin-catalyzed activation of chymotrypsinogen (Northrop *et al.*, 1948), as well as that of Linderstrøm-Lang and Ottesen on the subtilisin-mediated conversion of ovalbumin to plakalbumin (1947), paved the way for the concept that "limited proteolysis" might be a rather general vehicle for changing protein A into protein B.

Within the context of enzyme controlled post-translational assemblies, the clotting of fibrinogen by thrombin highlights the biological potential of tailoring proteins by processes of "limited proteolysis", and this is the problem I worked on after I left Albert Szent-Györgyi's institute in Budapest and joined the department of his friend, W. T. Astbury in Leeds. We were able to show that only about four peptide bonds per mole of fibrinogen were cleaved by thrombin (Bailey *et al.*, 1951; Lorand and Middlebrook, 1952, 1953) and that approximately 3% of the parent protein was removed from its N-terminal portion in the form of well-defined by-products which I thought deserved the special name of fibrinopeptide (Lorand, 1951, 1952). The α-chain in fibrinogen $[(\alpha\beta\gamma)_2]$ is attacked first, followed by a clip in the same region of the β-chain (Blomback and Yamashina, 1958):

$$\underset{\text{Fibrinogen}}{(\alpha\beta\gamma)_2} \xrightarrow[\text{2 Fibrinopeptide A}]{\text{Thrombin}} (\alpha'\beta\gamma)_2 \xrightarrow[\text{2 Fibrinopeptide B}]{\text{Thrombin}} \underset{\text{Fibrin}}{(\alpha'\beta'\gamma)_2}$$

Clearly, the ability for self-association is encoded within the fibrinogen molecule as such; yet, at the pH, temperature and ionic strength of blood plasma this protein shows little tendency for clotting. The expression to do so is acquired through the excision by thrombin of the negatively charged fibrinopeptide patches from the N-terminal portion of the protein which appears to form a tight disulfide-bridged knot (Blomback *et al.*, 1968):

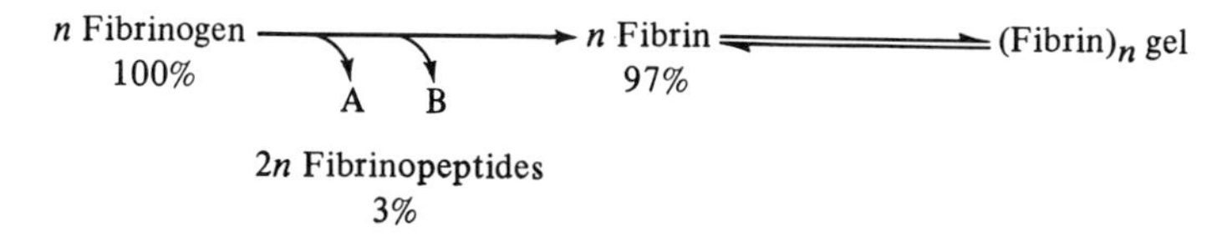

It is interesting to note that the $(\alpha'\beta\gamma)_2$ fibrin intermediate is already endowed with clot forming ability and this is the situation with clotting enzymes of snake venoms which do not attack the β-chain of fibrinogen (Blomback, 1958; Holleman and Coen, 1970; Markland and Damus, 1971).

The limited hydrolytic modification not only alters the solubility of the fibrinogen molecule, causing it to precipitate in the form of ordered aggregates, but also produces an unmasking of cross-linking sites which must be ready for the next phase in blood clotting. The latter involves the introduction of intermolecular γ-glutamyl-ε-lysine peptide bridges into the fibrin gel by a specific transamidating reaction (see, e.g., Lorand, 1972; Lorand *et al.*, 1972c). Although the actual number of such covalent protein-to-protein bridges is small (ca. 1 per

550 amino acid residues), they still contribute considerably to the physical stiffening of the fibrin network (Roberts *et al.*, 1973; Mockros *et al.*, 1974) and to its resistance to lytic enzymes (Lorand *et al.*, 1966). In the present discussion, I shall focus on enzymes which cross-link proteins by such γ–ϵ isopeptide linkages. I fully realize, of course, that this is only one of several possible covalent side chain interactions in biological cross-linking processes (see, e.g., Gallop and Paz, 1975). Interestingly enough, in his Lane Medical Lectures, Linderstrøm-Lang speculated about the possible remodeling role of transamidating enzymes, but he concerned himself with exchanges in the peptide backbone rather than with the reactions of protein side chains (Linderstrøm-Lang, 1952).

"*ENDO*-γ-GLUTAMINE:ϵ-LYSINE TRANSFERASE" ENZYMES WHICH CROSS-LINK PROTEINS

There is a group of transamidating enzymes, collectively best characterized by the name "*endo*-γ-glutamine:ϵ-lysine transferase" (Lorand and Stenberg, 1976), which have the function of modifying specific native proteins to yield covalently linked polymeric assemblies. These enzymes, widely distributed in nature, catalyze the nucleophilic displacement reaction depicted in Fig. 1. Depending on the particular physiological situation, the reaction may lead (A) to the formation of specific precipitates [e.g., lobster fibrinogen → clot (Lorand *et al.*, 1963; Fuller and Doolittle, 1971); seminal vesicle secretion protein → copulation plug (Williams-Ashman *et al.*, 1972)]; (B) to the covalent fusion ("ligation") of individual molecules within an already existing ordered protein aggregate [e.g., vertebrate fibrin gel → cross-linked or ligated fibrin (Lorand *et al.*, 1968a; Matacic and Loewy, 1968; Pisano *et al.*, 1968); ligation of keratin (Asquith *et al.*, 1970; Harding and Rogers, 1972)]; (C) to a possible rearrangement of the cytoskeleton, as in the transamidase-catalyzed cross-linking of cell membrane proteins

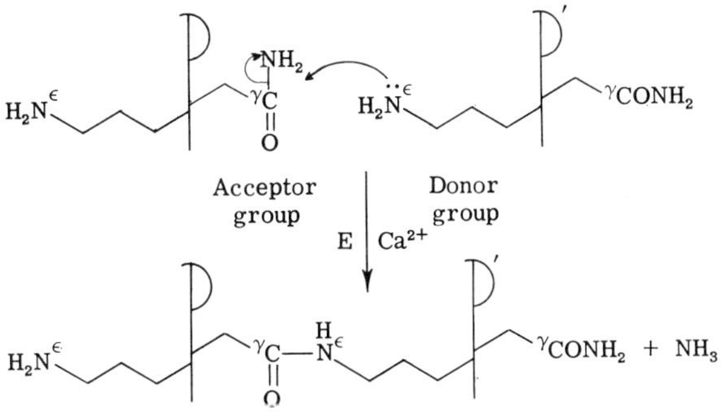

Fig. 1. Scheme for a transamidase (E)-catalyzed protein assembly. Dimerization by forming a single intermolecular γ-glutamyl-ϵ-lysine peptide bridge is shown between proteins P and P′.

which is known to occur in red blood cells upon the influx of Ca^{2+} (Lorand *et al.,* 1976). In these processes, the breaking of the γ-amide of a glutamine residue provides the energy for the synthesis of the protein-to-protein γ–ϵ peptide bond. There might be situations in which the γ-glutamyl-ϵ-lysine peptide forms in a reversible manner; in other cases, the precipitation of the polymeric protein product might pull the reaction irreversibly to the right (Fig. 1).

Characteristically transamidating enzymes of this class contain cysteine-thiol active centers and require calcium ions for activity; they are usually inhibited by iodoacetamide or by calcium chelators such as EDTA (Lorand and Stenberg, 1976). The fact that a biological reaction involves cross-linking by such enzymes can often be best recognized by the inhibitory actions of suitable synthetic amines which simultaneously become incorporated into the protein substrate. The competition between the amine donor group of a protein substrate and a pseudo (or substitute) donor is illustrated in Fig. 2 for the biological cross-linking of fibrin (Lorand, 1972). Typically, the pseudo-donor molecule (H_2NR) is a primary amine which contains an alkyl residue resembling that of the lysine donor side chains in proteins, anchored to some bulky hydrophobic nucleus; dansylcadaverine [*N*-(5-aminopentyl)-5-dimethylaminonaphthalenesulfonamide] would be a good example (Lorand *et al.*, 1968c). It is, of course, advantageous to have such molecules carry fluorescence or radioactive labels.

The transamidating activity of tissues is measured by the calcium ion and sulfhydryl-dependent incorporation of labeled amines into nonspecific protein acceptors (e.g., casein, β-lactoglobulin or their derivatives), and a number of good methods are available for this purpose (Lorand *et al.*, 1969, 1971, 1972b; Lorand and Campbell, 1971; Curtis and Lorand, 1976). Heinrich Waelsch pio-

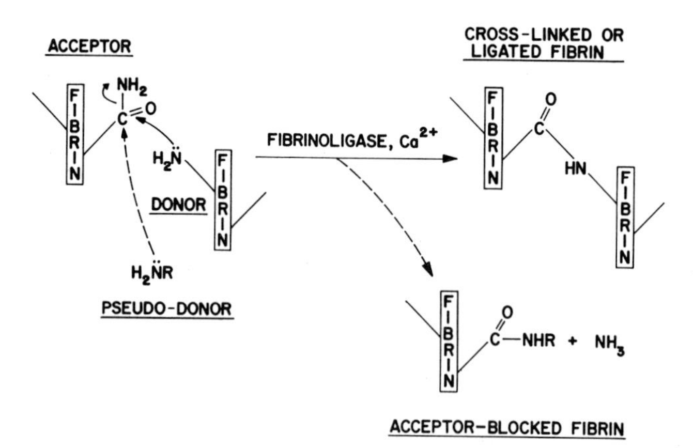

Fig. 2. Inhibition of a transamidase-catalyzed protein assembly reaction by primary amines. Inhibition of cross-linking goes together with incorporation of the pseudo or substitute donor into the parent protein, as illustrated for fibrin.

neered in the area of intracellular transamidases with his work on liver transglutaminase (Clarke *et al.*, 1959). This is an enzyme of still unknown function and its relationship to the more specific enzymes such as the one generated in blood for the cross-linking of fibrin (i.e., fibrinoligase) is about the same as that of trypsin to thrombin. Just as there are many serine proteases with varying nuances of affinities for arginyl or lysyl bonds [playing exquisitely specific roles in controlled biological sequences in the clotting cascade, kinin activation or the complement pathway (Davie and Fujikawa, 1975; Reich *et al.*, 1975; Lorand, 1976a)] one finds a whole range of variations for the transamidases. The specificity of fibrinoligase, for example, is so tuned toward fibrin that it can only effect a much slower incorporation of amines into the very similar fibrinogen molecule (Lorand and Ong, 1966; Lorand *et al.*, 1972c). Further, when acting on fibrin, it reacts in a singularly ordered manner first with glutamines in γ-chain sites, then with those in the α-chains (Lorand and Chenoweth, 1969; Lorand *et al.*, 1972c). We may assume that these remodeling enzymes co-evolved with the protein substrates which they are to "embroider upon" (Lorand *et al.*, 1972c).

Historically, the cross-linking of fibrin was the first physiological example of transamidase-catalyzed protein associations, and it is through this system that I became interested in such reactions. As a biochemist, my immediate objective was to purify the essential components from blood plasma and to study the isolated system with the attendant enzymology. From time to time I recall Szent-Györgyi's saying that "it doesn't matter where one starts digging, one is bound to progress toward the center of the globe," which to him always meant bigger and more challenging problems. It is obvious that it would not be satisfying to keep on digging in a single hole without searching for the jewels one may find in shafts sent out in different directions. (Szent-Györgyi once told me that he discovered vitamin C by looking into one of these side channels which at the time was only of minor interest to him).

First I shall summarize our knowledge of the protein and enzyme chemical background on the fibrin stabilizing system in blood plasma; then I shall indicate the multifaceted ramifications, including the clinical aspects, which flowed from the basic research.

FIBRINOLIGASE GENERATION IN BLOOD PLASMA

Of all transamidase controlled assemblies, the one operating in blood appears to be the most complex. For obvious reasons, the cross-linking enzyme cannot circulate in plasma under normal conditions in the active form but must be rapidly generated at the time clotting is required. Indeed, we have long known that the fibrin stabilizing factor (also named by an International Committee

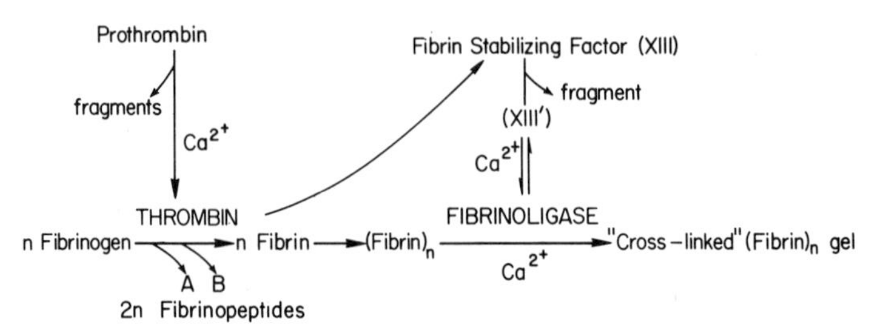

Fig. 3. Outline of the clotting reaction in the plasma of vertebrates. Thrombin, a hydrolytic enzyme of great specificity, brings about the conversion of fibrinogen to fibrin and it also regulates the rate of formation of fibrinoligase (activated Factor XIII) which cross-links fibrins (Lorand, 1975, 1976a).

(1963) as coagulation Factor XIII) was a zymogen, needing both thrombin and calcium ions for activation (Lorand and Konishi, 1964). As shown in Fig. 3, thrombin has a dual regulatory role in clotting because it controls the rate of formation of fibrin as well as that of the transamidase (called fibrinoligase or activated Factor XIII) which is responsible for the cross-linking of this protein substrate. Fibrinoligase is the last enzyme to form in the coagulation cascade and the only one with a cysteine-thiol active center (Curtis *et al.*, 1973, 1974b); all the others belong to the serine proteases of the trypsin variety (Davie and Fujikawa, 1975; Lorand, 1976a). In fact, as far as I know, fibrinoligase may be the only example of an extracellular thiol enzyme.

During the past decade, we have isolated the fibrin stabilizing factor zymogen to a high degree of purity from human as well as from bovine blood plasma (Lorand and Gotoh, 1970), and showed that activation by thrombin and Ca^{2+} proceeds in two independent and consecutive stages (Curtis *et al.*, 1974a,b):

$$\text{Zymogen} \xrightarrow{\text{Thrombin}} \text{Zymogen}' \xrightarrow{Ca^{2+}} \text{Enzyme}$$

The plasma zymogen has an (ab) subunit structure with a protomer weight of about 160,000 (Lorand *et al.*, 1968a; Schwartz *et al.*, 1971) [a = 75,000 (Schwartz *et al.*, 1971); b = 88,000 (Schwartz *et al.*, 1971)] and activation by thrombin consists of the removal of an N-terminal peptide fragment of 37 amino acids from the (a) subunit (Takagi and Doolittle, 1974). Next, the hydrolytically modified zymogen (a′b), under the specific influence of calcium ions, undergoes a heterologous dissociation of its subunits (Curtis *et al.*, 1974b; Lorand *et al.*, 1974) which permits the unmasking of the active center thiol in (a′) (Curtis *et al.*, 1974b; Lorand *et al.*, 1974). A glance at Fig. 4 shows that of the two Ca^{2+}-regulated events (i.e., dissociation and unmasking) the former is the slower step. The zymogen found in platelets contains no regulatory (b) subunit (Schwartz *et al.*, 1971) and addition of Ca^{2+} to its thrombin modified form, in

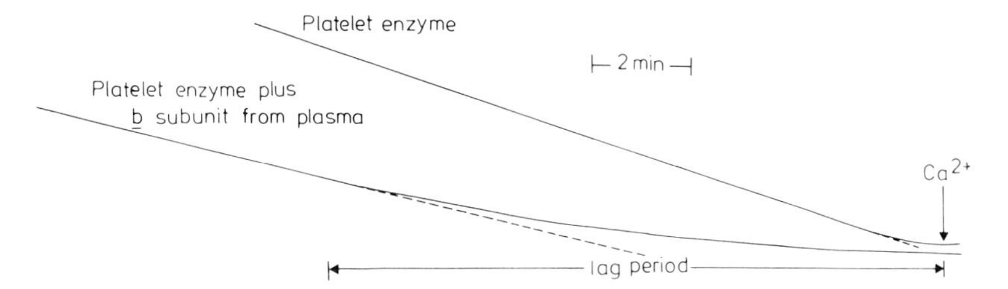

Fig. 4. Progression curves (from right to left) for the formation of the amide product in the reaction between β-phenylpropionylthiocholine iodide and dansylcadaverine, for the platelet enzyme with and without hybridization with the noncatalytic (b) subunit from plasma (Lorand *et al.*, 1974).

contrast to the plasma zymogen, produces enzyme activity without an appreciable lag phase (Lorand *et al.*, 1974). Also, the dissociation step appears to require a considerably higher concentration of Ca^{2+} than the conformational change which leads to the unmasking of the active center thiol in (a′) (see, e.g., Stenberg *et al.*, 1975). The conversion of the plasma zymogen to fibrinoligase can be summarized as follows:

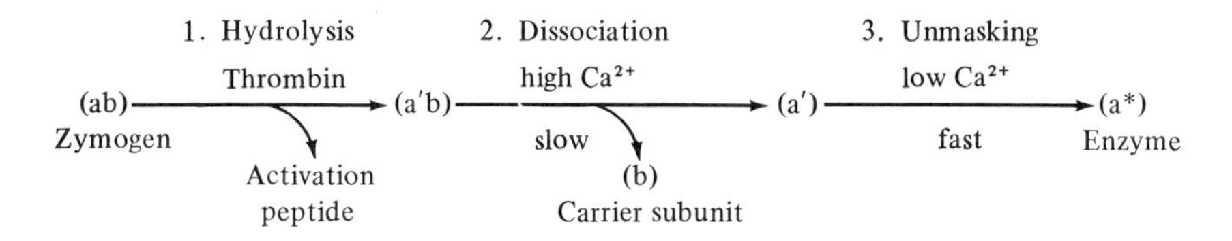

Both Ca^{2+}-dependent steps are seemingly reversible; addition of EDTA causes a reburying of the active center cysteine of (a*) and a reassociation of (a′) with (b) (Curtis *et al.*, 1974b; Lorand *et al.*, 1974). The dual role of Ca^{2+} in the activation of the hydrolytically altered zymogen can also be seen from our recent finding that Zn^{2+} competes against Ca^{2+} in inhibiting enzyme activity effectively (K_i for $Zn^{2+} \approx 6 \times 10^{-6}$ *M*), but that Zn^{2+} has no effect on the Ca^{2+}-dependent heterologous dissociation of the subunits, per se (Credo *et al.*, 1976).

CATALYSIS BY FIBRINOLIGASE

Of approximately 300 glutamine side chains (Henschen, 1964) in the native fibrin molecule (MW 330,000) only about six can react with fibrinoligase (Lorand and Chenoweth, 1969; Lorand *et al.*, 1972c) which points to the importance of secondary binding sites outside of the active center of the enzyme. As yet, we do not know how this aspect of specificity is regulated. By contrast, using simple synthetic substrates rather than fibrin, we made good headway in elucidating the pathway of catalysis which involves the primary binding sites of the enzyme. Acyl portions of certain thiocholine esters (Lorand *et al.*, 1972d; Curtis *et al.*, 1974a; Stenberg *et al.*, 1975) apparently fit rather well into the cleft or channel on the enzyme surface which complements the dimensions of the glutamine side chain of the native substrate. These esters offer numerous advantages for the kineticist because they are sufficiently water-soluble and rather stable near neutral pH's even at 37°C. Attack by the enzyme may be studied in terms of simple hydrolysis or, in the presence of an added amine, in terms of aminolysis with concomitant amide formation. As indicated in Fig. 5 for the β-phenylpropionylthiocholine (RCOSR′) and dansylcadaverine (H_2NR'') substrate pair, we have developed a variety of continuous methods for studying the fibrinoligase catalyzed reaction. The steady-state analysis shows that the enzymatic reaction proceeds by an acylation–deacylation pathway in which the enzyme (E-SH) gives rise to an acyl-enzyme intermediate (Curtis *et al.*, 1974a; Stenberg *et al.*, 1975), and the latter may undergo hydrolysis as well as aminolysis (Fig. 6). In general, this pathway is not unlike that of papain (Bender, 1971), and perhaps there is a mechanistic reason for the utilization of a thiol enzyme for promoting an efficient amide exchange in aqueous solutions. A striking feature of fibrinoligase is the great specificity of the acyl-enzyme intermediate for the amine substrate (Lorand *et al.*, 1968c; Lorand and Nilsson,

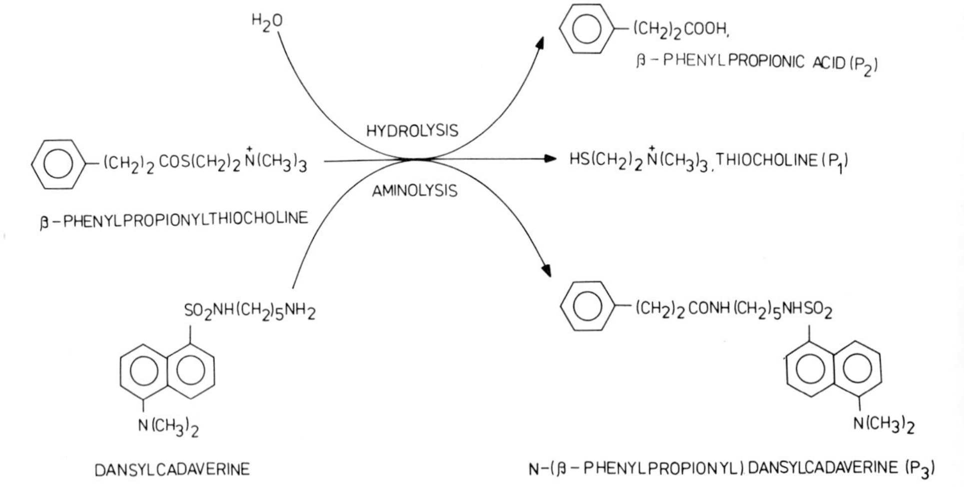

Fig. 5. Reaction of β-phenylpropionylthiocholine (S_1) with dansylcadaverine (S_2). Formation of thiocholine (P_1) is measured by reaction with 5,5′-dithio-bis(2-nitrobenzoic acid) (Curtis *et al.*, 1974a); production of acid (P_2) can be followed in a pH'stat and the water-insoluble coupling product (P_3) is quantitated by fluorescence after extraction into a heptane phase (Stenberg *et al.*, 1975).

$$R_2COSR_1 + HS\text{-}E \underset{}{\overset{K_a}{\rightleftharpoons}} [R_2COSR_1;\ HS\text{-}E] \xrightarrow[\text{acylation}]{k_2} R_2COS\text{-}E + HS\text{-}R_1 \quad (1)$$

$$R_2COS\text{-}E + H_2O \xrightarrow[\text{hydrolysis}]{k_3} HS\text{-}E + R_2COOH \quad (2)$$

$$R_2COS\text{-}E + H_2NR_3 \underset{}{\overset{K_b}{\rightleftharpoons}} [R_2COS\text{-}E;\ H_2NR_3] \xrightarrow[\text{aminolysis}]{k_5} HS\text{-}E + R_2CONHR_3 \quad (3)$$

Fig. 6. Pathway for the fibrinoligase-catalyzed hydrolysis and aminolysis of thiolesters.

1972). Whereas many amines at saturating concentrations for the enzyme produce the same maximal reaction velocity, the efficacy of amine binding to the enzyme intermediate is highly specific (Curtis *et al.*, 1974a). The acyl-enzyme surface may have a crevice into which alkylamines of a length similar to an ε-lysine side chain (e.g., dansylcadaverine) fit best. Furthermore, the fit must be rather tight because, in collaboration with Dr. Ake Jönsson, we found dansyl-

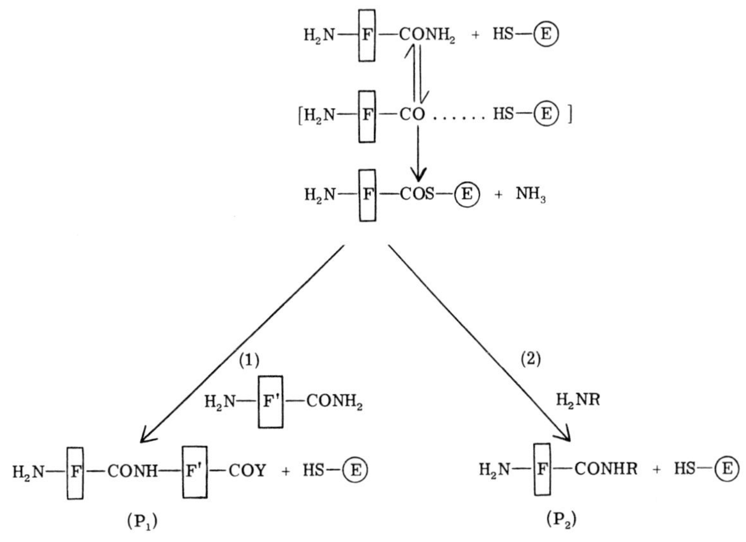

Fig. 7. Proposed pathway of fibrin (F) cross-linking by formation of an acyl (fibrinyl)-enzyme intermediate. The enzyme (E) is shown to form a thiolester with the fibrin substrate. The intermediate could be aminolyzed either by reacting with the natural ε-lysine donor groups of another fibrin molecule or by reacting with a synthetic pseudo-donor amine (H_2NR). The former would lead to a cross-linked fibrin product (P_1), the latter to an acceptor-blocked fibrin (P_2) species. This scheme explains the specific inhibitory effect of amines on cross-linking and also the simultaneous incorporation into fibrin.

cadaverine (I) to be a far better substrate than a bulkier branched analogue (II):

$$\text{DansylNHCH}_2\text{CH}_2\text{CH}_2\text{CH}_2\text{CH}_2\text{NH}_2 \quad \text{(I)}$$

$$\text{DansylNHCH}_2\text{CH}_2\overset{\overset{\text{CH}_3}{|}}{\underset{\underset{\text{CH}_3}{|}}{\text{C}}}\text{CH}_2\text{CH}_2\text{NH}_2 \quad \text{(II)}$$

These model studies with simple substrates suggest that the pathway of catalysis in the cross-linking of fibrin must also be quite similar (Lorand *et al.*, 1972c) (Fig. 7).

HYBRID SYSTEMS

Though fibrin is a paramount constituent, it is not the only component of clots and thrombi; thus covalent hybrid associations of fibrin with other proteins, including those of cell membranes, may be of great significance. It is already known that "cold-insoluble globulin", a protein somewhat larger than fibrinogen but present in plasma only at about one-tenth of the concentration of the latter, copolymerizes with fibrin in the presence of the transamidase (Mosher, 1975, 1976). Interestingly, the outside surface of fibroblasts is covered with a protein which by immunological tests is similar to or identical with "cold-insoluble globulin."

Dr. Isaac Cohen of Tel-Aviv University Medical School has recently joined our laboratory to investigate the question of whether fibrin forms hybrid γ-glutamyl-ϵ-lysine links with proteins on the surface of platelets and/or with contractile proteins of these "mini-smooth muscles." His experiments with Dr. Joyce Bruner-Lorand already show that myosin and actin from platelets as well as skeletal muscle are definitely substrates for fibrinoligase, both in terms of incorporation of amines and cross-linking (Cohen *et al.*, 1977). It is probably not an accident that myosin is among the rather few native proteins which react with fibrinoligase, and the existence of hybrid copolymers between fibrin and the platelet contractile proteins emerges as a strong possibility.

In mentioning the reactions of myosin with fibrinoligase and with transamidases of similar type, attention must also be drawn to the remarkable susceptibility of "spectrins", the myosin-like polypeptides of cell membranes, towards these enzymes (Dutton and Singer, 1975; Lorand *et al.*, 1975; Shishido *et al.*, 1976). As shown in Figs. 8 and 9, for example, the treatment of human red cell ghosts with transglutaminase produces cross-linking of "spectrins" and incorporation of amines into these proteins. Since red cells (and many other cells) are known to contain at least latent transamidase activities, the question arises whether such enzymes could somehow regulate the elastic properties of cell membranes just as transamidation between fibrin molecules stiffens clot

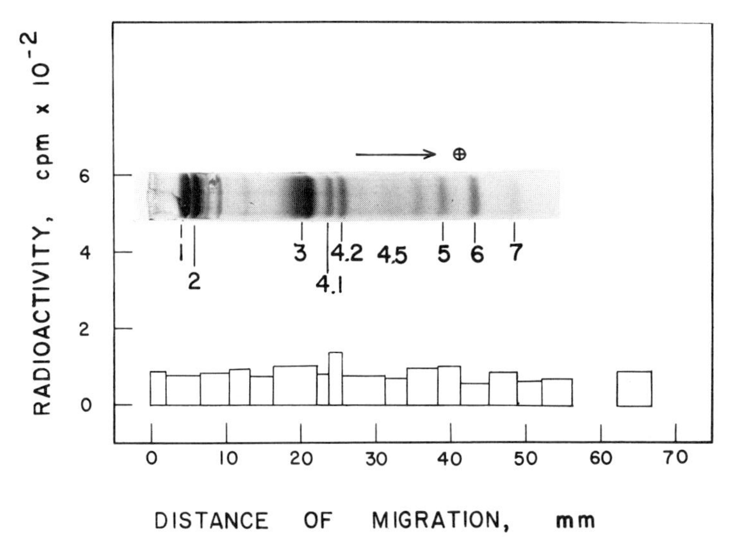

Fig. 8. Electrophoretic profiles of human erythrocyte ghosts after exposure to transglutaminase and ^{14}C-putrescine in the absence of calcium ions. Open bars denote radioactivity found in gel slices. The photographic insert shows the protein pattern after staining with Coomassie blue; numbering of bands corresponds to that given in the literature (Steck, 1974).

structure. The intracellular transamidase could play a variety of roles. It could catalyze the reversible cross-linking of the cytoskeletal framework in a dynamic fashion; alternately, irreversible cross-linking could lead to pathological shape changes and to the stiffening of cell membrane which would affect the life span of the cell. In specific terms we can now ask if the shorter survival of erythrocytes in sickle cell disease is due to the fact that the intrinsic transglutaminase is switched on by the influx of Ca^{2+} into these cells. We have actually been able to show that the Ca^{2+}-mediated polymerization of erythrocyte membrane proteins is due to the turning on of the intrinsic transamidase (Lorand *et al.*, 1976). Thus it appears that the functioning of this enzyme transmits the signal for the stiffening of cell membrane and the elimination of the cell by the spleen.

CLINICAL ASPECTS OF FIBRIN STABILIZATION

There seem to be few branches of "molecular medicine" where fundamental findings have clinical applications as direct as those relating to blood coagulation. In this field, biochemical discoveries often provide an immediate impetus for medical advances.

The development of meaningful tests applicable to human plasma opened up

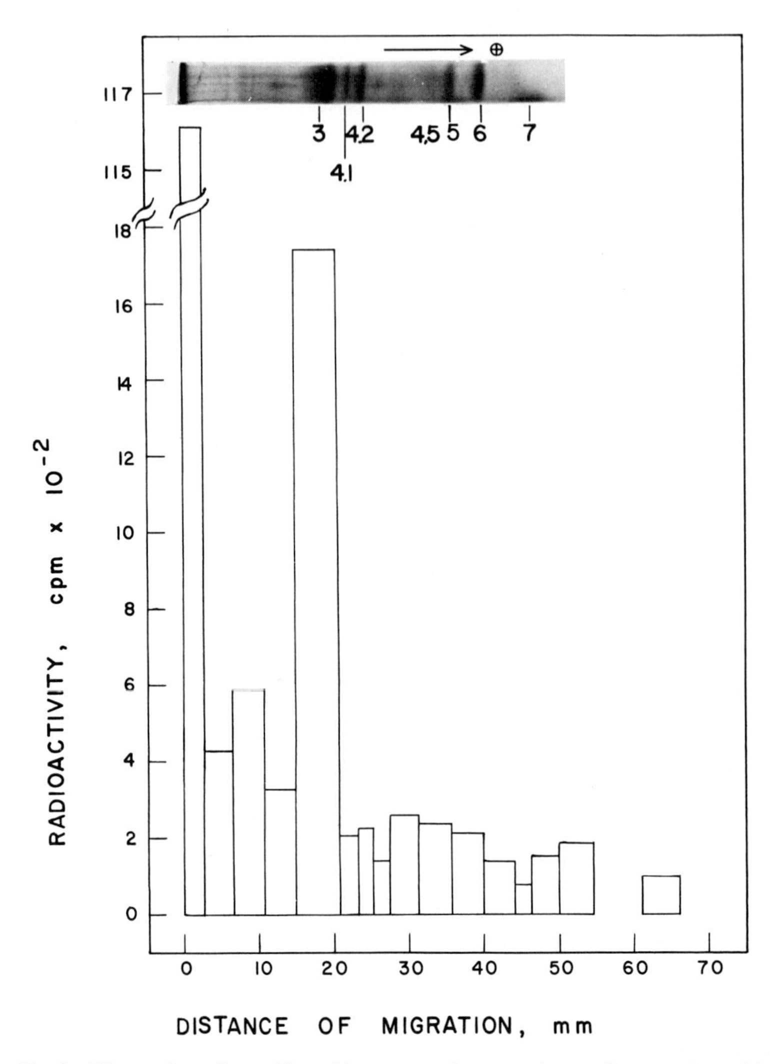

Fig. 9. Electrophoretic profiles of human erythrocyte ghosts after reaction with transglutaminase and ^{14}C-putrescine in the presence of calcium ions. Presentation as in Fig. 8.

new avenues for the differential diagnosis of hitherto uncategorized disorders of fibrin stabilization. Working primarily with the incorporation of synthetic amines (such as dansylcadaverine or putrescine) into casein (Lorand *et al.*, 1969, 1972b), we were able to analyze quantitatively the mode of inheritance for the genetic lack of the fibrin stabilizing factor zymogen (Lorand *et al.*, 1970; Henriksson *et al.*, 1974) (Factor XIII) (Fig. 10) and to evaluate the efficacy of transfusions given as therapy to the deficient individual (Lorand *et al.*, 1969). In

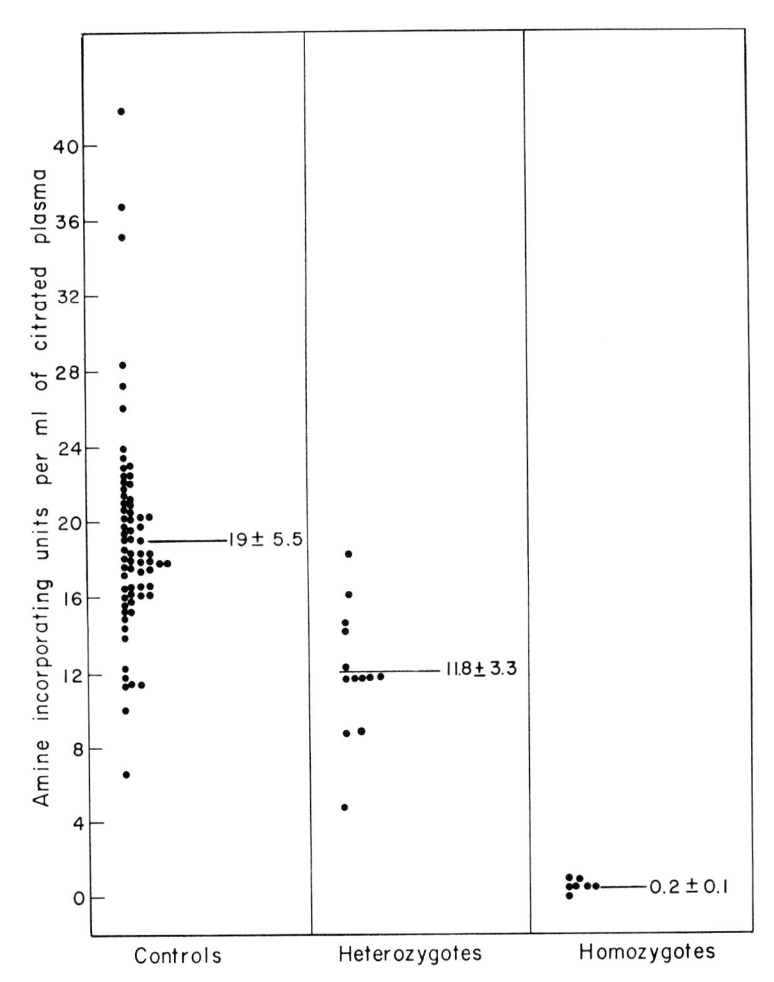

Fig. 10. Summary of data for control individuals for homozygote bleeders with deficiency of fibrin stabilizing factor zymogen and for their heterozygote parents, using the dansylcadaverine method of Lorand *et al.* (1969, 1970).

addition, we succeeded in defining three different types of bleeding abnormalities due to the sudden appearance of inhibitors in the patient's circulation (Lorand, 1972, 1976b). Characterization of prototypes of acquired inhibitors offers a real challenge to the molecular biologist. As shown in Fig. 11, the inhibitors interfere with different reactions in the pathway of fibrin stabilization. Type I specifically prevents activation of the zymogen, but does not inhibit transamidase activity (Lorand *et al.*, 1972e). By contrast, the Type II inhibitor interferes with transamidase activity as such (Lorand *et al.*, 1968b). In the Type

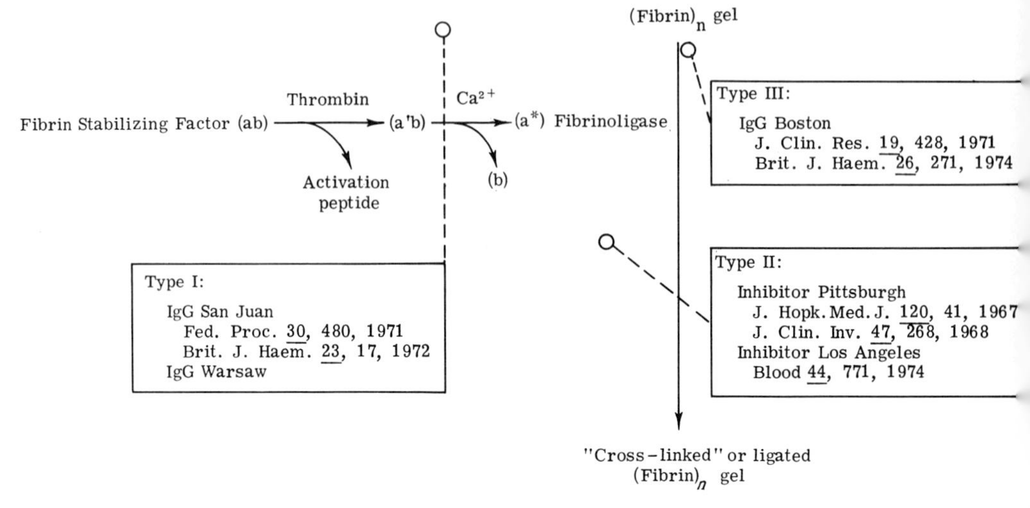

Fig. 11. Classification of acquired inhibitors in hemorrhagic disorders of fibrin stabilization (Lorand, 1972, 1976b).

III situation, the inhibitor combines with fibrin, rendering the cross-linking sites of this substrate unreactive towards the transamidase (Rosenberg *et al.*, 1974). The Type I and III inhibitors turned out to be IgG antibodies, and one could surmise that "autoimmunization" of the patients may have been set in motion by the administration of isoniazid and penicillin. Various molecular mechanisms may account for the breakdown of immunological tolerance following the use of these drugs. It is particularly noteworthy, however, that isoniazid can actually be shown (Lorand, 1971; Lorand *et al.*, 1972a) to be a substrate for transamidating enzymes and to become covalently incorporated into (presumably γ-glutaminyl side chains) of proteins as:

P -CONHNHC(=O)-⟨pyridyl⟩N

This type of modification could conceivably trigger antibody formation not only against the bound isoniazid haptene but also against neighboring determinants on the carrier protein (P). The native protein could become a cross-reacting antigen and could continue, by itself, to stimulate antibody production even after the elimination of the originally modified protein. Antibodies against the haptenic modifier might even disappear. Penicillin could react with the active center sulfhydryl group of fibrinoligase, as it does with the transpeptidase in bacterial cell walls (Tipper and Strominger, 1965; Wise and Park, 1965), or it might be a less specific nonenzymatic modifier of proteins (Schneider, 1970).

CONCLUSION

After I finished medical school and went to England in 1948, I thought I started research which would move me further away from medicine. Now, I realize that I only made a scientifically necessary, albeit quite a long detour. In offering this article as a token of my affection to Albert Szent-Györgyi, I fondly remember the early stimuli he and his associate, Koloman Laki, gave me. If I hadn't qualified for the departmental volleyball team, I might have missed all the fun of the research I recalled here.

ACKNOWLEDGMENTS

Research work pertaining to this article was supported by a USPHS Career Award (5 K06 HL-03512) and by grants from the National Institutes of Health, National Heart and Lung Institute (HL-02212 and HL-16346).

REFERENCES

Asquith, R. S., Otterburn, M. S., Buchanan, J. H., Cole, M., Fletcher, J. C., and Gardner, K. L. (1970). *Biochim. Biophys. Acta* **221**, 342.

Bailey, K., Bettelheim, F. R., Lorand, L., and Middlebrook, W. R. (1951). *Nature (London)* **167**, 233.

Bender, M. L. (1971). "Mechanisms of Homogeneous Catalysis from Protons to Proteins." Wiley, New York.

Blomback, B. (1958). *Ark. Kemi* **12**, 321.

Blomback, B., and Yamashina, I. (1958). *Ark. Kemi* **12**, 299.

Blomback, B., Blomback, M., Henschen, A., Hessel, B., Iwanaga, S., and Woods, K. R. (1968). *Nature (London)* **218**, 130.

Clarke, D. D., Mycek, M. J., Neidle, A., and Waelsch, H. (1959). *Arch. Biochem. Biophys.* **79**, 338.

Cohen, I., Bruner-Lorand, J., Blankenberg, T. A., and Lorand, L. (1977). *J. Biol. Chem.*, submitted for publication.

Credo, R. B., Stenberg, P., Tong, Y. S., and Lorand, L. (1976). *Fed. Proc., Fed. Am. Soc. Exp. Biol.* **35**, 1631 (Abstr. No. 1390).

Curtis, C. G., and Lorand, L. (1976). *In* "Methods of Enzymology" (L. Lorand, ed.), Vol. 45, p. 177. Academic Press, New York.

Curtis, C. G., Stenberg, P., Chou, C.-H. J., Gray, A., Brown, K. L., and Lorand, L. (1973). *Biochem. Biophys. Res. Commun.* **52**, 51.

Curtis, C. G., Stenberg, P., Brown, K. L., Baron, A., Chen, K., Gray, A., Simpson, I., and Lorand, L. (1974a). *Biochemistry* **13**, 3257.

Curtis, C. G., Brown, K. L., Credo, R. B., Domanik, R. A., Gray, A., Stenberg, P., and Lorand, L. (1974b). *Biochemistry* **13**, 3774.

Davie, E. W., and Fujikawa, K. (1975). *Annu. Rev. Biochem.* **44**, 799.

Dutton, A., and Singer, S. J. (1975). *Proc. Natl. Acad. Sci. U.S.A.* **72**, 2568.
Fuller, G., and Doolittle, R. F. (1971). *Biochemistry* **10**, 1311.
Gallop, P. M., and Paz, M. A. (1975). *Physiol. Rev.* **55**, 418.
Harding, H. W. J., and Rogers, G. E. (1972). *Biochim. Biophys. Acta* **257**, 37.
Henschen, A. (1964). *Ark. Kemi* **22**, 1.
Henriksson, P., Hedner, U., Nilsson, I. M., Boehm, J., Robertson, B., and Lorand, L. (1974). *Pediatr. Res.* **8**, 789.
Holleman, W. H., and Coen, L. J. (1970). *Biochim. Biophys. Acta* **200**, 587.
International Committee on Blood Clotting Factors (1963). *Thromb. Diath. Haemorrh. Suppl.* **13**, 419.
Klotz, I. M., Darnall, D. W., and Langerman, N. R. (1975). *In* "The Proteins" (H. Neurath and R. L. Hill, eds.), 3rd ed., Vol. I, p. 294. Academic Press, New York.
Linderstrøm-Lang, K. (1952). *Stanford Univ. Publ., Univ. Ser., Med. Sci.* **6**, 1–115.
Linderstrøm-Lang, K., and Ottesen, M. (1947). *Nature (London)* **159**, 807.
Lorand, L. (1951). *Nature (London)* **167**, 992.
Lorand, L. (1952). *Biochem. J.* **52**, 200.
Lorand, L. (1971). *Thromb. Diath. Haemorrh., Suppl.* **45**, 243.
Lorand, L. (1972). *Ann. N. Y. Acad. Sci.* **202**, 6.
Lorand, L. (1975). *In* "Proteases and Biological Control" (E. Reich, D. B. Rifkin, and E. Shaw, eds.), p. 79. Cold Spring Harbor Lab., Cold Spring Harbor, New York.
Lorand, L. (1976a). *In* "Methods in Enzymology" (L. Lorand, ed.), Vol. 45, p. 31. Academic Press, New York.
Lorand, L. (1976b). *In* "Biochemistry, Physiology and Pathology of Haemostasis" (D. Ogston and B. Bennett, eds.). Wiley, New York (in press).
Lorand, L., and Campbell, L. K. (1971). *Anal. Biochem.* **44**, 207.
Lorand, L., and Chenoweth, D. (1969). *Proc. Natl. Acad. Sci. U.S.A.* **63**, 1247.
Lorand, L., and Gotoh, T. (1970). *In* "Methods in Enzymology" (G. E. Perlmann and L. Lorand, eds.), Vol. 19, p. 770. Academic Press, New York.
Lorand, L., and Konishi, K. (1964). *Arch. Biochem. Biophys.* **105**, 58.
Lorand, L., and Middlebrook, W. R. (1952). *Biochem. J.* **52**, 196.
Lorand, L., and Middlebrook, W. R. (1953). *Science* **118**, 515.
Lorand, L., and Nilsson, J. L. G. (1972). *In* "Drug Design" (E. J. Ariens, ed.), Vol. 3, p. 415. Academic Press, New York.
Lorand, L., and Ong, H. H. (1966). *Biochemistry* **5**, 1747.
Lorand, L., and Stenberg, P. (1976). *In* "Handbook of Biochemistry and Molecular Biology, Proteins Vol. II" (G. D. Fasman, ed.), 3rd ed., p. 669. CRC Press, Cleveland, Ohio.
Lorand, L., Doolittle, R. F., Konishi, K., and Riggs, S. K. (1963). *Arch. Biochem. Biophys.* **102**, 171.
Lorand, L., Bruner-Lorand, J., and Pilkington, T. R. E. (1966). *Nature (London)* **210**, 1273.
Lorand, L., Downey, J., Gotoh, T., Jacobsen, A., and Tokura, S. (1968a). *Biochem. Biophys. Res. Commun.* **31**, 222.
Lorand, L., Jacobsen, A., and Bruner-Lorand, J. (1968b). *J. Clin. Invest.* **47**, 268.
Lorand, L., Rule, N. G., Ong, H. H., Furlanetto, R., Jacobsen, A., Downey, J., Oner, N., and Bruner-Lorand, J. (1968c). *Biochemistry* **7**, 1214.
Lorand, L., Urayama, T., deKiewiet, J., and Nossel, H. A. (1969). *J. Clin. Invest.* **48**, 1054.
Lorand, L., Urayama, T., Atencio, A., and Hsia, D. Y. Y. 1970). *Am. J. Hum. Genet.* **22**, 89.
Lorand, L., Lockridge, O. M., Campbell, L. K., Myhrman, R., and Bruner-Lorand, J. (1971). *Anal. Biochem.* **44**, 221.
Lorand, L., Campbell, L. K., and Robertson, B. (1972a). *Biochemistry* **11**, 434.
Lorand, L., Campbell-Wilkes, L. K., and Cooperstein, L. (1972b). *Anal. Biochem.* **50**, 623.

Lorand, L., Chenoweth, D., and Gray, A. (1972c). *Ann. N. Y. Acad. Sci.* **202**, 155.
Lorand, L., Chou, C.-H. J., and Simpson, I. (1972d). *Proc. Natl. Acad. Sci. U.S.A.* **69**, 2645.
Lorand, L., Maldonado, N., Fradera, J., Atencio, A. C., Robertson, B., and Urayama, T. (1972e). *Br. J. Haematol.* **23**, 17.
Lorand, L., Gray, A. J., Brown, K., Credo, R. B., Curtis, C. G., Domanik, R. A., and Stenberg, P. (1974). *Biochem. Biophys. Res. Commun.* **56**, 914.
Lorand, L., Shishido, R., Parameswaran, K. N., and Steck, T. L. (1975). *Biochem. Biophys. Res. Commun.* **67**, 1158.
Lorand, L., Weissmann, L. B., Epel, D. L., and Bruner-Lorand, J. (1976). *Proc. Natl. Acad. Sci. U.S.A.* **73**, 4479.
Markland, F. S., and Damus, P. S. (1971). *J. Biol. Chem.* **246**, 6460.
Matacic, S., and Loewy, A. G. (1968). *Biochem. Biophys. Res. Commun.* **30**, 356.
Mockros, L. F., Roberts, W. W., and Lorand, L. (1974). *Biophys. Chem.* **2**, 164.
Mosher, D. F. (1975). *J. Biol. Chem.* **250**, 6614.
Mosher, D. F. (1976). *J. Biol. Chem.* **251**, 1639.
Northrop, J. H., Kunitz, M., and Herriott, R. M. (1948). "Crystalline Enzymes," 2d ed. Columbia Univ. Press, New York.
Pisano, J. J., Finlayson, J. S., and Peyton, M. P. (1968). *Science* **160**, 892.
Roberts, W. W., Lorand, L., and Mockros, L. F. (1973). *Biorheology* **10**, 29.
Rosenberg, R., Colman, R. W., and Lorand, L. (1974). *Br. J. Haematol.* **26**, 271.
Schneider, C. H. (1970). *In* "Penicillin Allergy, Clinical and Immunological Aspects" (G. T. Stewart and J. P. McGovern, eds.), pp. 23–58. Thomas, Springfield, Illinois.
Schwartz, M. L., Pizzo, S. V., Hill, R. L., and McKee, P. A. (1971). *J. Biol. Chem.* **246**, 5851.
Shishido, R. Y., Parameswaran, K. N., and Lorand, L. (1976). *Fed. Proc., Fed. Am. Soc. Exp. Biol.* **35**, 475 (Abstr. No. 1488).
Stenberg, P., Curtis, C. G., Wing, D., Tong, Y. S., Credo, R. B., Gray, A., and Lorand, L. (1975). *Biochem. J.* **147**, 155.
Takagi, T., and Doolittle, R. F. (1974). *Biochemistry* **13**, 750.
Tipper, D. J., and Strominger, J. L. (1965). *Proc. Natl. Acad. Sci. U.S.A.* **54**, 1133.
Williams-Ashman, H. G., Notides, A. C., Pabalan, S. S., and Lorand, L. (1972). *Proc. Natl. Acad. Sci. U.S.A.* **69**, 2322.
Wise, E. M., Jr., and Park, J. T. (1965). *Proc. Natl. Acad. Sci. U.S.A.* **54**, 75.

15

Physical Properties of the Inner Histones (H2a, H2b, H3, H4)

IRVIN ISENBERG

Among the many things I learned in Albert Szent-Györgyi's Institute was the importance of *complexes* for biology. Two molecules come together, form a structure whose properties are not simple sums of the properties of the constituents, and a new entity is built. Nowhere is this shown better, or in a more simple fashion, than in the histones.

INTRODUCTION

In 1956, Luck, Cook, Eldridge, Haley, Kupke, and Rasmussen stated, without fear of contradiction, that "The histones are commonly regarded as unpleasant proteins for rigorous studies." Until recently there was no reason to argue against this opinion.

Histones have been known and studied for many years. As early as 1884, Kossel separated a basic substance combined with nucleic acid in goose erythrocytes, and gave this substance the name, histone. Later (Kossel, 1928) he realized that there was not a single substance called histone; histones formed a class of proteins. However only in 1950 did the Stedmans show in a much more definitive manner that there were indeed different types of histone molecules, and only in 1951 was it reported that histones were found in every type of cell (Stedman and Stedman, 1951).

What made the histones such unpleasant proteins? Fundamentally it was because pure histones had to be prepared under conditions that commonly denatured proteins and, furthermore, they aggregated. Starting with Kossel (1884) everyone used acids to extract histones. This has continued to the present day. The widely used methods of Johns (1964), for example, use acids, sometimes mixed with ethanol.

It is not a defect to use acids. It is a perfectly satisfactory method to obtain pure histones, but it yields proteins that are unfolded–in a more or less random coil state. Such a state is not native.*

Much more serious than the denaturation was the problem of aggregation.† It will be seen later that the aggregation occurs at a much slower rate than histone refolding; it is unlikely therefore that the histones aggregate *because* they are denatured. Aggregation was the reason that histones were "commonly regarded as unpleasant proteins." In general, biochemists consider it to be a bad omen when the proteins they are studying aggregate, but, for histones, aggregation has been a *special* curse. Aggregation has interfered with the interpretation of the physical properties of histones from the earliest work down through some of the most recent papers. If we go back through the history of the field we find that almost all of the workers studying the physical properties of histones were riveting their attention on the aggregation, often being misled by their results. The Stedmans, for example, proposed that histones were cell specific and even showed differences between normal and malignant cells. Looking back at their work one sees that most of the evidence supporting this viewpoint stemmed from studies of aggregation; the only substantive support was the difference observed between histone f2c of bird erythrocytes and the protamines of sperm. Numerous other instances could be given, in which investigators were misled by aggregation phenomena, but it would serve no useful function to detail these.

It does not follow that studies of aggregation are uninteresting. The recent work of Sperling and Bustin (1974, 1975), to quote just one example, has shown

*Treatment with acid, under certain conditions, may remove phosphates from histidines or lysines (Chen *et al.,* 1974). Phosphorylation occurs at particular times during the cell cycle. If a dephosphorylation were to occur during preparation, the histones would be characteristic of a particular physiological state, but would not be "damaged". As will be seen, acid-extracted histones are denatured, but not denatured *irreversibly*; they may be refolded.

†In this paper the term aggregation will be restricted to the formation of large units. It will not refer to the formation of small oligomers such as dimers and tetramers.

that histones aggregate to very definite large structures: bent rods or fibers. These form in the manner of a self-assembled structure and this formation is certainly a remarkable property.

Nevertheless, as far as we know now, it is evident that the most important things about histones are what they do *other* than aggregate. As will be discussed, the essential features of histones are that the unfolded molecules can refold, and that these refolded molecules can interact with one another to form dimers, tetramers and octamers. Recent work on chromatin has shown that such oligomers are fundamental constituents of chromatin.

As has become evident recently, the histone oligomers can be obtained in two ways: they can be reformed from separated pure histones and they can be extracted directly from chromatin. The knowledge derived from both methods of study complement one another. As will be seen, the methods give different, though partly overlapping, insights into histone properties, and have provided important elements of our new knowledge of chromatin.

THE PRIMARY SEQUENCES OF THE HISTONES

There are five common classes of histones. Table I gives the names that have been used for them. In this paper we shall use the CIBA nomenclature: H2a, H2b, H3, H4 and H1.

As has become clear in recent years, H1 occupies a very different position and undoubtedly has a very different function from the others. Histones H2a, H2b, H3 and H4 are found in the chromatin particles, named ν bodies by Olins and Olins (1974) and nucleosomes by Oudet *et al.* (1975), and it is desirable to classify these histones differently from H1. Because they are present inside the nucleosomes I shall call them the *inner histones.* This paper will be limited to a review of the inner histones. By an obvious extension, the name *outer histones*

TABLE I
Classes of Histones[a]

Histone names	Molecular weight
H1, f1, I, KAP	About 21,000
H2b, f2b, IIb2, KAS	13,774
H2a, f2a2, IIb1, LAK	13,960
H3, f3, III, ARE	15,273
H4, f2a1, H4, IV, GRK	11,236

[a]The most commonly used names and molecular weights of the histones are given. In this paper the CIBA nomenclature is used. This is shown first in each listing.

appears appropriate for the H1 histones and any others that may also be outside the nucleosome, even if they bind to it.

Figure 1 shows a representation of the distribution of residues of these proteins. Upward bars show the positions of lysine and arginine; downward bars show glutamic and aspartic acid, and dots show the hydrophobic residues, valine, methionine, isoleucine, leucine, tyrosine, phenylalanine, proline and alanine.

One property is immediately evident from the figure: the charge distribution is uneven. All of the amino terminal halves of the inner histones are heavily basic, a fact which has been noted ever since the first histone, H4, was sequenced (DeLange *et al.*, 1969; Ogawa *et al.*, 1969). Indeed, based upon the determination of the H4 sequence, it was suggested (DeLange *et al.*, 1969) that the N- and C-terminal regions played two different roles in chromatin; the N region because of its basicity might bind to DNA, while the C region could intereact with other chromatin elements. This has been a fruitful concept.

The carboxy-terminal halves have been called either hydrophobic or apolar, but an additional useful characterization is also possible. We have compared the distribution of amino acids in the C-terminal halves with the distribution in a number of globular proteins. They are very similar in nature. A good descriptive word for the C-terminal region might then be "ordinary" or "globular."

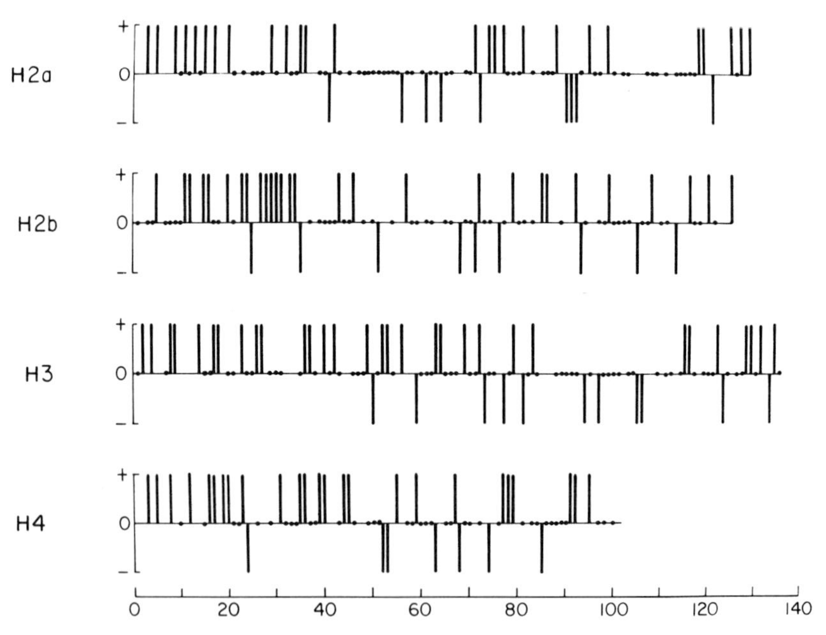

Fig. 1. Distribution of amino acid residues of the inner histones. Upward bars are lysine and arginine, downward bars are aspartic and glutamic acids and dots arc Val, Mct, Ilc, Lcu, Tyr, Phe, Pro, and Ala.

This choice of words is not simply a matter of semantics. We know a great deal about globular proteins, and what we know globular proteins to be able to do, we might suspect that a histone C-terminal region could do. Denatured globular proteins can be refolded; perhaps histones can too. The refolding of globular proteins, especially small globular proteins, is usually cooperative; perhaps that of histones is too. As will be seen, histones do indeed undergo a conformational change that does appear to be a cooperative folding. This folding occurs upon the addition of salt to histones in water at neutral pH.

We might also ask what might be expected of the part that does not look globular: the N-terminal region of the molecule. There is little that one can gain here by examining the primary structure, other than to note its basicity. However, there have been some suggestions and, in addition, interesting nuclear magnetic resource data.

Sung and Dixon (1970) have shown that models of the amino terminal portion of H4, if folded in an α-helical array, will fit nicely into the major groove of DNA, with four lysyl residues making close contact with four consecutive DNA phosphates. This is interesting, but, of course, is a proposed model only.

Bradbury *et al.* (1971) cleaved H2b at methionines 59 and 62. Using optical rotatory dispersion they then measured Moffit b_0 values for the two polypeptides, 1 to 58, and 63 to 125. With increasing salt concentrations the amino terminal portion showed little change of b_0 while the carboxy part showed large changes. If one makes the assumption, which appears reasonable, that studies on such separated parts of H2b can help elucidate the behavior of the entire molecule, and, in addition, that b_0 measures the existence of definite secondary structure of any type, then it would follow that the N-terminal portion is mainly in a random coil conformation even after the salt-induced change.

Studies of severe line broadening of proton magnetic resonances (Bradbury *et al.*, 1971, 1973; Bradbury and Rattle, 1972) have indicated which parts of the molecule are involved in aggregation and which are not, but these studies have not yet elucidated properties of the folding of the inner histones as such. However, it is of considerable interest that all such studies have indicated that the N-terminal regions have a high mobility and must be considered to be in a more or less random coil arrangement.

Recent magnetic resonance data has added to our knowledge. Studies of histone complexes both by proton magnetic resonance (E. M. Bradbury and C. Crane-Robinson, personal communication) and by ^{13}C resonance (D. J. J. Lilley, personal communication) have shown that the N-terminal regions are highly mobile and must be regarded as unfolded. Such tails provide in a natural manner the possibility of links with the DNA of a chromatin particle, perhaps in the manner suggested by Van Holde *et al.* (1974). In this model the histone tails extend from the histone core and wrap themselves about the DNA, thus forming a compact particle.

The histone tails may have a different structure when bound to DNA than they do in free solution. When bound, they may fold, perhaps in a manner like that suggested by Sung and Dixon (1970) although nothing is known about this at the present time.

HISTONE REFOLDING

The study of histone refolding was made possible by the observation that a conformational change could occur in histone H4 which was *not* accompanied by aggregation (Li *et al.*, 1972).

When salt is added to a dilute solution (10^{-5} *M*) of H4 in water two different changes occur. The changes are distinguishable from one another in several ways. In the first place they occur on completely different time scales. There is a fast change and a slow change and, at low concentrations, these differ by orders of magnitude in time (Li *et al.*, 1972; Wickett *et al.*, 1972). Similar phenomena are observed for H3 (D'Anna and Isenberg, 1974d).

Slow changes of state, when salt is added to histone solutions, had been observed previously (Cruft *et al.*, 1958; Laurence, 1966), but the fast change had not been noticed.

Our laboratory has used three principal techniques in studying fast and slow changes: fluorescence polarization, circular dichroism and light scattering. Figure 2 shows some typical results for H3; similar data has been obtained for the other inner histones. At zero time, the anisotropy is a function of the salt concentration. The conformational change that is being monitored at zero time has been named the fast step. Above a critical salt concentration, the fast step is followed by a slower change taking minutes to hours, depending on the salt concentration, the histone concentration, and the temperature.

It may be noted that the light scattering data has a completely different character: only the slow step is evident; there is no scattering change with salt at zero time.

The same point is demonstrated in Fig. 3. The anisotropy and light scatter are shown for both zero and infinite (in practice 24 hr) times. Below the critical salt concentration $r_\infty = r_0$ and the scattered light E/L is a constant.* Only after r_∞ diverges from r_0 is there an increase in the scattered light.

The conclusion is clear. There is aggregation *only* during the slow step, *not* during the fast step.

Lewis *et al.* (1975) verified the existence of both a fast and slow step, but reported that there was aggregation during what they defined in their work as

*The subscripts on the anisotropy denote infinite time or zero time. E/L is the scattered light in arbitrary units.

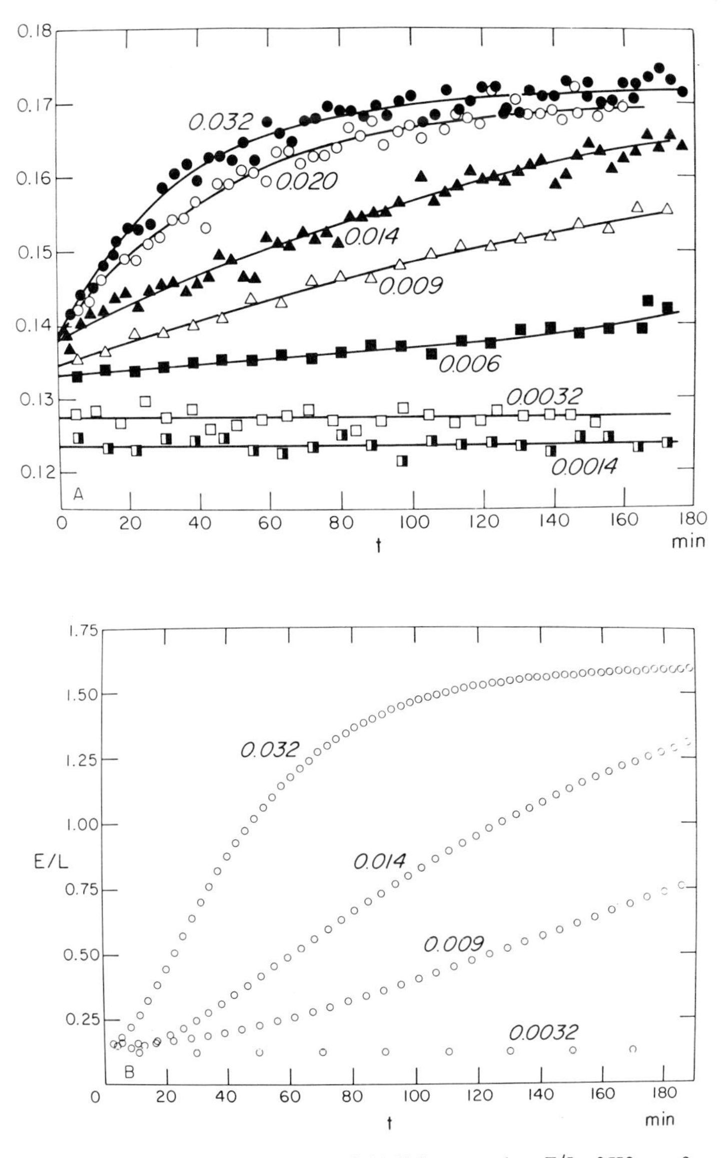

Fig. 2. (a) Fluorescence anistropy, r, and (b) light scattering, E/L of H3 as a function of time. The phosphate concentration is given on the curves. pH was 7.0. The anisotropy of salt-free solutions was 0.100. The histone concentration was $1.0 \times 10^{-5} M$. Reprinted with permission from D'Anna and Isenberg, *Biochemistry* **13**, 4987 (1974d). Copyright by the American Chemical Society.

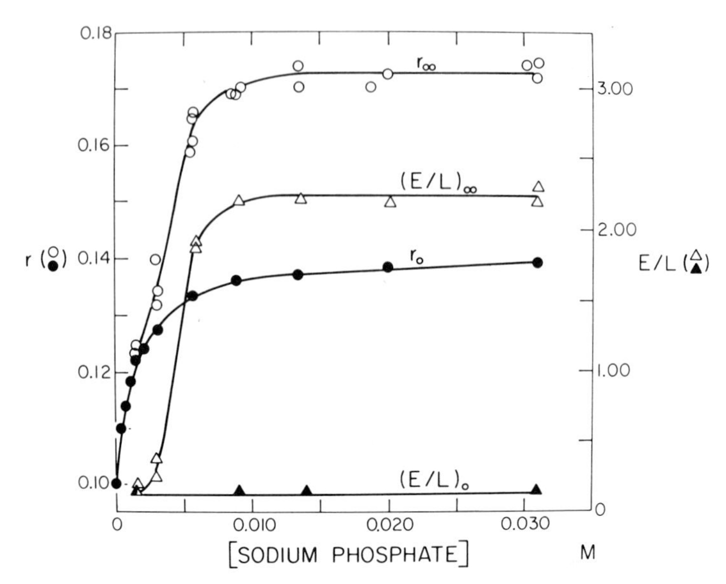

Fig. 3. Anisotropy and light scatter of H3 at zero time and infinite time (in practice, 24 hr). Note that there is no increase in light scatter in the region for which $r_\infty = r_0$. Aggregation occurs only during the slow step. Solution conditions are the same as in Fig. 2. Reprinted with permission from D'Anna and Isenberg, *Biochemistry* **13,** 4987 (1974d). Copyright by the American Chemical Society.

the fast step. However, these workers used an H4 concentration of 10^{-3} *M,* which is some 100 times larger than that used by Li *et al.* (1972) and Wickett *et al.* (1972). At these high histone concentrations, there is considerable line broadening of proton magnetic resonances at very early times, implying aggregation at these early times. It is known, however (Li *et al.,* 1972; Wickett *et al.,* 1972), and in any case reasonable on *a priori* grounds, that the aggregation is a sensitive function of the histone concentration. It is easily possible, and perhaps likely, that an appreciable amount of aggregation could have taken place during an initial measurement. It is clear that the studies of Lewis *et al.* (1975) do not permit one to observe the clean distinction between folding and aggregation that work at dilute concentrations does.

The change in secondary structure that occurs during the fast step is very different from that occurring in the slow step. During the fast change in H4 there is α-helix, but little or no β-sheet formation; during the slow change there is β-sheet but little or no α-helix formation (Li *et al.,* 1972; Wickett *et al.,* 1972). For H3 there is again α-helix, but no β-sheet, formation during the fast change but in the slow change there is not only β-sheet formation but also a decrease in α-helical content (D'Anna and Isenberg, 1974d). The structure of an H3 or H4 molecule, when in an aggregate, is completely different from the structure in the folded state before aggregation.

While the arginine-rich histones, H4 and H3, show both fast and slow steps, only a fast change has been observed in H2b solutions at dilute concentrations (D'Anna and Isenberg, 1972). Histone H2a does show a slow change but it is one with a very long time constant (D'Anna and Isenberg, 1974a). There is no appreciable slow change in many hours and, for most purposes this may be ignored. Again it must be emphasized that this holds true only at low concentrations of the order of 10^{-5} *M*. Bradbury *et al.* (1975) reported that folding and aggregation occurred "simultaneously" in histone H2a, but they used histone concentrations of 10^{-3} *M*.

Globular proteins generally, but not always, fold in a cooperative manner (Baldwin, 1975) and it is of some interest to examine the cooperativity of the fast change.

As of the date of this writing there is only one type of experimental evidence bearing on this point. Figure 4 shows the fractional change in the fast folding of H3 as a function of sodium phosphate. The important thing about the figure is that the fraction changing is measured in two different ways. In one method, what is observed is the anisotropy and the intensity of fluorescence of the tyrosines of H3; in the other, one measures the circular dichroism of the peptide absorbance. These techniques are sensitive to completely different properties of the molecule. In the first, we observe a parameter that is sensitive to the rotatory diffusion of the tyrosine transition moment and, perhaps, energy transfer be-

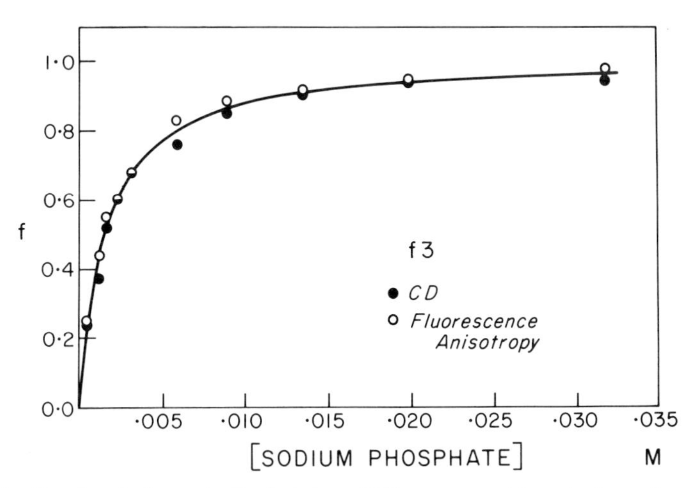

Fig. 4. Fractional change in the fast step of H3 as measured by anisotropy (open circle) and CD (closed circle).

tween tyrosines; in the second, we examine a property that is sensitive to the exciton splitting due to the interaction of the peptide transition moments. These bear no necessary relationship to one another. The fact that the fractional change is the same when measured by both techniques means that each technique is looking at a different aspect of the same event, the cooperative folding of an entire molecule, otherwise the curves would be different. As the salt concentration changes, the number of molecules in each of the two states, folded and unfolded, alters. At any particular salt concentration there is a dynamic equilibrium between the folded and unfolded forms, just as there is in any chemical equilibrium.

A contrary interpretation has been given (Boublik *et al.*, 1970; Lewis *et al.*, 1975; Bradbury and Rattle, 1972) in which the folding of a histone molecule is a function of the salt concentration. This interpretation envisions the folding as essentially noncooperative.

Other workers have used the idea of looking at protein folding by different techniques to judge the cooperativity of the conformational changes. Holcomb and Van Holde (1962) and Ginsburg and Carroll (1965) used it to study ribonuclease and Anfinsen *et al.* (1972) adopted it in work on staphylococcal nuclease.

It would be valuable to have additional and independent evidence on the cooperativity of the histone foldings. Nevertheless, in its absence, it must be recognized that what exists supports the idea that all of the histones fold cooperatively (Li *et al.*, 1972; Wickett *et al.*, 1972; D'Anna and Isenberg, 1972, 1974a,d). There is one fundamental fast change upon salt addition and this change bears all of the earmarks of the renaturation of a denatured globular protein. The data supporting this is particularly extensive for H4 (Wickett *et al.*, 1972). Folding was induced by seven different salts. The effectiveness in inducing folding was a function of the salt, more specifically the anion, and varied from one salt to another by two orders of magnitude. Yet with the exception of $NaClO_4$ (a result which is not understood), each salt showed the same effectiveness in inducing folding when viewed by CD or by anisotropy of emission, despite the wide variation in effectiveness from one salt to another.

HISTONE AGGREGATION

The aggregation of histones has been extensively investigated by a wide variety of techniques (Cruft *et al.*, 1954, 1957, 1958; Laurence, 1966; Davison and Shooter, 1956; Mauritzen and Stedman, 1959; Phillips, 1965, 1967; Johns, 1968, 1971; Fambrough and Bonner, 1968; Edwards and Shooter, 1969; Boublik *et al.*, 1970; Diggle and Peacocke, 1971; Bradbury *et al.*, 1973, 1975; Bradbury and Rattle, 1972; Lewis *et al.*, 1975; Li *et al.*, 1972; Wickett *et al.*,

1972; D'Anna and Isenberg, 1974d; Sperling and Bustin, 1974, 1975; Smerdon and Isenberg, 1973, 1974).

In contrast to the fast folding, the aggregation of histones is noncooperative. This is clearly shown by the nuclear magnetic resonance studies reported by Bradbury and co-workers (1971, 1973, 1975; Bradbury and Rattle, 1972; Lewis *et al.*, 1975). These studies show that the amount of severe proton magnetic resonance line broadening, and hence aggregation, is a function of the salt concentration in the histone solutions. The aggregation is not a simple cooperative process and, in fact, there may be a number of significant intermediate conformational states in the aggregation phenomenon; how many is not clear. Of considerable additional interest is the finding that the histones show *selective* line broadening; some parts of the molecule have resonances that broaden a great deal compared to other parts. The aggregation is therefore selective. Furthermore, the protein resonances of DNA–histone complexes also show selective broadening. A comparison of the two studies show that the regions of the molecule that aggregate are different from the regions that broaden in the presence of DNA.

The histones aggregate to large but definite structures. These are bent rods with a diameter of 22 Å and fibers of diameter 44 Å (Sperling and Bustin, 1974, 1975). These results are not only interesting in their own right, but help explain the selectivity of the resonance broadening reported by Bradbury and co-workers. It would be very difficult to see how only a part of a histone molecule could participate in a large aggregate if the aggregate were essentially globular. However, if the aggregate were assembled in a linear array then part of the molecule could participate in the aggregation, the remainder being free to stick out of the sides. Such linear arrays have been postulated as models to explain the resonance data (Bradbury *et al.*, 1975; Lewis *et al.*, 1975) and it is therefore satisfying that such linear arrays have been observed in electron micrographs (Sperling and Bustin, 1974, 1975).

A considerable amount of β sheet is found in the aggregate while little or no β sheet is formed in the fast folding (Li *et al.*, 1972; Wickett *et al.*, 1972; D'Anna and Isenberg, 1974a,d; Baker and Isenberg, 1976).

The aggregation of histones is a sensitive function of temperature. This temperature dependence has been known for many years (Ui, 1957; Cruft *et al.*, 1958), but only recently has the phenomenon been studied in depth and clarified (Small *et al.*, 1973; Smerdon and Isenberg, 1973, 1974). The aggregation is a very sensitive function of the temperature; at 0°C, the slow step can be blocked for many days (Smerdon and Isenberg, 1973).

The sharp temperature dependence may be related to the fact that the aggregation is slow. If activation barriers must be surmounted for aggregation to occur, then the barriers will be more difficult to overcome as the temperature is lowered.

HISTONE–HISTONE INTERACTIONS

The first hint that different histones complexed with one another came many years ago when Cruft *et al.* (1958) mentioned, in passing, that what they were calling "subsidiary histones" inhibited the aggregation of an arginine-rich fraction that they were calling "β histone." Somewhat clearer evidence was obtained by Laurence (1966) who, it is interesting to note, knew that H3 and "fraction f2a" changed their properties with time after salt was added to a previously salt-free solution. Laurence showed that H2b inhibited the aggregation of fraction f2a, clearly suggesting complex formation.

Shih and Bonner (1970) reported melting studies of nucleohistones that had been reconstituted with mixtures of histones, and concluded that the individual histones were distributed among one another in the complex.

Skandrani *et al.* (1972) ran calf thymus "fraction f2b" on Amberlite IRC-50 columns and reported that two fractions were obtained. One was a serine-rich, lysine-rich histone and another was an equimolar complex of the serine-rich, lysine-rich histone and an arginine-rich, lysine-rich histone.

D'Anna and Isenberg (1973) reported that H2b and H4 formed a strong one-to-one complex. In a one-to-one mixture of the two histones, the slow step of H4 was completely inhibited by H2b.

Figure 5 shows the anisotropy as a function of time for H4 plus H2b, the expected anisotropy if the histones did not interact, and the measured anisotropy of a one-to-one mixture. The anisotropy of the mixture is independent of time and is higher than the zero-time anisotropies of the individual histones. The presence of H2b inhibits the aggregation of H4 and it is clear that the two histones interact strongly.

Figure 6 shows a similar phenomenon for circular dichroism data. The slow change in CD of H4 is abolished by the presence of H2b. In addition, however, it should be noted that the CD of the interacting proteins is appreciably higher than that of the noninteracting proteins. Furthermore if one analyzes a CD spectrum rather than the CD at a particular wavelength, one finds that the α-helical content of the complex is greater than that of the individual folded histones. A recently developed method of analyzing CD spectra (Baker and Isenberg, 1976) shows that there are 15 additional α-helical residues induced by complex formation, somewhat larger than the prior estimate of 8 residues (D'Anna and Isenberg, 1973, 1974d) based on difference spectra calculations.

Figure 7 shows data on the formation of the H3–H4 complex. Whereas both histones have slow aggregations individually, in one-to-one mixtures each histone quenches the slow aggregation of the other.

The H3–H4 complex is a tetramer as shown by sedimentation analysis (D'Anna and Isenberg, 1974c). Kornberg and Thomas (1974) and Roark *et al.* (1974) obtained this tetramer using a completely different procedure, and it is

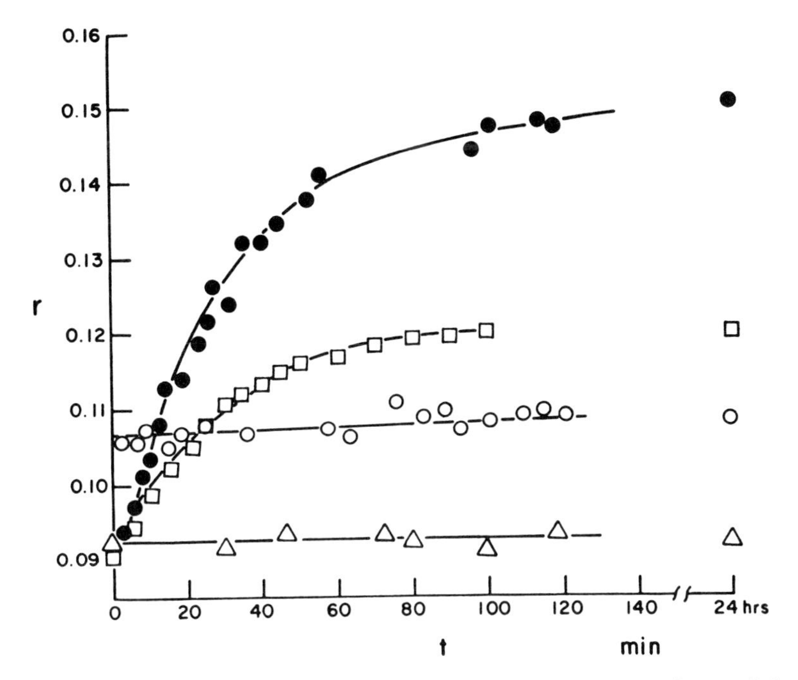

Fig. 5. Anisotropy of H4 (closed circle), H2b (triangle), and a 1:1 mixture of the two (open circle). The curve with the square symbol is what would be obtained for a noninteracting mixture. Concentrations were 0.90×10^{-5} *M* for each histone in 0.016 *M* phosphate pH 7.0. Reprinted with permission from D'Anna and Isenberg, *Biochemistry* **12**, 1035 (1973). Copyright by the American Chemical Society.

of considerable interest to contrast the method of D'Anna and Isenberg with that of Kornberg and Thomas (1974) and Roark *et al.* (1974). The method used by D'Anna and Isenberg may be called a renaturation procedure. The individual, pure histones are added to each other and the complex is formed upon the addition of salt. Kornberg and Thomas and Roark *et al.* used what may be termed a mild extraction procedure. The tetramer was obtained by salt extraction of chromatin at neutral pH.

Neither procedure, in and of itself, demonstrates that the tetramer exists as such in chromatin. Reconstitution obviously cannot show this, but neither can extraction methods since if both H3 and H4 had existed separately in chromatin, but were extracted simultaneously, they could then form the tetramer.

As with the dimer (H2b)(H4), the $(H3)_2(H4)_2$ tetramer also has a structure which is different from the simple juxtaposition of its constituents. Again, when the molecules come together there is a structural alteration induced in the process of complexing.

Histones H2a and H2b also form a strong complex (Kelley, 1973; D'Anna and

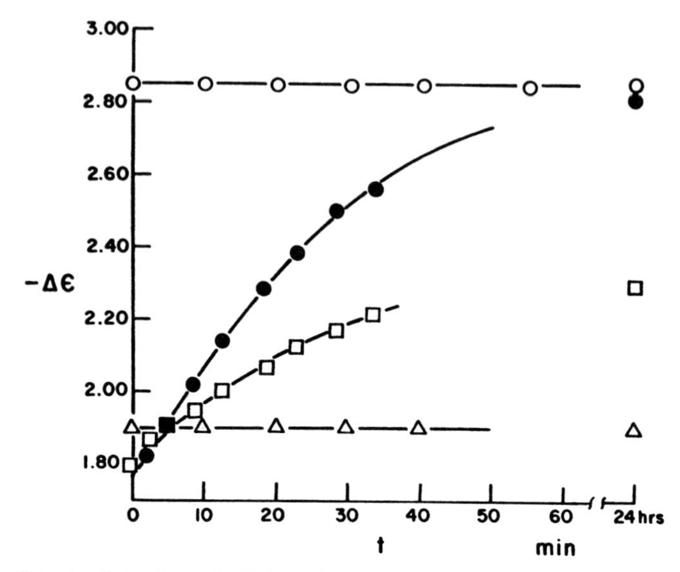

Fig. 6. Circular dichroism of solutions described in Fig. 5. Symbols are the same as for Fig. 5. Units are cm^{-1} l/mole of residue. Reprinted by permission from D'Anna and Isenberg, *Biochemistry* **12,** 1035 (1973). Copyright by the American Chemical Society.

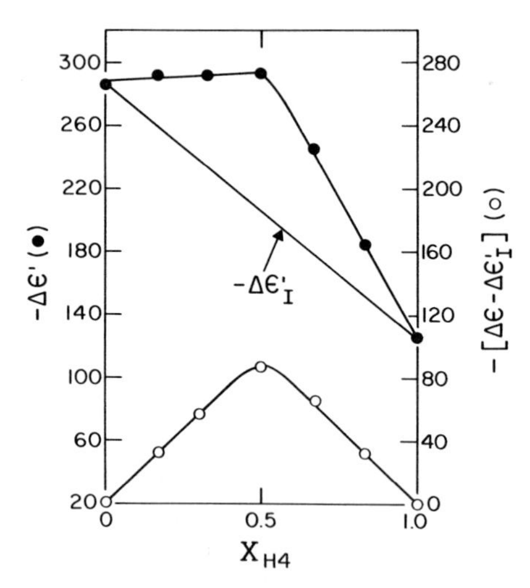

Fig. 7. Continuous variation curve of histones H4 and H3 CD at 220 nm vs. mole fraction of H4. CD units are cm^{-1} l/mole of protein. The total protein concentration was 0.5×10^{-5} *M* in 0.0085 *M* phosphate, pH 7.0. The subscript, I, means the expected value of noninteracting pairs. Reprinted by permission from D'Anna and Isenberg, *Biochemistry* **13,** 4992 (1974e). Copyright by the American Chemical Society.

Isenberg, 1974b; Kornberg and Thomas, 1974). The complex is a dimer and again there is a marked enhancement of the α-helical content upon complexing.

The binding of these complexes is very strong. The value of the binding constant varies with salt, so only representative values may be given. However, such a value for H2b, H4 and H2a, H2b is 10^6 M^{-1} and for $(H3)_2(H4)_2$ is 10^{21} M^{-3}. Recent work by D'Anna and Isenberg shows that even the high value of 10^{21} M^{-3} may be exceeded at physiological salt concentrations.

Aside from these three strong interactions there are three weak ones, so that any of the four inner histones interacts with any of the others. Histone H2a interacts with H4 (D'Anna and Isenberg, 1974b; Clark *et al.*, 1974) and with H3 (D'Anna and Isenberg, 1974e) and H3 interacts weakly with H2b (D'Anna and Isenberg, 1974e). The stoichiometries of the weak complexes are not yet known.

The separation of the histone–histone interactions into strong and weak classes does not, of course, imply that only the strong interactions should, or could, occur in chromatin. We know, if fact, that both weak and strong interactions do occur, perhaps serving different functions.

HISTONE–HISTONE INTERACTIONS IN CHROMATIN

Recent developments have shown that chromatin has a subunit structure (Olins and Olins, 1973, 1974; Woodcock, 1973; Sahasrabuddhe and Van Holde, 1974; Van Holde *et al.*, 1974; Noll, 1974; Kornberg, 1974; Kornberg and Thomas, 1974; Oudet *et al.*, 1975). Although at the time that this review is being written, there are some features of the chromatin subunit structure that are either in doubt or controversial, it appears as if there is general agreement that a major feature of the subunit is a particle, termed a *ν* body by Olins and Olins (1974), a PS particle by Rill and Van Holde (1973) and a nucleosome by Oudet *et al.* (1975). The core of the chromatin particle appears to be a set of eight histones with probably two of each of the inner histones. The DNA is wrapped around the histone core.

The present paper cannot attempt a survey of these recent findings on chromatin. Two reviews have recently been written (Van Holde and Isenberg, 1975; Elgin and Weintraub, 1975) and the field is now developing at a very rapid pace. This section will therefore limit itself to only those aspects of the structure of chromatin that are directly related to histone-histone interactions.

The principal and certainly the most important tool used to date to study histone arrangements in chromatin is linking by bifunctional reagents. Before proceeding to discuss the results of such studies, however, it is necessary to comment on both the power and limitations of the method.

Cross-linking is a means of studying the *geometry* of a structure. If two molecules are close to one another then, at least in principle, they can be cross-linked by a choice of the proper bifunctional reagent. The molecules do not have to interact to any significant extent to be cross-linked; they must simply be sufficiently close to one another.

Now, if two molecules interact they, of course, are indeed close and can be cross-linked, and therefore a search for interacting species can be carried out by cross-linking techniques. In addition, if the cross-linking reagent is short, the likelihood becomes high that cross-linking will pick out pairs that are complexed. However, fundamentally, complexing and cross-linking measure two different properties: one thermodynamic, the other geometric.

It follows that cross-linking will not be able to distinguish between strong and weak interactions. We have seen that between the histones there are both strong and weak forces and this may have functional significance. In other areas of biochemistry there are both strong and weak interactions. The distinction is important, and strong and weak interactions have quite different functions. Different roles may exist also for the strong and weak interactions of histones, but it would be premature at the present time to attempt to say what these roles might be.

Another comment should be made. Chemical systems are almost always in a state of dynamic equilibrium between different species; biological systems are too. This means that there is some danger that misleading interpretations are possible if cross-linking data are interpreted in what, at first glance, might be considered a straightforward way. Consider, for example, the system

$$A + B \underset{k_2}{\overset{k_1}{\rightleftharpoons}} AB$$

The complex, AB, can be cross-linked. Suppose now that the dissociation constant, $K = k_2/k_1$ is very large, so that practically all of the molecules are in the separated form A and B, and there are extremely few complexes, AB. A reaction cross-linking A and B will proceed slowly, if only because there are so few complexes. However, as soon as some complexed molecules are cross-linked, they will be pulled out of the equilibrium and, by Le Chatelier's principle, more molecules will be complexed. These, in turn, will also be cross-linked and the reaction could, in principle at least, proceed until all of the molecules existed as cross-linked complexes. If we then were to analyze our sample, and find that all of the molecules are cross-linked, we could not then conclude that the predominant species in the solution before cross-linking was AB. A rather vicious uncertainty principle is operative: the cross-linking itself alters what is being measured.

In addition, one must recognize that cross-linking can not only cause more complexes to form but can also break up pre-existing structures by interference. Thus, while cross-linking methods are extremely valuable and, indeed, have been of unique importance so far, one should recognize the possibility of being misled by them. It is possible, for example, that the report that H2a–H2b exists in solution as a polymer (Kornberg and Thomas, 1974) rather than a dimer, was due to the cross-linking of a small polymer fraction in the solution and the subsequent distortion of the equilibrium as described above.

With these precautionary statements made, however, let us turn to the cross-linking experiments themselves.

Martinson and McCarthy (1975) used tetranitromethane (TNM) to cross-link histones in chromatin. They obtained unambiguous evidence for a cross-linked H4–H2b dimer. Furthermore, they found the same dimer in growing HeLa cells treated with TNM. In addition, however, Martinson and McCarthy cross-linked nucleohistones reconstituted in a number of different ways.

Reconstitution with all five histones yielded an H4–H2b product. Histones H1 and H3 could be eliminated and the same dimer still obtained. However, if H2a were not present, the H4–H2b product was also absent. Histone H2a is needed to enable H4 to cross-link to H2b.

TNM is a zero-length cross-linking reagent. It is highly likely, therefore, that by its use complexing as well as close proximity is being detected.

The results of Martinson and McCarthy provide evidence, therefore, for an H2b–H4 complex and, in addition, the complexing of H2a with one or both of the other histones.

Using formaldehyde and gluteraldehyde, Van Lente *et al.* (1975) also found a cross-linked dimer of H4–H2b but, in addition, reported an H2a–H2b dimer. Bonner and Pollard (1975) reported a carbodiimide cross-linked H4–H3 dimer.

Thomas and Kornberg (1975) cross-linked chromatin with dimethyl suberimidate. By this means they were able to demonstrate the octamer $(H4)_2(H3)_2(H2b)_2(H2a)_2$. This important step was the first direct demonstration given for the existence of such an octamer even though such an octamer had been previously postulated on the existence of indirect evidence (Kornberg, 1974; Van Holde *et al.*, 1974). Thomas and Kornberg also reported that they could obtain such an octamer free in histone solutions formed by the elution of histones from chromatin at an ionic strength of 2.0 at pH 9. However, the method used to detect this octamer in free solution was cross-linking. It would be helpful to have more direct studies on the salt-extracted material to see if an octamer could be identified without cross-linking.

The cross-linking studies of Thomas and Kornberg also led to the identification of various smaller cross-linked products: $(H3)_2$, H2a–H2b, H4–H3, H4–H2a, and H4–H2b.

Weintraub *et al.* (1975) have reported that, in solutions of all of the inner histones in 2 *M* NaCl at pH 7.1, the majority of the histones are in a tetramer, probably containing one each of H4, H2b, H3, and H2a. This work is an apparent contradiction to that of Thomas and Kornberg (1975), referred to in the previous paragraph. At first glance the implication might be that at pH 7.1 the inner histones form a tetramer and at pH 9, two tetramers join to form an octamer. However, such an important issue deserves some direct experimental resolution.

The finding by Weintraub *et al.* (1975) of a tetramer of all four inner histones, which might, under proper conditions, dimerize to form an octamer, is consistent with the model proposed by D'Anna and Isenberg (1974e). These authors noted that H2a, H2b, H3, and H4 could form a linear chain by pair-wise strong interactions. This chain has the feature that all of the histones are different; hence the chain has a direction. If such a chain were to be associated with one of the DNA strands, another such chain could be associated in an opposite sense with the other DNA strand. The weak histone–histone interactions might hold the two chains to one another, thus increasing the DNA stability.

This model is, of course, still untested and may be wrong. It provides, however, for a simple picture of replication. During replication each histone tetramer could remain associated with its own DNA strand, with the newly synthesized histones binding as a new histone tetramer to the new DNA. Tsanev and Russev (1974) have reported that histone replication is semiconservative in just this sense.

In any event, regardless of specific models, or of detailed interpretations of cross-linking experiments, it must be concluded that all such experiments are at least consistent with, and lend support to the idea that the major structure of eukaryotic chromatin is determined by histone-histone interactions. These are all specific. Both weak and strong interactions can occur, but most of them appear to be strong. The meaning of such interactions for the functioning of eukaryotic chromatin is not yet clear, but if the present state of vigorous progress in the field continues, one can hope that such an understanding may begin soon.

SUMMARY OF THE KNOWN PHYSICAL PROPERTIES OF THE INNER HISTONES (H2a, H2b, H3, H4)*

1. The inner histones have a very heterogeneous distribution of amino acid residues. The amino terminal regions are very basic; the carboxy halves (very approximately) have distributions that are typical of globular proteins.

*No references are given in the summary. They are, of course, in the body of the text. The references have been omitted to facilitate the reading of the summary. Many laboratories, of course, have contributed to our present knowledge of histone physical properties and the author hopes that he has given adequate reference to many of these contributions in the main part of this review.

2. In salt-free aqueous solutions the histones are in a more or less random coil form. There may or may not be definite residual secondary structure, but the dominant feature is the random coil nature of the molecule.

3. At low concentrations of a histone (10^{-5} *M*) there is a very rapid folding of the molecule upon the addition of salt.

4. The folding is cooperative. It is not yet known how much of each molecule folds.

5. Above a critical salt concentration, after the folding, histones H3 and H4 aggregate slowly. At low concentrations H2b and H2a either do not aggregate or do so exceedingly slowly so that for many hours they are in an unaggregated form.

6. The secondary structure, and hence the tertiary structure, of a histone molecule in an aggregate is different from that of a folded, fast step product.

7. At high histone concentrations (10^{-3} *M*) all of the inner histones aggregate. The aggregation involves only parts of the molecules: other parts do not aggregate. The aggregation may be visualized as a linear array with tails that are free of aggregation.

8. The aggregates are definite structures: bent rods of 22 Å diameter and fibers of 44 Å diameter.

9. The aggregation is very temperature dependent. Lowering the temperature markedly decreases the rate of aggregation and, at 0°C and 10^{-5} *M* concentration, aggregation is negligible for days.

10. All of the inner histones complex with one another. The complexing is specific, but may be strong or weak.

11. The strong complexes are $(H4)_2$ $(H3)_2$, (H2b)(H4), and (H2a)(H2b). The histones change their structure upon complexing.

12. The weak complexes are H2a–H3, H2a–H4, and H2b–H3. The stoichiometry of the weak complexes is not yet known.

13. Two of the strong complexes, $(H4)_2(H3)_2$ and (H2a)(H2b) may be obtained from chromatin by mild, salt extraction procedures.

14. Cross-linking experiments on nuclei or chromatin yield linked molecules of all of the strong complexes, one of the weak complexes, the dimer (H2a)(H4), the homodimer $(H3)_2$ and the octamer $(H4)_2(H3)_2(H2a)_2(H2b)_2$.

ACKNOWLEDGMENTS

The work of this paper was supported by Public Health Service Grant CA-10872. It is a pleasure to thank the many associates and students who were associated with various phases of the work reviewed here: Carl C. Baker, A. Morrie Craig, Joseph A. D'Anna, Jr., Roswitha Hopkins, J. H. Li, Enoch W. Small, Michael J. Smerdon, and R. Randall Wickett.

REFERENCES

Anfinsen, C. B., Schechter, A. N., and Taniuchi, H. (1972). *Cold Spring Harbor Symp. Quant. Biol.* **36**, 249.
Baker, C. C., and Isenberg, I. (1976). *Biochemistry* **15**, 629.
Baldwin, R. L. (1975). *Annu. Rev. Biochem.* **44**, 453.
Bonner, W. M., and Pollard, H. B. (1975). *Biochem. Biophys. Res. Commun.* **64**, 282.
Boublik, M., Bradbury, E. M., and Crane-Robinson, C. (1970). *Eur. J. Biochem.* **14**, 486.
Bradbury, E. M., and Rattle, H. W. E. (1972). *Eur. J. Biochem.* **27**, 270.
Bradbury, E. M., Cary, P. D., Crane-Robinson, C., Riches, P. L., and Johns, E. W. (1971). *Nature (London) New Biol.* **233**, 265.
Bradbury, E. M., Cary, P. D., Crane-Robinson, C., and Rattle, H. W. E. (1973). *Ann. N.Y. Acad. Sci.* **222**, 266.
Bradbury, E. M., Cary, P. D., Crane-Robinson, C., Rattle, H. W. E., Boublik, M., and Sautiere, P. (1975). *Biochemistry* **14**, 1876.
Chen, C. C., Smith, D. L., Bruegger, B. B., Halpern, R. M., and Smith, R. A. (1974). *Biochemistry* **13**, 3785.
Clark, V. M., Lilley, D. J. J., Howarth, O. W., Richards, B. M., and Pardon, J. F. (1974). *Nucleic Acids Res.* **1**, 865.
Cruft, H. J., Mauritzen, C. M., and Stedman, E. (1954). *Nature (London)* **174**, 580.
Cruft, H. J., Mauritzen, C. M., and Stedman, E. (1957). *Philos. Trans. R. Soc. London, Ser. B* **241**, 93.
Cruft, H. J., Mauritzen, C. M., and Stedman, E. (1958). *Proc. R. Soc. London, Ser. B* **149**, 21.
D'Anna, J. A., Jr., and Isenberg, I. (1972). *Biochemistry* **11**, 4017.
D'Anna, J. A., Jr., and Isenberg, I. (1973). *Biochemistry* **12**, 1035.
D'Anna, J. A., Jr., and Isenberg, I. (1974a). *Biochemistry* **13**, 2093.
D'Anna, J. A., Jr., and Isenberg, I. (1974b). *Biochemistry* **13**, 2098.
D'Anna, J. A., Jr., and Isenberg, I. (1974c). *Biochem. Biophys. Res. Commun.* **61**, 343.
D'Anna, J. A., Jr., and Isenberg, I. (1974d). *Biochemistry* **13**, 4987.
D'Anna, J. A., Jr., and Isenberg, I. (1974e). *Biochemistry* **13**, 4992.
Davison, P. F., and Shooter, K. V. (1956). *Bull. Soc. Chim. Belg.* **65**, 85.
DeLange, R. J., Fambrough, D. M., Smith, E. L., and Bonner, J. (1969). *J. Biol. Chem.* **244**, 319.
Diggle, J. H., and Peacocke, A. R. (1971). *FEBS Lett.* **18**, 138.
Edwards, P. A., and Shooter, K. V. (1969). *Biochem. J.* **114**, 227.
Elgin, S. C. R., and Weintraub, H. (1975). *Annu. Rev. Biochem.* **44**, 725.
Fambrough, D. M., and Bonner, J. (1968). *J. Biol. Chem.* **243**, 4434.
Ginsburg, A., and Carroll, W. R. (1965). *Biochemistry* **4**, 2159.
Holcomb, D. N., and Van Holde, K. E. (1962). *J. Phys. Chem.* **66**, 1999.
Johns, E. W. (1964). *Biochem. J.* **92**, 55.
Johns, E. W. (1968). *J. Chromatogr.* **33**, 563.
Johns, E. W. (1971). *In* "Histones and Nucleohistones" (D. M. P. Phillips, ed.), pp. 1–45.
Kelley, R. I. (1973). *Biochem. Biophys. Res. Commun.* **43**, 1588.
Kornberg, R. D. (1974). *Science* **184**, 868.
Kornberg, R. D., and Thomas, J. O. (1974). *Science* **184**, 865.
Kossel, A. (1884). *Hoppe-Seyler's Z. Physiol. Chem.* **8**, 511.
Kossel, A. (1928), "Protamines and Histones." Longmans, Green, New York.
Laurence, D. J. R. (1966). *Biochem. J.* **99**, 419.
Lewis, P. N., Bradbury, E. M., and Crane-Robinson, C. (1975). *Biochemistry* **14**, 3391.

Li, H. J., Wickett, R., Craig, A. M., and Isenberg, I. (1972). *Biopolymers* **11**, 375.
Luck, J. M., Cook, H. A., Eldridge, N. J., Haley, M. I., Kupke, D. W., and Rasmussen, P. S. (1956). *Arch. Biochem. Biophys.* **65**, 449.
Martinson, H. G., and McCarthy, B. J. (1975). *Biochemistry* **14**, 1073.
Mauritzen, C. M., and Stedman, E. (1959). *Proc. R. Soc. London, Ser. B* **150**, 299.
Noll, M. (1974). *Nature (London)* **251**, 249.
Ogawa, Y., Quagliarotti, G., Jordan, J., Taylor, C. W., Starbuck, W. C., and Busch, H. (1969). *J. Biol. Chem.* **244**, 4387.
Olins, A. L., and Olins, D. E. (1973). *J. Cell Biol.* **59**, 252a.
Olins, D. E., and Olins, A. L. (1974). *Science* **183**, 330.
Oudet, P., Cross-Bellard, M., and Chambon, P. (1975). *Cell* **4**, 281.
Phillips, D. M. P. (1965). *Biochem. J.* **97**, 669.
Phillips, D. M. P. (1967). *Biochem. J.* **105**, 46P.
Rill, R., and Van Holde, K. E. (1973). *J. Biol. Chem.* **248**, 1080.
Roark, D. E., Geoghegan, T. E., and Keller, G. H. (1974). *Biochem. Biophys. Res. Commun.* **59**, 542.
Sahasrabuddhe, C. G., and Van Holde, K. E. (1974). *J. Biol. Chem.* **249**, 152.
Shih, T. Y., and Bonner, J. (1970). *J. Mol. Biol.* **48**, 469.
Skandrani, E., Mizon, J., Sautiere, P., and Biserte, G. (1972). *Biochemie* **54**, 1267.
Small, E. W., Craig, A. M., and Isenberg, I. (1973). *Biopolymers* **12**, 1149.
Smerdon, M. J., and Isenberg, I. (1973). *Biochem. Biophys. Res. Commun.* **55**, 1029.
Smerdon, M. J., and Isenberg, I. (1974). *Biochemistry* **13**, 4046.
Sperling, R., and Bustin, M. (1974). *Proc. Natl. Acad. Sci. U.S.A.* **71**, 4625.
Sperling, R., and Bustin, M. (1975). *Biochemistry* (in press).
Stedman, E., and Stedman, E. (1950). *Nature (London)* **166**, 780.
Stedman, E., and Stedman, E. (1951). *Philos. Trans. R. Soc. London, Ser. B* **235**, 565.
Sung, M. T., and Dixon, G. H. (1970). *Proc. Natl. Acad. Sci.* **67**, 1616.
Thomas, J. O., and Kornberg, R. D. (1975). *Proc. Natl. Acad. Sci. U.S.A.* (in press).
Tsanev, R., and Russev, G. (1974). *Eur. J. Biochem.* **43**, 257.
Ui, N. (1957). *Biochim. Biophys. Acta* **22**, 205.
Van Holde, K. E., and Isenberg, I. (1975). *Acc. Chem. Res.* **8**, 327.
Van Holde, K. E., Sahasrabuddhe, C. G., and Shaw, B. R. (1974). *Nucleic Acids Res.* **1**, 1579.
Van Lente, F., Jackson, J. F., and Weintraub, H. (1975). *Cell* **5**, 45.
Weintraub, H., Palter, K., and Van Lente, F. (1975). *Cell* **6**, 85.
Wickett, R., Li, H. J., and Isenberg, I. (1972). *Biochemistry* **11**, 2952.
Woodcock, C. L. F. (1973). *J. Cell Biol.* **59**, 360a.

SUBMOLECULAR BIOLOGY

16

Molecular Dimensionality in Relation to Phenomenology

From the Atom to the Living Cell*

MICHAEL KASHA

The biological mysteries of Life have long engaged scientific discussions of vitalism versus mechanism. The questions which remain today center on, largely, in what manifestation will the laws of physics and chemistry apply. The development of modern quantum mechanics led atomic physicists (Schroedinger, 1944; Bohr, 1958; Elsasser, 1958) steeped in the restrictions on observation in the atomic realm to speculate on possible analogues in the biological realm. To a physical chemist interested in molecular phenomena and the phenomena in complex molecular aggregates, the jump from the atom to the living cell seems too gross to permit the detailed mechanistic connection required for a deeper understanding of life processes.

In this paper, as a tribute to Albert Szent-Györgyi on his 82nd birthday, I shall present a physical chemist's view of molecular dimensionality in relation to observable phenomena from the atom to the living cell. Szent-Györgyi has

*This work was supported by a contract between the Division of Biomedical and Environmental Research, U. S. Energy Research and Development Administration, and the Florida State University.

himself devoted many years of research exploring the application of various molecular-electronic mechanisms as a means of enhancing the understanding of the biochemistry of biological phenomena. He has inspired a number of researches exemplified in the applications given in this review. He has brought excitement and joy to research!

CORRELATION OF MOLECULAR DIMENSIONALITY WITH PHENOMENOLOGY

The thesis which we shall develop here is that the mechanistic realms from the atom to the living cell are separated by discontinuities or boundaries. For each change in order of molecular complexity, there is a corresponding change in the complexity of typical phenomena for that order of dimensionality.

The correlation diagram of Fig. 1 has been developed to illustrate this thesis. We take our clue from the behavior of the "atomic realm" with respect to the macroscopic realm. The anthropocentric concepts of classical behavior in the macroscopic world are abruptly modified and in the atomic world we are introduced to typically quantal phenomena. Among these are wave-particle duality, uncertainty relations on energy and time, or momentum and position, penetration or tunneling through potential barriers, and zero-point energy.

The "diatomic realm" abruptly brings in characteristically molecular behavior: vibrations and rotations, and novel electronic states. However the simplicity of the molecular structure restricts the full generation of molecular phenomena.

TYPICALLY POLYATOMIC BEHAVIOR

The "polyatomic realm" brings forth changes in behavior, compared with atomic and diatomic entities, as dramatically abrupt as the change from the macroscopic to the atomic realm. We may call these typically polyatomic phenomena. There are numerous examples of polyatomic molecule behavior deriving from their more complex structure which does not appear in simpler systems. The most familiar of these may be the "stereoisomerism" characteristic of some polyatomic structures. More closely related to biological function is "molecular chirality" and its relation to specificity. Spectroscopists are acquainted with the role of "antisymmetric vibrations", especially in facilitating electronic processes forbidden in the simplest molecules and atoms or ions. Radiationless electronic transitions, largely excluded in atomic and diatomic systems, become a common, most probable event in polyatomic molecules, both in electronic excitation phenomena as well as in reaction mechanisms. Thus, the

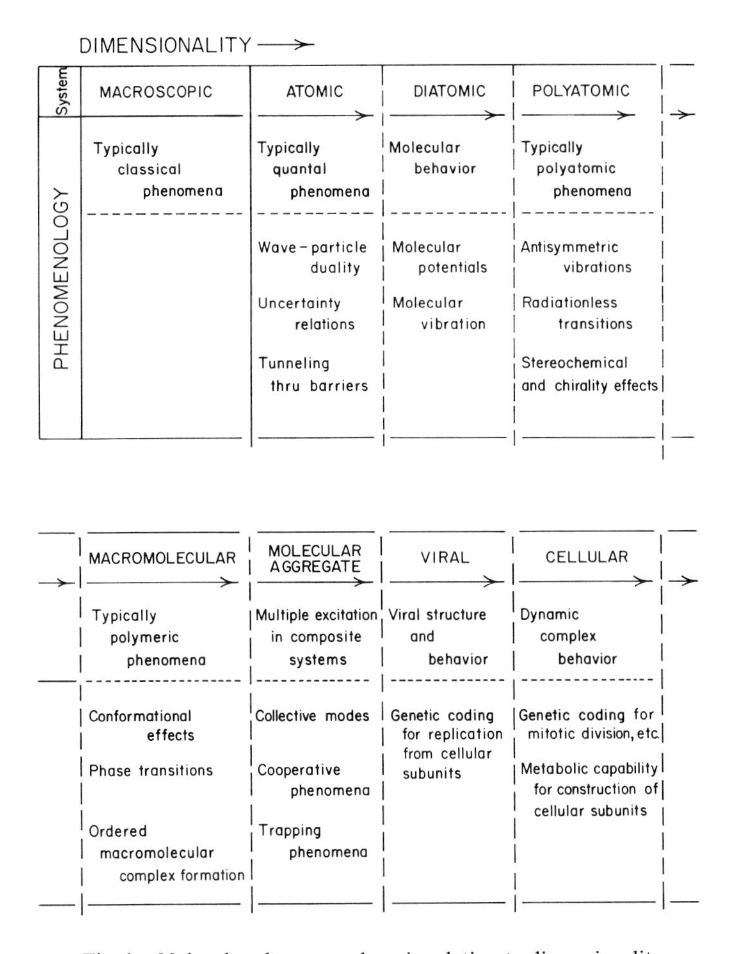

Fig. 1. Molecular phenomenology in relation to dimensionality.

polyatomic molecule behavior brings forth new manifestations of the quantal laws, essential for bridging the gap to still more complex systems.

ELECTRONIC PROCESSES IN MOLECULAR AGGREGATES

The "macromolecular realm" marks the next abrupt jump in complexity from the simpler polyatomic realm (Fig. 1). The characteristic features of macromolecular behavior include properties which dominate the phenomena of these large entities and have direct applications to biological function. John Platt

(1961) has developed this subject as an expansion of our conversations on the general thesis crystallized in this paper.

The "molecular aggregate realm" generalized to include homogeneous and heterogeneous clusters (crystalline and noncrystalline) of molecules and macromolecules once again leads to a whole order of increased complexity of electronic phenomena. Characteristic features are collective modes and trapping phenomena. Intermolecular coupling can feature electron transfer and exchange, as well as electron localization with excitation exchange. Figure 2 correlates (Kasha and El-Bayoumi, 1974) some multiple excitation events observed for molecular pairs, clusters, lamellae, and crystals. It is instructive to note that all of the phenomena correlated in this diagram go beyond the simple concept of single-photon, single-molecule interaction. Yet, biological systems always involve complex aggregates, and the knowledge of the phenomena involved in the latter constitute stepping stones to the understanding of complex events in photobiology and molecular biology in general.

Of course it is known to everyone that biological structures and processes involve molecular entities of various dimensionalities. It is not widely recognized, however, that the phenomenology of these molecular entities abruptly change as the molecular dimensionality changes. Although the pace of research in molecular biology is highly accelerated by the current evolution of new lines of thought and new instrumentation techniques, the middle ground of fundamental molecular phenomenology is still evolving and needs to be explored in much greater depth.

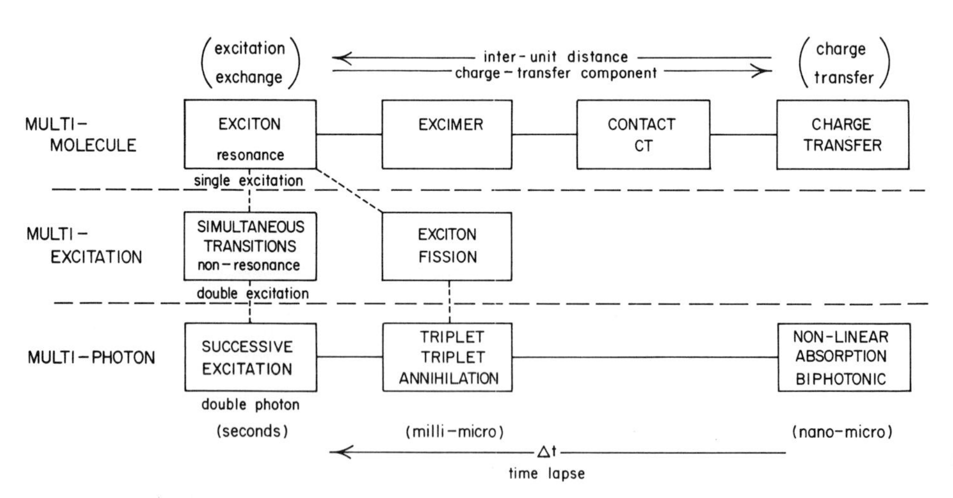

Fig. 2. Multiple excitation phenomena in composite molecules (Kasha and El-Bayoumi, 1974).

In the next sections four specific examples will be described as examples of novel molecular phenomena under recent and current development.

SIMULTANEOUS TRANSITIONS IN SINGLET MOLECULAR OXYGEN PAIRS

Molecular oxygen is one of the most ubiquitous chemical species in nature, and yet it is only in the last decade that the metastable excited form of oxygen has been recognized to be of importance chemically and biologically (Kasha and Khan, 1970; Wilson and Hastings, 1970; Foote, 1976). Normal molecular oxygen exists in a triplet electronic spin state, an almost unique example of a nondiamagnetic ground state of a stable even-electron molecule (Kasha and Khan, 1970). G. N. Lewis speculated on the low rate of oxygen reaction with unsaturated hydrocarbons on the basis of magnetic restriction, a topic taken up again recently. Singlet molecular oxygen, previously known largely to astrophysicists, is a diamagnetic excited form of oxygen of extreme metastability in the gas phase. Fortunately, it has been shown to be readily produced in aqueous systems in chemical oxidation of hydrogen peroxide, and by other means (Khan and Kasha, 1963; Browne and Ogryzlo, 1964; Kasha and Khan, 1970).

Singlet molecular oxygen is characterized by several features which make it of exceptional spectroscopic, chemical, and biological interest. Spectroscopically, the presence of singlet oxygen is made prominent by the appearance of a simultaneous electronic transition arising from pairs of singlet molecular oxygen. Figure 3 illustrates an example of one of the three possibilities observed in singlet oxygen systems. The ground triplet state to excited singlet state transition is exceedingly highly forbidden, so that in effect the natural lifetime of the lowest excited singlet state of molecular oxygen ($^1\Delta_g$) is 45 min. But in the presence of other oxygen (unexcited) molecules, a pair state is produced (S+T), so the lifetime becomes pressure-dependent and greatly shortened. More interesting still, a pair of excited singlet states can undergo a simultaneous transition to the ground state, that is, a collision of two excited molecules can result in the emission of one photon, leaving two ground state molecules (Khan and Kasha, 1970). As Fig. 3 indicates, the very highly forbidden $S \rightarrow T$ emission process becomes an $(S+S) \rightarrow (S+T+Q)$ process in the simultaneous transition, i.e., the singlet component of the singlet–triplet–quintet ensemble (for the pair of ground triplet oxygens) effectively removes the spin restriction on the emission.

Another description of the simultaneous transition is that it is tantamount to a photon doubling process. Thus, in Fig. 3, the 7620 Å photon corresponding to $^1\Sigma_g^+$ excitation appears as a 3810 Å ultraviolet photon for the molecular pair state $(^1\Sigma_g^+)(^1\Sigma_g^+)$.

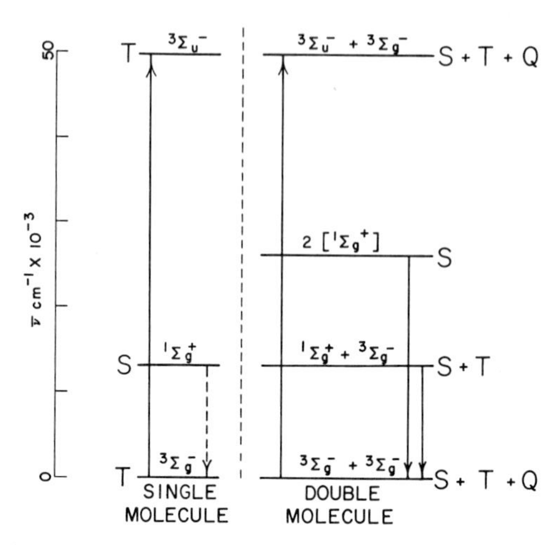

Fig. 3. Simultaneous transitions in molecular oxygen pairs (the $^3\Sigma_u^-$ is the allowed perturbing state).

Since it has been shown that the singlet excited states of molecular oxygen are chemically produced, have characteristic reactivities, and also participate in energy transfer processes, there has been an explosive development of the research on the chemistry and biological effects of singlet molecular oxygen.

MOLECULAR EXCITONS IN SMALL AGGREGATES

The subject of "molecular excitons" was developed by A. S. Davydov (1962) to explain excitation delocalization in molecular crystals and the consequent band sharpening and splittings observed. Eion McRae and I were asked by Albert Szent-Györgyi to explain why frozen water media caused dye molecules to phosphoresce while the same dyes dissolved in alcohol glasses were fluorescent. Following Beukers and Behrends' work on thymine photo dimers produced by irradiating frozen water solutions of thymine, Szent-Györgyi thought that the crystal structure of ice could induce intersystem crossing to the triplet state of the dye, thereby enhancing phosphorescence.

McRae and Kasha (1958, 1964) then showed that the molecular exciton model explained the Szent-Györgyi experiments as molecular aggregate (dimer and polymer) effects. In brief, the formation of molecular pairs or clusters leads to energy delocalization, and the stationary exciton states which are present in the cluster can lead to enhanced triplet state excitation.

The molecular exciton model has now been applied to a wide range of molecular cluster geometries, from dimers to linear and helical polymers, to lamellae of various geometries, and, qualitatively, to spherical aggregates (cf. Kasha, 1976).

The lamellar exciton studies by Hochstrasser and Kasha (1964) will serve to illustrate the general operation of the molecular exciton model. In the upper part of Fig. 4 is diagrammed a "pin-cushion" array of chromophores: the transition dipole vectors aligned normal to the lamellar plane. This simple square lattice has, effectively, one molecule per unit cell, so by the Davydov rule, of the N different exciton states for the array of N molecules, only one state will be allowed for light absorption (by electric dipole radiation). This allowed exciton state can be quickly identified as corresponding to the transition dipole array with all dipoles in phase (as diagrammed in Fig. 4). All other dipole arrays properly constructed for the overall symmetry for the infinite lamella correspond to forbidden exciton states. For example, in the array with dipoles half up and half down, the resultant vector for light absorption is zero.

Energetically, the highest energy exciton state is the in-phase array, as it is the most repulsive for the dipole–dipole interaction. The lowest energy state is that corresponding to the alternate *up* versus *down* orientation of transition dipoles;

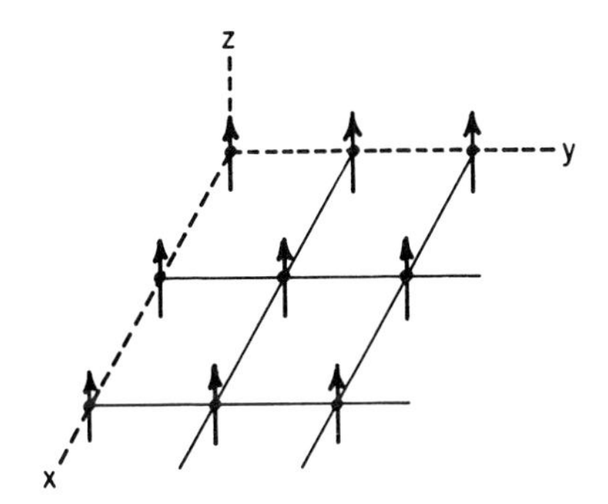

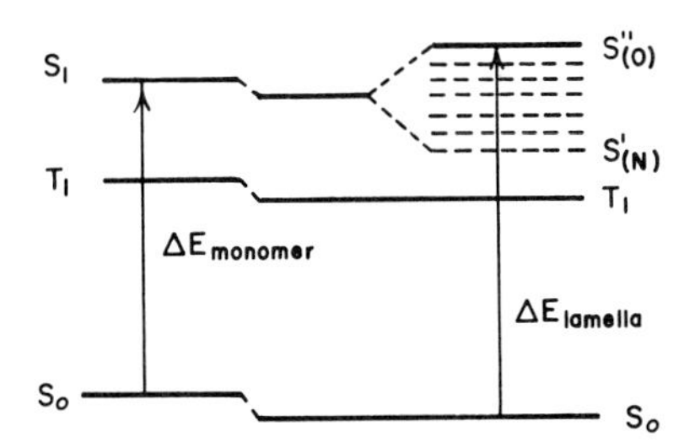

Fig. 4. Molecular excitons as a simple square lamellar array. *Upper*: in-phase arrangement of transition dipoles; *lower*: energy level diagram for the simple square lamella relative to single molecule electronic states.

of course, this lowest state is a forbidden exciton state as far as (electric dipole) light absorption is involved.

The triplet state in the first order Davydov model is little affected by the exciton interaction.

Phenomenologically, the allowed (singlet–singlet) absorption band of the lamella is found to be strongly blue-shifted (to shorter wavelengths) compared with the monomer absorption, a "metachromasy." The fluorescence of the monomer (which in dyes dominates intersystem crossing to the triplet state) is quenched in the lamella, and intersystem crossing from the metastable lowest exciton $S'_{(N)}$ state to the triplet state now results in a powerful phosphorescence.

The molecular exciton model represents a resonance interaction model for excited states of weakly coupled molecular aggregates, with a specific geometry dependence and inverse cube of distance dependence. The model offers semi-quantitative calculation by simple formulas of the spectroscopic states of molecular aggregates, using the states of the monomer units as a starting point. Specific applications in numerous photobiological systems remain to be made. The molecular (Davydov) exciton model offers no charge delocalization in contrast to the atomic (Wannier) exciton model (Kasha, 1959).

BIPROTONIC PHOTOTAUTOMERISM AND WATER STRUCTURE

Water as a medium in biological systems makes its interaction with polyatomic molecular species of great interest. In our laboratory we have discovered a luminescence probe for monomeric water. Our first studies (Taylor *et al.*, 1969) of the doubly H-bonded base pairs of 7-azaindole indicated that the change from the violet-fluorescing monomer to the green-fluorescing dimer was the result of a fast cooperative transfer of the two hydrogen-bonding protons in the excited state (Fig. 5).

The special effects of solvation of 7-azaindole by H-bonding solvents proved to be more puzzling: ethanol solvent for 7-azaindole resulted in a green-fluorescing tautomer, while water solutions of 7-azaindole showed only violet fluorescence. Recently we discovered how to find monomeric H_2O in water. Simpson and Kasha (1977) found that if 7-azaindole (in the concentration range for monomer stability, 10^{-5} *M*) is dissolved in methylene chloride or toluene, a violet fluorescence is observed as expected. Upon addition of small amounts of water to these solutions, comparable to the 7-azaindole concentration (or small multiples thereof), the green luminescence of the 7-azaindole appears. These results indicate clearly that the water clusters in methylene chloride and toluene

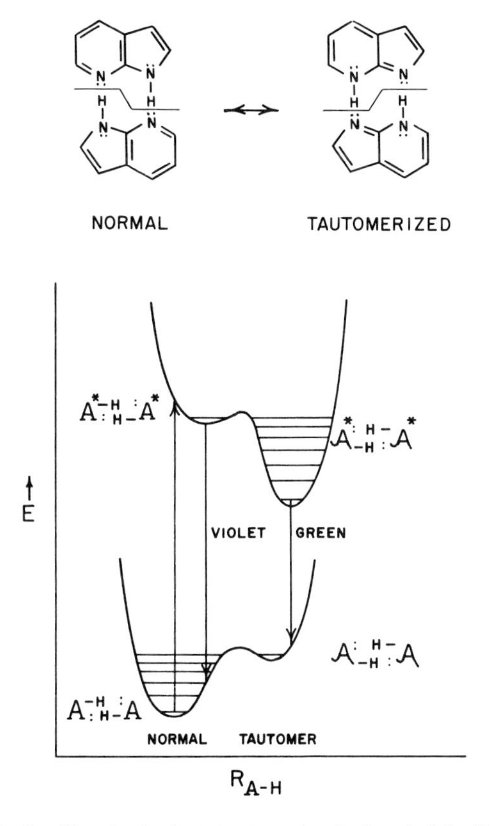

Fig. 5. Biprotonic phototautomerism in 7-azaindole dimers.

solvent dissociate, and that monomeric water is finally available to form a 1:1 hydrogen-bonded complex with 7-azaindole.

Figure 6 illustrates the two models for the interaction of 7-azaindole with water. On the *left* is shown a cyclical H-bonded complex involving monomeric water. In the excited state, as the large electronic charge rearrangement changes the relative base strength of the two N-atoms, the protons can flip cooperatively in unison if they are related to the same complexing molecule. But on the *right* side is shown the H-bonding complex of 7-azaindole with two independent water chains, which must exist in liquid water as solvent; the cooperative transfer of the two protons cannot occur during the lifetime of the excited state, so the tautomerization of the 7-azaindole is prevented.

Since 7-azaindole is a molecule which could be described as the "universal solute," dissolving in hydrocarbons, ethers, aromatics, alcohols, water, etc., it

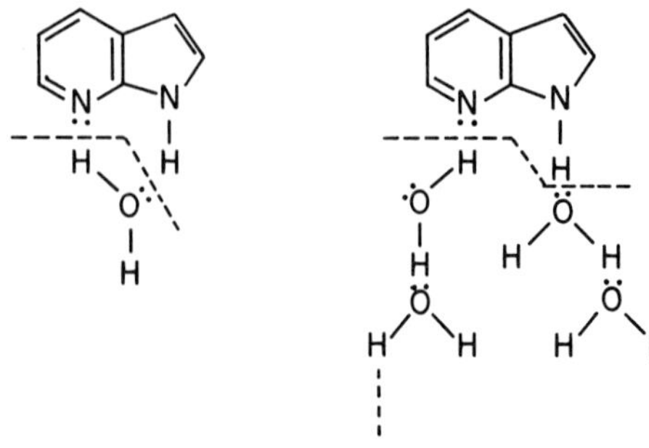

Fig. 6. Solvates of 7-azaindole with monomeric water and water chains.

may serve as a useful probe for hydrophilic and hydrophobic regions of macromolecular complexes and proteins.

POTENTIAL MODEL OF THE SOLVENT CAGE

One common circumstance for molecule behavior in complex systems is the restriction of molecular motion by a solvent cage. The cage may consist of a liquid or rigid matrix solvent surrounding the solute molecule, or a surface to which a molecule is adsorbed, or a viscous macromolecular environment for an occluded molecule. The classical solvent cage has been treated by kinetic methods as a kinetic effect. Recently Dellinger and Kasha (1975, 1976) have translated the solvent cage theory into spectroscopic language by using the Born-Oppenheimer resolution of the electronic wave functions, and the wave functions for solvent cage relaxation and the solute vibrational relaxation.

Thus, the solvent cage is developed as a potential barrier perturbation (about Q_0) of the solute molecule potential (about Q_e). Figure 7 illustrates one of the many molecular cases considered (Morse vibrational, repulsive, double-bond torsional, double-minimum, pseudo Jahn-Teller potentials). The ground state S_0 is represented by a hydrogen-bonding potential for a heteroaromatic molecule in an H-bonding solvent. For the n,π^* excited S_1 state (Kasha, 1961) the potential will be repulsive, and the solvate dissociates. The consequence of this normal behavior is that the absorption band is blue-shifted in H-bonding solvents (water, alcohols) relative to the absorption spectrum of a hydrocarbon solvent.

The fluorescence spectrum of H-bonded heteroaromatics is, however, identical with the fluorescence spectrum of the solvate in hydrocarbon solvents. Thus, even though the ground state is stabilized by H-bonding, the repulsive excited state leads to fluorescence (F_ℓ) from the uncomplexed molecule in liquid solvent. Dellinger and Kasha have found, however, that in a rigid glass solvent

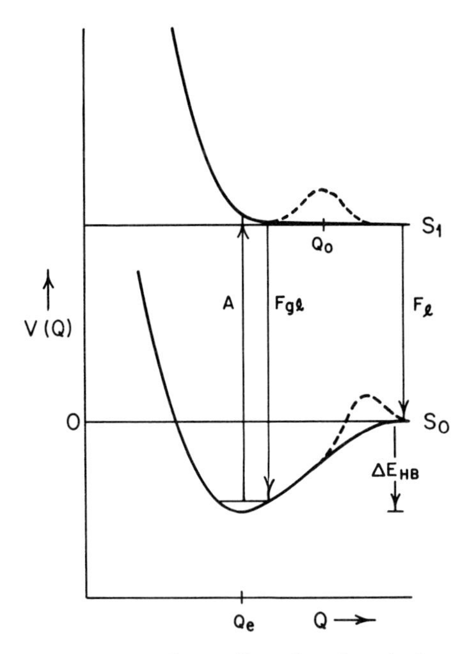

Fig. 7. Potential curves for ground state S_0 and n,π^* excited state S_1 in an H-bonding coordinate. Dashed curves indicate solvent cage perturbation of the heteroaromatic solute potentials.

matrix, the fluorescence is now blue-shifted ($F_{g\ell}$) and corresponds to the blue-shifted absorption band. As indicated in Fig. 7 by dashed contours, the rigid glass matrix interposes a potential barrier to dissociation of the bulky H-bonded complex (e.g., 4-methylcinnoline with perfluoro-*tert*-butanol in 3-methylpentane glass).

The viscosity and adsorption effects on dye molecule fluorescence are common manifestations of potential perturbations by the solvent cage. The study of inhibition of molecular motion and change in spectral phenomena by solute-cage interaction may be expected to have many biological applications as molecular aggregate phenomena.

SUMMARY

The thesis is developed that the mechanistic realms from the atom to the living cell are separated by boundaries or discontinuities. For each change in order of molecular complexity, there is a corresponding change in the complexity of typical phenomena for that order of dimensionality. Empnasis is given to typically polyatomic behavior, and the phenomenology of molecular aggregates. Examples are given from current research on singlet molecular oxygen, molecular excitons in small aggregates, biprotonic phototautomerism as a luminescence probe for water, and a potential model for solvent cage effects.

ACKNOWLEDGMENT

The thesis developed in this paper was inspired by rich conversations with William T. Simpson on typically polyatomic behavior and its significance with increasing order of complexity. It is a great pleasure for the author to acknowledge these early discussions and the developments which followed.

REFERENCES

Bohr, N. (1958). "Atomic Physics and Human Knowledge." Wiley, New York.

Browne, R. J., and Ogryzlo, E. A. (1964). *Proc. Chem. Soc., London* p. 117.

Davydov, A. S. (1962). "Theory of Molecular Excitons." McGraw-Hill, New York.

Dellinger, B., and Kasha, M. (1975). *Chem. Phys. Lett.* **36**, 410.

Dellinger, B., and Kasha, M. (1976). *Chem. Phys. Lett.* **38**, 9.

Elsasser, W. M. (1958). "The Physical Foundation of Biology." Pergamon, Oxford.

Foote, C. (1976). *In* "Free Radicals in Biology" (W. A. Pryor, ed.), Vol. 2, p. 85. Academic Press, New York.

Hochstrasser, R. M., and Kasha, M. (1964). *Photochem. Photobiol.* **3**, 317.

Kasha, M. (1959). *Rev. Mod. Phys.* **31**, 162; cf. "Biophysical Science" (J. L. Oncley *et al.*, eds.), pp. 162–169. Wiley, New York.

Kasha, M. (1961). *In* "Light and Life" (W. D. McElroy and B. Glass, eds.), pp. 31–64. Johns Hopkins Press, Baltimore, Maryland.

Kasha, M. (1976). *In* "Spectroscopy of the Excited State" (B. Di Bartolo, ed.), p. 337. Plenum, New York.

Kasha, M., and El-Bayoumi, M. A. (1974). *In* "Physical Mechanisms in Radiation Biology" (R. D. Cooper and R. W. Wood, eds.), Bull. CONF-721001. Tech. Inf. Cent., US At. Energy Comm., Washington, D. C.

Kasha, M., and Khan, A. U. (1970). *Ann. N. Y. Acad. Sci.* **171**, 5.

Khan, A. U., and Kasha, M. (1963). *J. Chem. Phys.* **39**, 2105.

Khan, A. U., and Kasha, M. (1970). *J. Am. Chem. Soc.* **92**, 3293.

McRae, E. G., and Kasha, M. (1958). *J. Chem. Phys.* **28**, 721.

McRae, E. G., and Kasha, M. (1964). *Phys. Processes Radiat. Biol.*, Academic Press, New York, pp. 23–42.

Platt, J. R. (1961). *J. Theor. Biol.* **1**, 342.

Schroedinger, E. (1944). "What is Life?" Cambridge Univ. Press, London and New York.

Simpson, J., and Kasha, M. (1977). To be published.

Taylor, C. A., El-Bayoumi, M. A., and Kasha, M. (1969). *Proc. Natl. Acad. Sci. U. S. A.* **63**, 253.

Wilson, T., and Hastings, J. W. (1970). *Photophysiology* **5**, 49–95.

17

Bound Water in Biological Systems

A Quantum-Mechanical Investigation

ALBERTE PULLMAN

INTRODUCTION

One of the most important recent developments in the application of quantum mechanical methods to the study of biological molecules is in the attempt to evaluate the effect of the environmental conditions on their structure and properties. Foremost among these applications is the investigation of the role of water, and a number of laboratories are presently engaged in such investigations.

Broadly speaking, these efforts are developing along two different lines. The majority of the authors proceed in the "traditional" way of dealing with the problem through the use of a "continuum" model which tries in one way or another to account for the bulk effect of the surrounding medium. Outstanding among these efforts are those of Sinanoglu (1974), Beveridge (1974), Hopfinger (1974), and Hylton *et al.* (1974).

One constant and essential limitation of these models is the absence of precision concerning the arrangement of the solvation layer(s) around the solute, which is considered as residing in a cavity (generally spherical) embedded in a polarizable dielectric. No information is obtainable in this way about the details of the short-range solute-medium interactions.

Recently, a different line of research started developing; it consists of a "discrete" treatment, in which one tries to establish the individual effect of the medium molecule(s) upon the system studied. This is achieved in general through the utilization of the "supermolecule" model which combines, in a unique exploration, the molecular entities in interaction, such as the solute and the solvent molecule(s). Such an approach became possible owing to the considerable, very recent developments in the applicability of the molecular orbital method; it is now possible to compute nonempirically a system solute and solvent of reasonable size, treating all the electrons in the field of all the nuclei. In this fashion one may evaluate the most probable sites of interaction and gather information about the nature of the binding (if any), the binding energy, the flexibility of the association and the mutual perturbations produced in the electronic structure of the partners by the interaction.

This second approach is obviously particularly well adapted for dealing with the problem of "bound" water in a broad sense. Its particular aim is the detection of the preferred modes of interaction of the solvent molecules with the receptor and the characterization of such interactions. The second stage of the research consists in determining the extent to which this primary interaction produces observable and measurable effects.

We have applied this approach in our laboratory to the exploration of the hydration scheme of a series of fundamental biological molecules, and, in particular, of the constituents of the biological polymers and related systems: formamide (Alagona *et al.*, 1973), *N*-methylacetamide (Pullman *et al.*, 1974), the hydrogen-bonded peptide linkages of proteins (Port and Pullman, 1974), the amino acid side chains of proteins (Port and Pullman, 1974), the purine and pyrimidine bases of the nucleic acids (Port and Pullman, 1973a), the ammonium and alkylammonium groups (Port and Pullman, 1973b; Pullman and Armbruster, 1974, 1975) and the phosphate group (A. Pullman *et al.*, 1975; B. Pullman *et al.*, 1975; Perahia *et al.*, 1975). The method utilized is, in most cases, the *ab initio* SCF procedure, generally using the STO 3G basis set (Hehre *et al.*, 1969, 1970). Applications have been made, in particular, for the evaluation of the solvent effect on the conformational properties of a series of biological and pharmacological molecules. General reviews may be found in A. Pullman and B. Pullman (1975), B. Pullman (1977), and B. Pullman (this volume).

HYDRATION SITES AND BINDING ENERGIES

The studies produce primarily two fundamental results. The first is on the preferred hydration sites and the corresponding energies of binding of the water

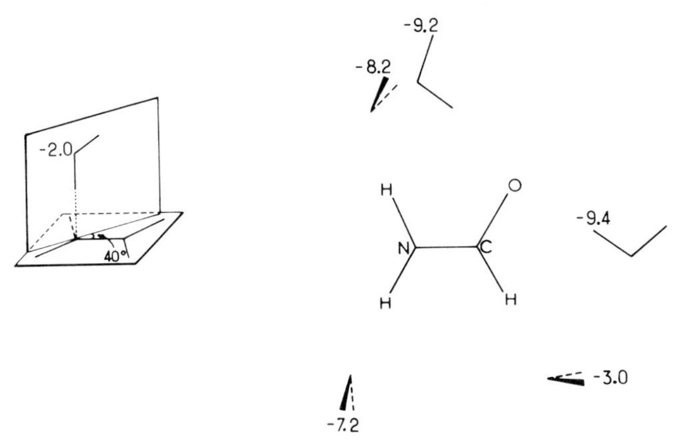

Fig. 1. Principal hydration sites in formamide and the corresponding energies for mono-adducts. (Energies in kcal/mole.) Water molecules with one full and one dotted line indicate that the water plane is perpendicular to the formamide plane (oxygen of water in the plane of formamide).

molecules to these sites. The results of such studies for the above-quoted biological systems or groups may be found in the original references and reviews. They refer, according to the case, to monohydration or polyhydration and extend, according to the case again, to hydration in the first shell only or beyond. We shall not comment on these results here but only give in Fig. 1–4 some typical examples pertinent to the situation prevalent in typical biomolecules.

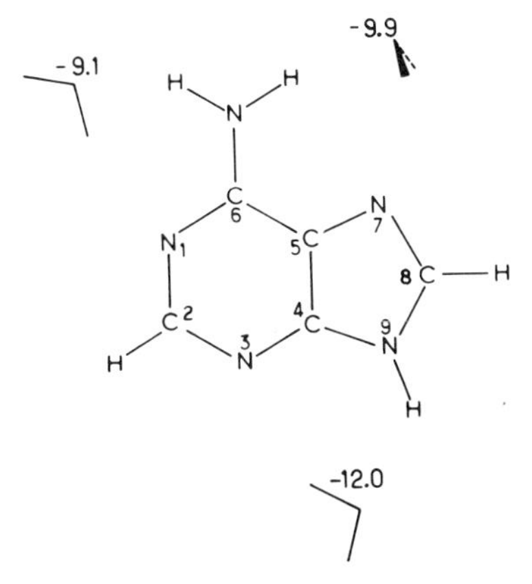

Fig. 2. Preferred hydration sites in adenine. Energies in kcal/mole. Full lines: co-planar arrangement of water and base. Half-dashed: perpendicular arrangement of water with respect to the plane of the base.

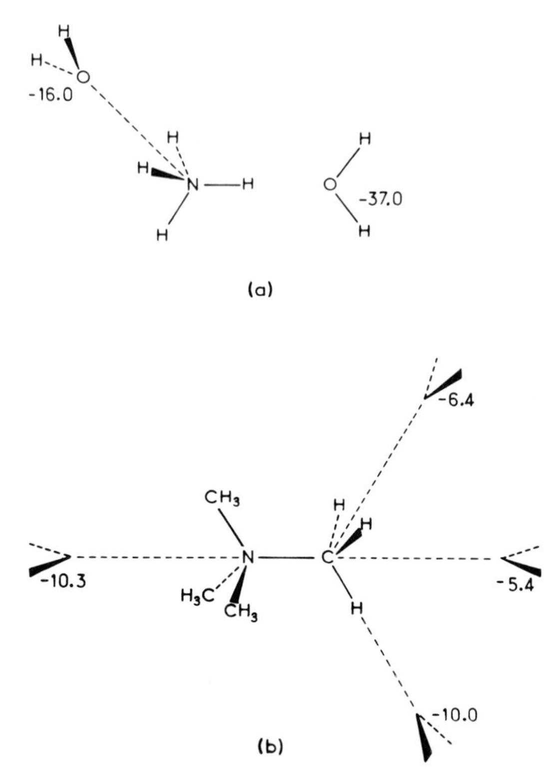

Fig. 3. Preferred sites of water binding to (a) NH_4^+ and (b) $N^+(CH_3)_4$. Energies in kcal/mole.

THE STRUCTURE OF "BOUND" WATER

The second result of primary interest to the biologist concerns the structure of the "bound" water or, more precisely, the effect produced by the formation of the hydration shells on the properties of the "free" water molecules, in particular, on their electronic distribution. This type of information for which biologists have been eagerly looking for a number of years is hardly obtainable by any experimental technique presently available. This makes the theoretical analysis of the problem, even if primitive and approximate, particularly desirable. It will be the main subject of this paper.

The procedure used will consist of visualizing the distribution of the electronic charges as obtained by a Mulliken population analysis (Mulliken, 1955) of the wave function. It must be borne in mind that the atomic charges defined in such an analysis do not correspond to an "observable" in the quantum mechanical sense, and represent only a conventional convenient partition of the exact

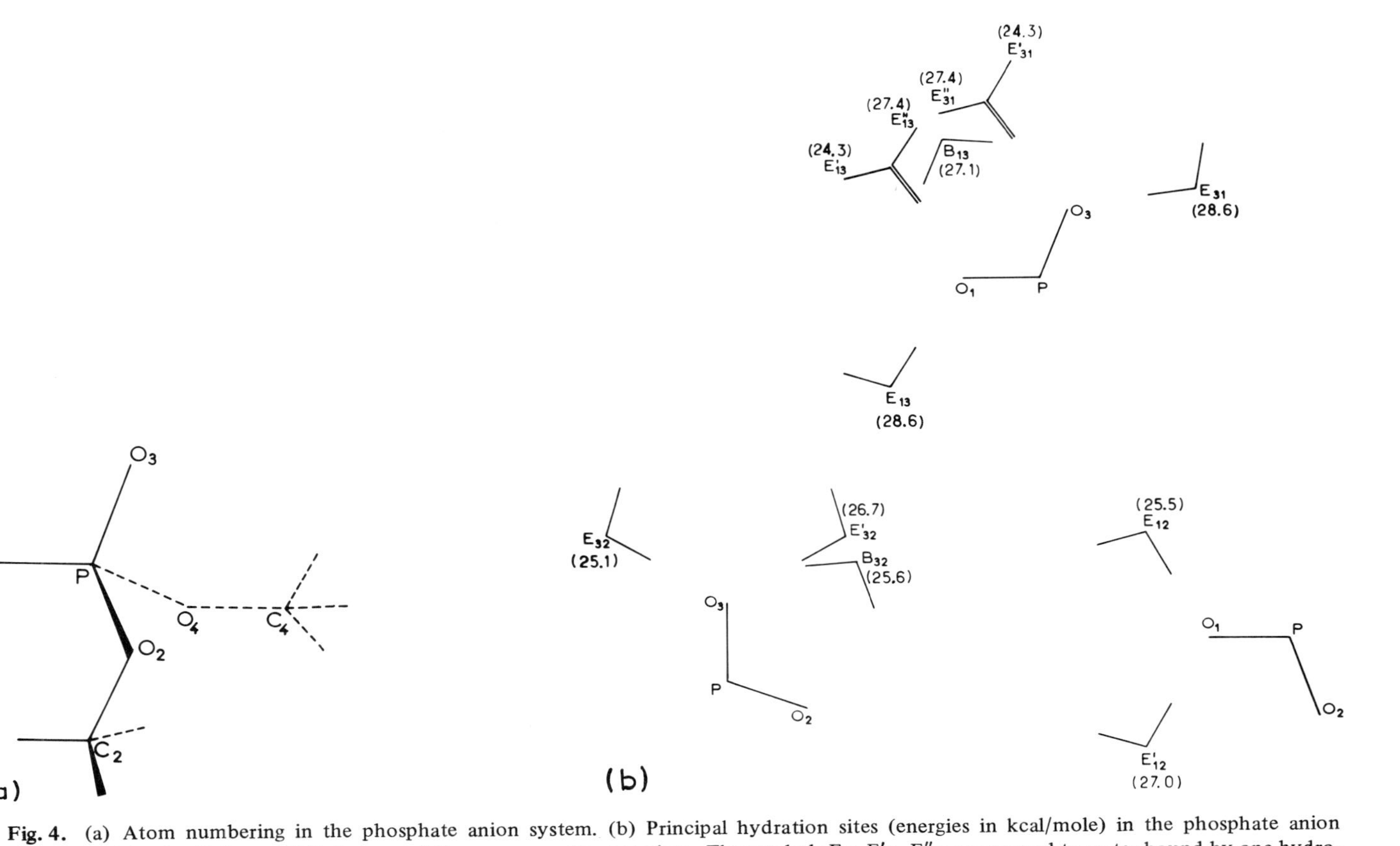

Fig. 4. (a) Atom numbering in the phosphate anion system. (b) Principal hydration sites (energies in kcal/mole) in the phosphate anion system, located in the different POP planes, and the corresponding notations. The symbols E_{ij}, E'_{ij}, E''_{ij} correspond to water bound by one hydrogen bond to oxygen i in the plane O_iPO_j. E stands for water external to the OPO angle, E′ and E″ for water internal to this angle. In the unprimed and primed E_{ij} positions the second hydrogen of water is turned toward the PO_i axis. In the double-primed position this second hydrogen is turned away from the PO_i axis. The symbol B_{ij} corresponds to water making a bridge between oxygens i and j.

electron-density distribution into atom-centered fractions (cf., for instance, A. Pullman, 1970; Jug, 1973; Politzer and Mulliken, 1971; Bader *et al.*, 1971). Moreover, the numerical values of the charges depend on the basis set used in the computations. Nevertheless, their relative values as well as their behavior under a given perturbation are generally sufficiently independent of the basis set for providing useful information in a problem such as the present one.

We shall distinguish three types of solute molecules: anionic, cationic and neutral.

Water Bound to an Anionic System

The system considered will be the phosphate group as represented by the dimethyl phosphate anion, DMP^- (B. Pullman *et al.*, 1975).

Figure 5 presents the distribution of the net electronic charges in free DMP^- and water. These computations have been performed by the SCF *ab initio* procedure, using an STO 3G basis set and neglecting the *d* orbitals on the

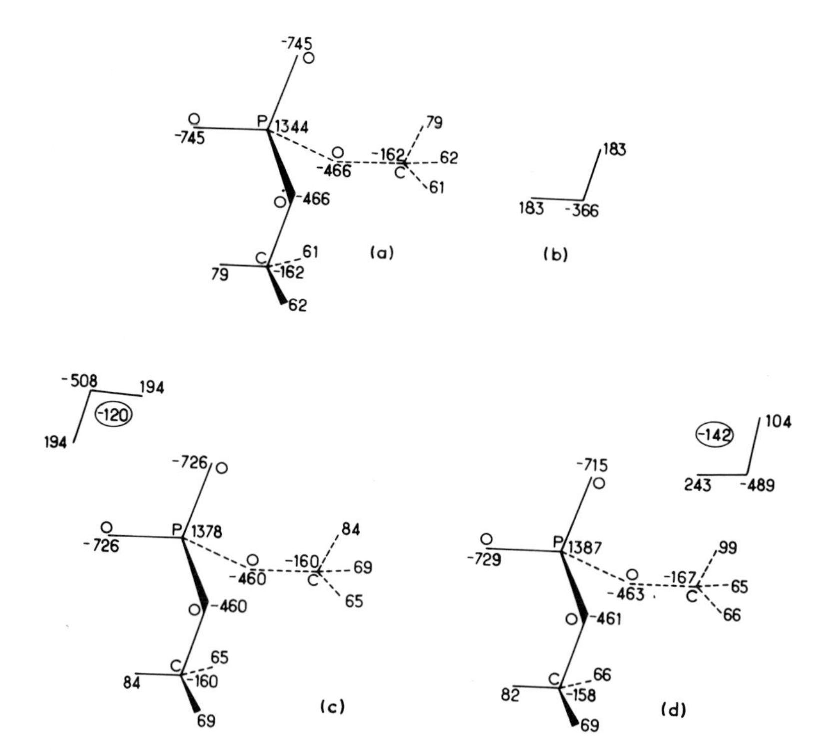

Fig. 5. Net charges (in millielectron units) in (a) DMP^-, (b) water, (c) B_{13} hydrate, (d) E_{31} hydrate. The numbers in circles indicate the global net charge on the water molecule.

phosphorus. As will be seen later, the absence of *d* functions on the phosphorus atom results in a strong exaggeration of the numerical values of the net charges on the phosphate ion. It is also known that the utilization of an STO 3G basis set leads to an overestimation of the charge transfer (Pullman and Armbruster, 1974). While these restrictions must be borne in mind in evaluating the *quantitative* significance of the results, they are of secondary importance in a comparative, qualitative study that will be our first aim. We shall see later the effect of including the *d* orbitals of phosphorus upon the results.

Figure 5 also contains the distribution of the electronic charges in the first hydration shell of the two mono-adducts B_{13} and E_{31} at their equilibrium O . . . H distances. It may be observed that the structure of the "bound" water is significantly perturbed with respect to that of free water: its oxygen atom has become appreciably more negative and its hydrogen(s) engaged in the hydrogen bond(s) with DMP^- has become slightly, but significantly more positive. Concomitantly, the phosphorus of DMP^- has increased its net positive charge while its anionic oxygens have undergone a decrease of their negative charges. *Altogether the binding of water to DMP^- produces a partial transfer of electrons from the latter to the former*; the transfer is greater (0.142 e) in case of the nonsymmetrical binding E_{31} than in the case of the symmetrical binding B_{13} (0.120 e). The bound water thus carries an excess of electronic density. The situation may be compared to the case of water dimers or polymers in which there is already a small charge transfer from the proton acceptor to the proton donor molecule (Del Bene and Pople, 1969, 1970) and to the case of the interaction of water with OH^- (Hankins *et al.*, 1970). The charge transfer in the case of $DMP^- \ldots H_2O$ [and similarly of $OH^- \ldots (H_2O)_n$] is more important than in the case of water dimer where an STO 3G calculation indicates a value of 0.034 e. This enhancement of the charge transfer is a result of the anionic nature of the solute which transmits a small partial anionic character to the bound water molecule of the first layer, thereby enhancing its attraction toward other water molecules.

Polyhydration in the first shell, while confirming these general features, indicates that the presence of a number of water molecules reduces in each of them the amount of charge transfer found for a single molecule. This is illustrated in Fig. 6 which presents, as an example, the distribution of the electronic charges in the trihydrate B_{13}, E'_{14}, E'_{32} of DMP^- (all water molecules in the first hydration shell).

New features appear in the structure of water when solvation is extended to the second hydration shell. They are illustrated in Figs. 7, 8 and 9.

In Fig. 7 there is one water molecule in the first shell and one in the second. In Fig. 8, the first shell is completed to six molecules of water, with one water molecule in the second shell. Figure 9 corresponds to three molecules in the first shell and three in the second shell. Let us denote them 1/1, 6/1, and 3/3, respectively.

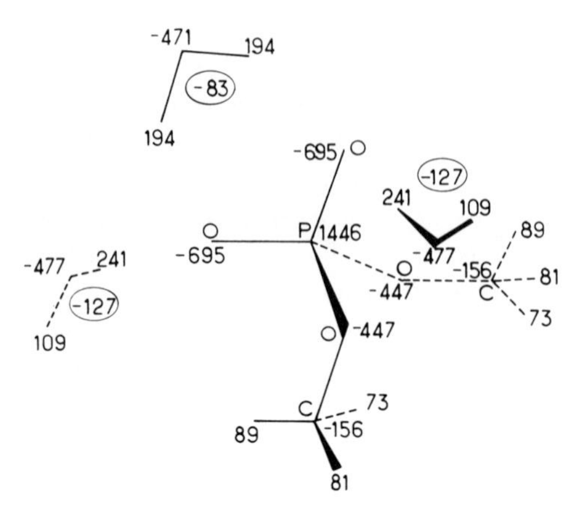

Fig. 6. Net charges (in millielectron units) in the B_{13}, E'_{14}, E'_{32} trihydrate of DMP^-. The numbers in circles indicate the global net charge on the water molecule.

The most conspicuous feature which appears when comparing Fig. 7 (1/1) with Fig. 5d is that the overall transfer of charge from DMP^- to the water molecules takes place predominantly to the benefit of the terminal molecule. Thus the excess of charge of the water molecule of the first hydration shell is only 0.041 e while that of the water molecule of the second hydration shell

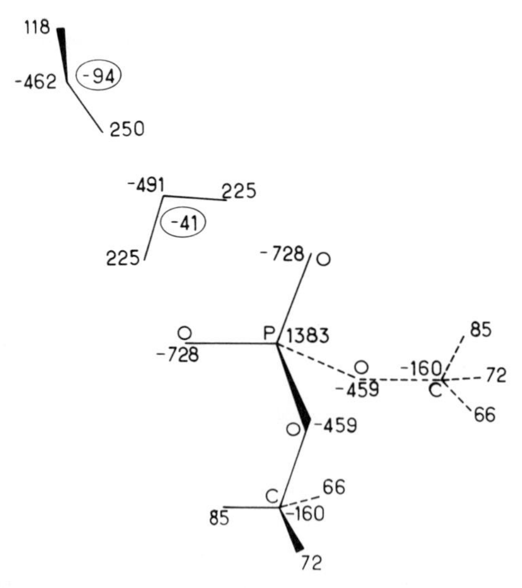

Fig. 7. Net charges (in millielectron units) in a dihydrate of DMP^- (one water molecule in the first hydration shell and one in the second). The numbers in circles indicate the global net charge on the water molecule.

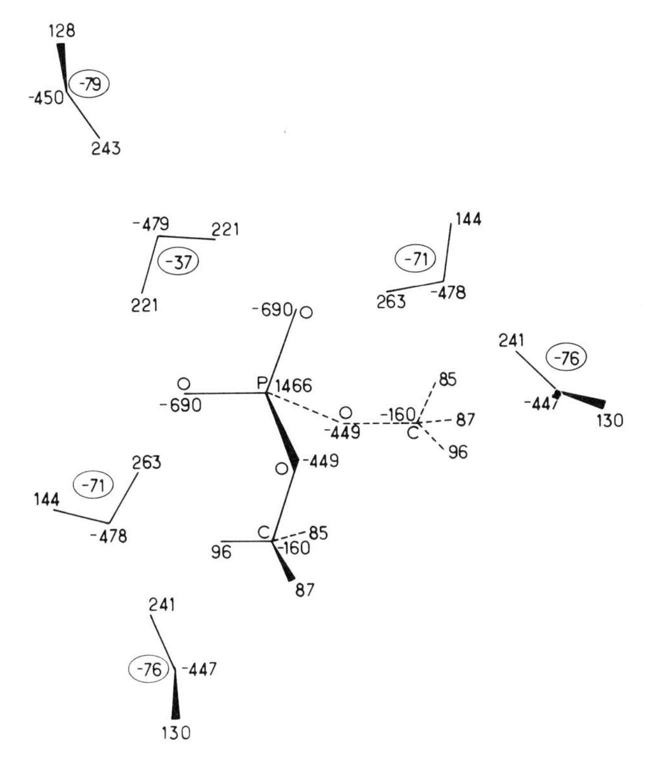

Fig. 8. Net charges (in millielectron units) in a heptahydrate of DMP⁻ (six water molecules in the first hydration shell and one in the second). The numbers in circles indicate the global net charge on the water molecule.

amounts to 0.094 e. This last number is nevertheless smaller than the 0.120 e transferred to the single water molecule of Fig. 5c, indicating a decreasing ability of the water molecules of the second shell for further binding. This is equally shown by the decrease in negative charge on the oxygen of the terminal water (−0.462 instead of −0.508 in Fig. 5c).

A more complete and significant picture may be observed in the heptahydrate (6/1) of Fig. 8. It appears here that the transfer of electronic charges leads to an excess of approximately 0.1 e at each water molecule of the first hydration shell, with the exception of the molecule in site E_{13} (bound to the water of the second shell), where the excess is smaller. Furthermore, on the water molecule of the second shell, the charge transfer amounts only to 0.057 e. In analogy to case 1/1, the negative charge on the oxygen atom of the water molecule of the second shell has decreased to −0.433. This redistribution of charge transfer over the successive layers of bound water is particularly evident in Fig. 9 (3/3).

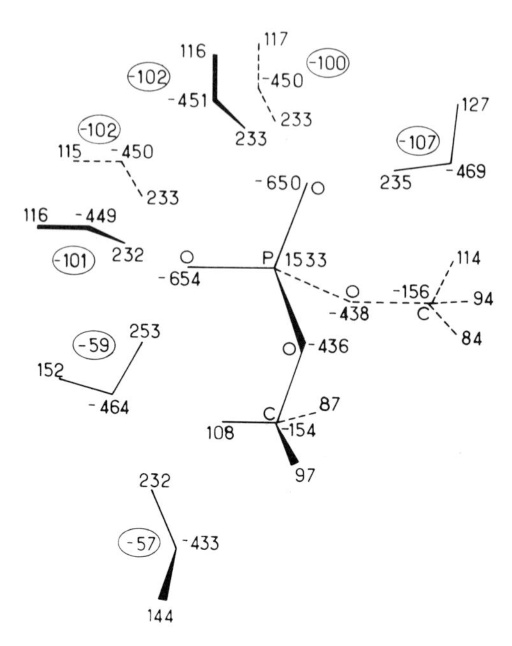

Fig. 9. Net charges (in millielectron units) in a hexahydrate of DMP$^-$ (three water molecules in the first hydration shell and three in the second). The numbers in circles indicate the global net charge on the water molecule.

It may thus be estimated that in the completion of rather large hydration shells, the polarization of the water molecules due to the effect of the perturbing anionic solute will decrease in successive solvation layers. The decrease is estimated as being relatively rapid and it seems probable that the water molecules of the third hydration shell around DMP$^-$ will be only very slightly perturbed, if at all. Compared with the 1/1 data of Fig. 7, this result shows the necessity of taking into account the totality or at least large parts of the first hydration shell in drawing conclusions about the characteristics of the second layer. It also agrees with the conclusion based on energy computation (B. Pullman *et al.*, 1975) that "bound" perturbed water around DMP$^-$ is essentially restricted to the first two hydration shells.

The preceding results have recently been reexamined by introducing the 3*d* orbitals of phosphorus into the computations (Perahia *et al.*, 1975). The consequences of this refinement appear different according to whether we consider its effect on the electronic structure of DMP$^-$ itself or on the electronic transfers in the water adducts.

Figure 10 presents the distribution of electronic charges in DMP$^-$ computed by including the 3*d* orbitals. By comparing the result with those of Fig. 5a it may be seen that the major effect of the 3*d* orbitals is to decrease substantially

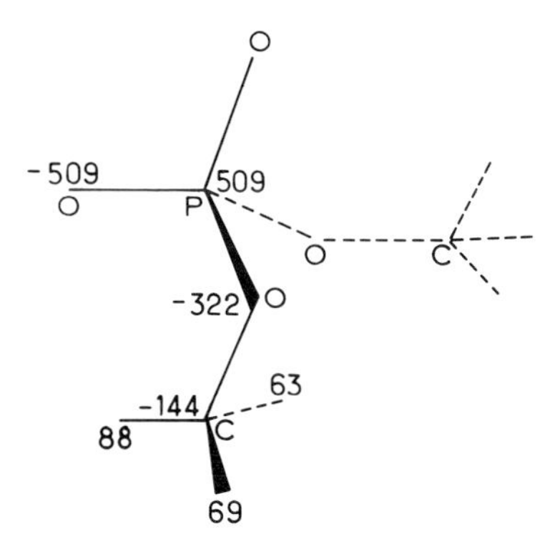

Fig. 10. Net charges in DMP^- (in millielectron units) obtained by *ab initio* computations including $3d$ orbitals on the phosphorus.

both the net positive charge on the phosphorus and the net negative charges on the oxygens. The carbon atoms become somewhat less negative. Only the methyl hydrogens increase their net charges. This overall smoothing out of the net charge distribution has been made possible by a concomitant, substantial accumulation of population in the $3d$ orbitals of phosphorus. These effects seem to be a general feature in compounds where the participation of the $3d$ orbitals is required to represent properly the valence of this atom in molecules (Guest *et al.*, 1972) as indicated also by other *ab initio* computations (Hillier and Saunders, 1970; Boyd, 1970; Demuyinck and Veillard, 1970; Johansen, 1974), CNDO (McAloon and Perkins, 1972; Alving and Laki, 1972) and even by EHT computations (Boyd, 1970; Boyd and Lipscomb, 1969).

Similarly, we may now investigate the influence of introducing $3d$ orbitals on the structure of the "bound" water. The net charges in the $DMP^- \ldots H_2O$ adducts with and without $3d$ orbitals may be compared. Typical examples for comparison are the B_{13} and E_{31} sites, with and without $3d$ orbitals, as seen in Figs. 11a and b and Figs. 5c and d, respectively. It is seen that the qualitative modifications of the charges of the two molecules on binding are similar. Moreover, the quantitative modifications are not very different in spite of the appreciable modifications brought about in the phosphate itself by the d orbitals. The global charge transfer is decreased from 0.145 to 0.125 e in E_{13} and from 0.120 to 0.100 e in the bridged site. The negative character of the oxygen of the bound water molecule remains very large, particularly in the bridge position. It seems likely that this partial anionic character acquired by water should enable it to bind further water molecules in a second hydration shell as was found in the computation without d functions. This state of affairs is verified by explicit computations on the energies of interactions (Perahia *et al.*, 1975).

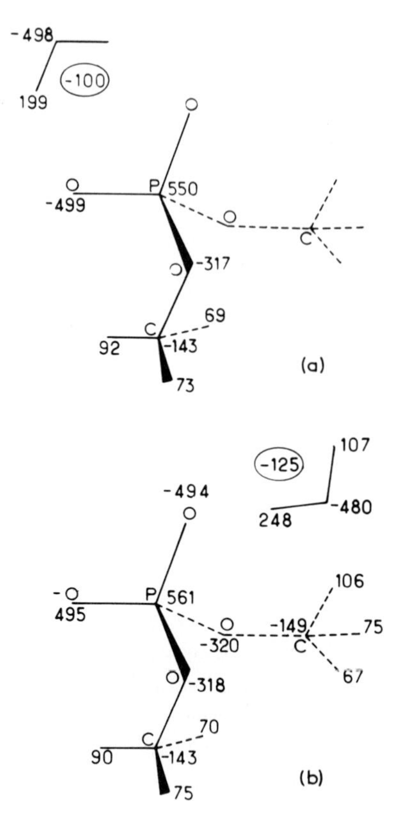

Fig. 11. Net charges in the B_{13} and E_{31} monohydrates of DMP^- (in millielectron units) obtained by *ab initio* computations including 3*d* orbitals on the phosphorus. The numbers in circles indicate the global net charges on the water molecules.

Water Bound to Cationic Systems

The system considered will be the ammonium group, NH_4^+ (Armbruster and Pullman, 1975).

Figure 12 presents the electronic charges in free NH_4^+ and H_2O computed by an SCF *ab initio* calculation with an STO 3G basis. It is worth stressing that, contrary to the usual chemical representation, the positive charge of the NH_4^+ ion is not localized on the N atom but spread over the four hydrogens to the point that the N atom carries in fact a formal negative charge. The essence of this situation has been confirmed experimentally by a combination of x-ray and neutron diffraction data pertaining to the related case of the $-ND_3^+$ end of perdeuterio-α-glycine (Griffin and Coppens, 1975).

Figure 13a–d presents the charge distribution in water adducts, from the

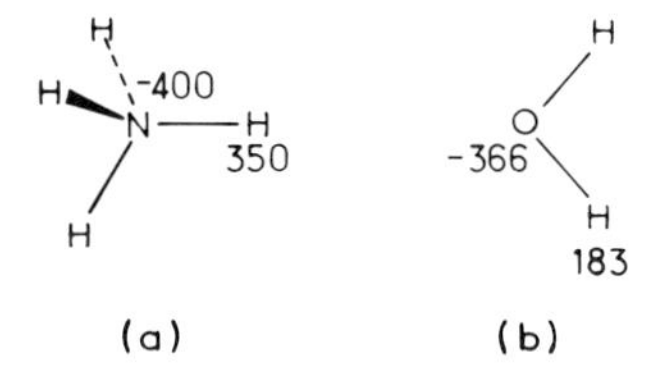

Fig. 12. Net charges (in millielectron units) in (a) NH_4^+, (b) water.

mono-adduct to the tetra-adduct, with the solvent molecules fixed along the N–H bonds of the solute, at their equilibrium distance, to the completion of the first hydration shell. Figure 14 presents the results for a penta-adduct, the fifth water molecule being placed in the second hydration layer. Naturally, while in the water adducts to DMP^-, the phosphate group functions as a proton acceptor and the NH_4^+ group acts as a proton donor. It is observed, in the NH_4^+–H_2O mono-adduct, that the bound proton becomes slightly more positive while both the nitrogen and the proton acceptor oxygen of water become more negative. Furthermore, the hydrogens of the proton acceptor (water) become more strongly positive than in the isolated state (+0.274 instead of +0.183). Globally a *partial transfer of the positive charge from the ammonium ion to the ligand is thus observed* so that the apparent global charge of the ion is decreased from unity to +0.837 and the bound water carries the concomitant excess of the positive charge.

Each time a new ligand molecule is added to one of the remaining free NH bonds of NH_4^+, the changes in the new hydrogen bond are of the same nature as those observed on the first binding. But a partial electronic rearrangement occurs whereby each of the H-bonded protons is now less positive than before (and even less positive than in NH_4^+); the central nitrogen becomes progressively more negative and the positive charge of the remaining free proton(s) is decreased. In fact similar displacements also occur when the fifth ligand is added on to a molecule of the first solvation layer.

The amount of electron transfer per molecule of ligand decreases progressively, following a parallel decrease of the binding energy per supplementary H–bond. The formal positive charge of the ammonium ion decreases to $n = 4$, becoming progressively dispersed on the ligand molecules; the positive charge on each of those decreases, however, with increasing n.

When the second solvation shell begins to form by the addition of one more molecule of H_2O to one of the already bound molecules of the complete first solvation layer, the central ion loses almost none of its supplementary positive charge. The three single ligands of the first shell are very little affected, while the fourth, which carries the molecule of the second layer, loses a part of its positive charge in favor of the terminal molecule. The positive charge of this second shell ligand is smaller, however, than that of the single molecules of the first shell. Its

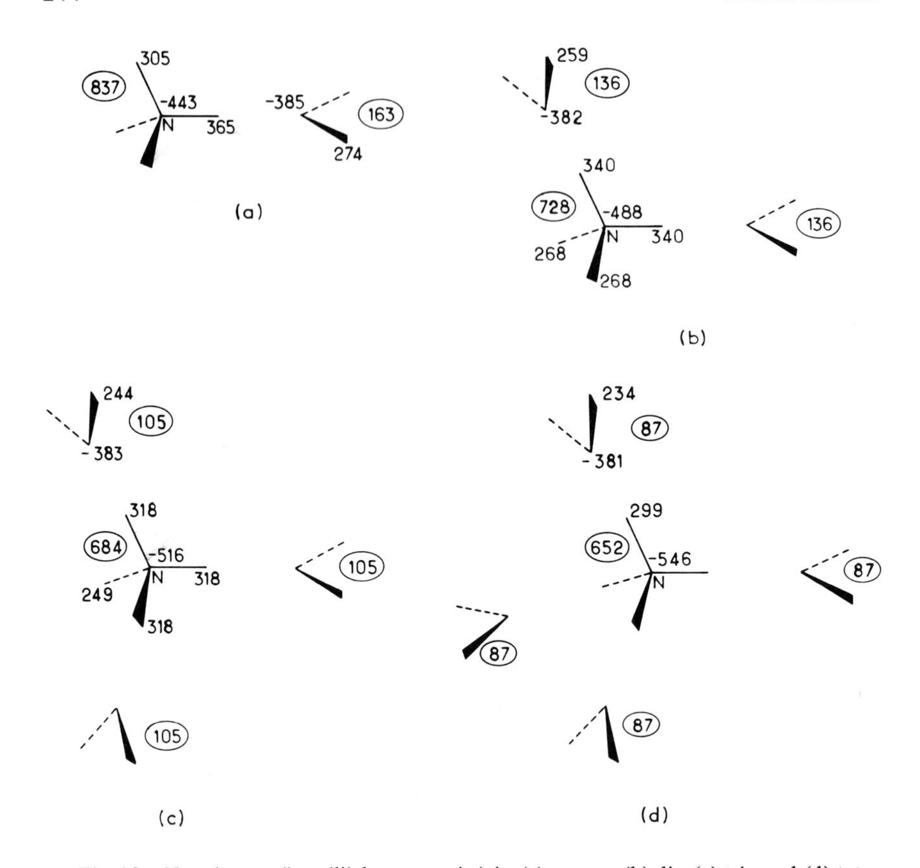

Fig. 13. Net charges (in millielectron units) in (a) mono-, (b) di-, (c) tri-, and (d) tetra-adducts of water to NH_4^+. The numbers in circles indicate the global net charges on the water molecules.

hydrogens are also less positive than those of the molecules of the first layer. Thus the perturbation (activation) of the ligand molecules decreases on their addition in successive layers. It may be inferred from the computed data that the completion of the second solvation shell will be accompanied by an evolution of the charge dispersion similar to that found on the building-up of the first layer. Thus there is a progressive decrease of the positive charge carried by the ligand molecules that will therefore become less and less capable of attracting further ligand molecules, making rather unlikely the structural formation of more than two solvent shells around the central ammonium ion.

These features are thus very similar to those found in the stepwise addition of water molecules to the phosphate ion where a progressive dispersion of negative charge was observed at the periphery of the solvation layer(s).

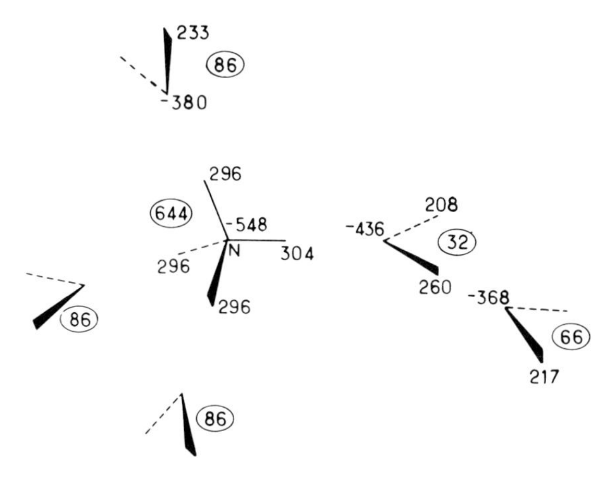

Fig. 14. Net charges (in millielectron units) in the pentahydrate of NH_4^+, with four molecules of H_2O in the first hydration shell and one in the second shell. The numbers in circles indicate the global net charges on the water molecules and on NH_4^+.

Water Bound to Neutral Systems

To illustrate the hydration of neutral species we have chosen formaldehyde as a model of a compound which can only act as a proton acceptor in its binding to water, thus being similar to the phosphate group. Figure 15 gives the distribution of the net charges in the mono-, di- and tetrahydrates compared to the isolated species. In the mono- and dihydrate with two water molecules in the first shell, the situation is qualitatively similar to that in Fig. 5, the displacements of charge being significantly smaller.

When water molecules are added in the second hydration shell, they take up electronic arrangements reminiscent of a linear water dimer (Fig. 16). The perturbation due to the formaldehyde substrate is visible in the decrease of the positive charge of the first water molecule and in the enhancement of the negative charge on the second water molecule. The overall loss of charge by the substrate increases slightly on going from $n = 2$ to $n = 4$.

Another example of a neutral solute is NH_3 (Fig. 17) which may act toward water both as a proton donor and acceptor. The progressive saturation of the three NH bonds by water molecules (Fig. 17c, d) is qualitatively analogous to the situation observed in the progressive hydration of NH_4^+, although the charge displacements are much smaller here.

Water acting as a proton donor (Fig. 17e) produces charge displacements analogous to those observed in the water dimer (Fig. 16). A particularly interesting scheme of charge displacements appears in Fig. 17f, which combines the two types of binding: the NH_3 molecule serves as a relay for electron-shift from the three water molecules attached to the NH bond toward the water

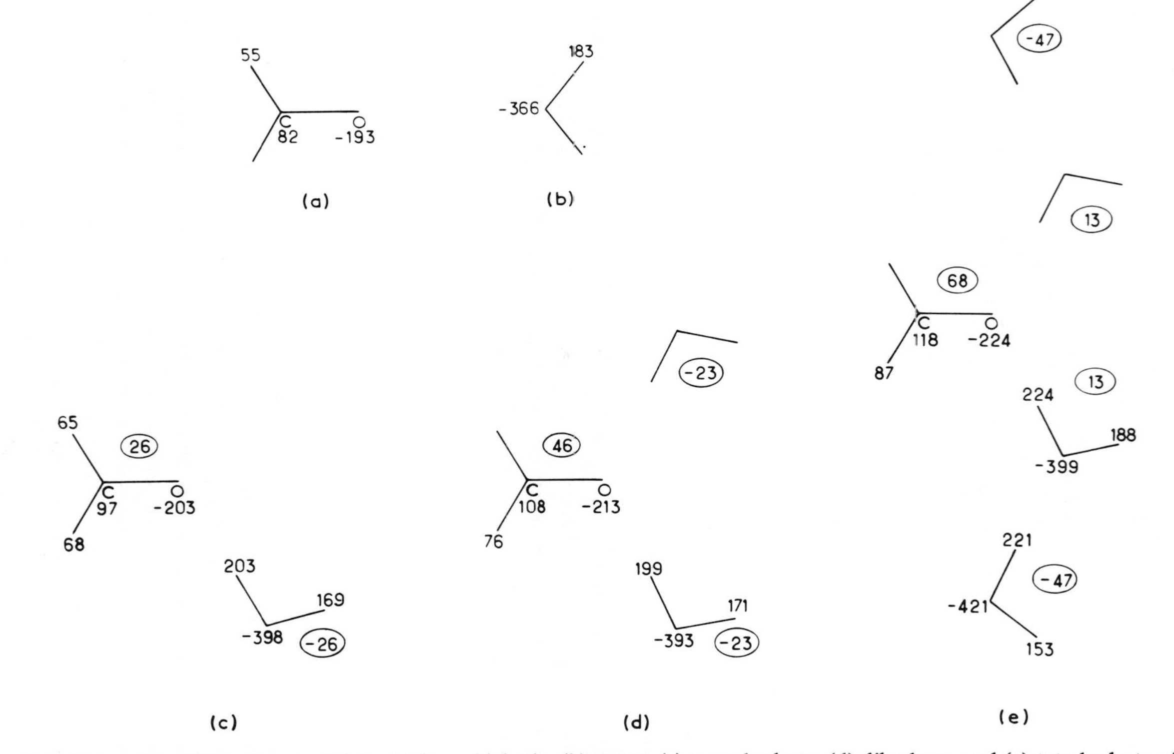

Fig. 15. Net charges (in millielectron units) in (a) formaldehyde, (b) water, (c) monohydrate, (d) dihydrate, and (e) tetrahydrate of formaldehyde. The numbers in circles indicate the global net charges on the water and on formaldehyde.

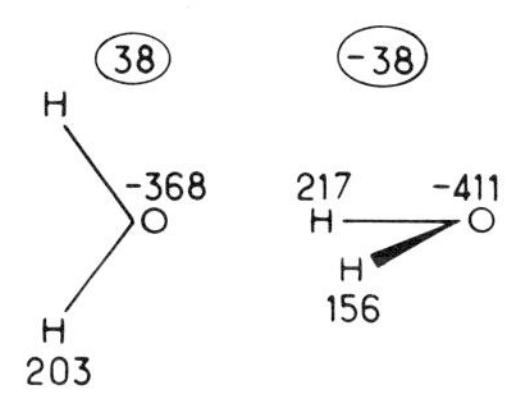

Fig. 16. Net charges (in millielectron units) in the linear water dimer. The numbers in circles indicate the global net charge on each water molecule.

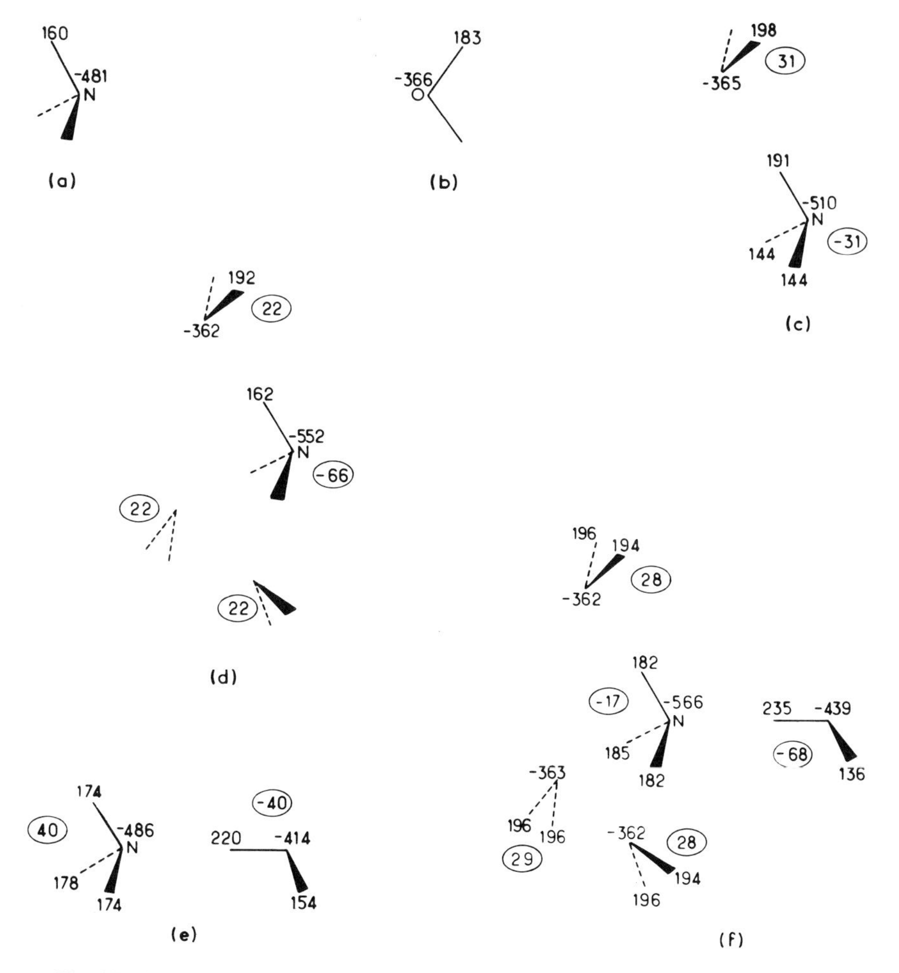

Fig. 17. Net charges (in millielectron units) in (a) ammonia, (b) water, (c) monohydrate on NH side, (d) trihydrate on NH side, (e) monohydrate on lone-pair side, and (f) tetrahydrate.

molecule linked to the nitrogen lone-pair, the loss of charge by the three proton-acceptor water molecules being enhanced by the presence of the proton-donor water molecule on the other side. It is manifest that this situation increases the ability to form a second hydration shell.

CONCLUSIONS

One of the main conclusions to be drawn from the above computations is the clear-cut indication of the existence around ionic substrates of strongly perturbed layers of bound water. Essentially, two such layers seem to be involved, the perturbation decreases appreciably from the first layer to the second and becoming small or negligible beyond the second layer. This conclusion seems to agree with the results of different experimental investigations on the bound water in the particularly abundantly studied case of phospholipids (for reference, see A. Pullman *et al.*, 1975; B. Pullman *et al.*, 1975). The nature of the electronic perturbation depends on the negative or positive character of the ionic center of hydration and corresponds to the transfer of electrons towards water in the former and to the transfer of the positive charge towards water in the latter. As a general rule, however, the region of the hydrogen bond itself becomes electron deficient. This situation which we have already indicated in our earlier studies on hydrogen bonding between peptide units (Dreyfus *et al.*, 1970) is today confirmed experimentally using different examples by combined x-ray and neutron diffraction studies (Griffin and Coppens, 1975; Thomas *et al.*, 1975).

When hydration occurs around ionic groups, the atoms of which carry appreciable net negative or positive charges, the amount of charge transferred is large. The effects are of a similar nature but much smaller when hydration occurs around neutral species.

REFERENCES

Alagona, G., Pullman, A., Scrocco, E., and Tomasi, J. (1973). *Int. J. Pept. Protein. Res.* **5**, 251.

Alving, R. E., and Laki, K. (1972). *J. Theor. Biol.* **34**, 199.

Armbruster, A. M., and Pullman, A. (1974). *Int. J. Quantum Chem., Symp.* **8**, 169.

Bader, R. F. W., Beddall, P. M., and Cade, P. E. (1971). *J. Am. Chem. Soc.* **93**, 3095.

Beveridge, D. L. (1974). *In* "Molecular and Quantum Pharmacology" (E. D. Bergmann and B. Pullman, eds.). Reidel Publ., Dordrecht, Netherlands.

Boyd, D. B. (1970). *Theor. Chim. Acta* **18**, 184.

Boyd, D. B., and Lipscomb, W. N. (1969). *J. Theor. Biol.* **25**, 403.

Del Bene, J., and Pople, J. A. (1969). *Chem. Phys. Lett.* **4**, 426.

Del Bene, J., and Pople, J. A. (1970). *J. Chem. Phys.* **52,** 4858.
Demuyinck, J., and Veillard, A. (1970). *Chem. Commun.* p. 873.
Dreyfus, M., Maigret, B., and Pullman, A. (1970). *Theor. Chim. Acta* **17**, 109.
Griffin, J. F., and Coppens, P. (1975). *J. Am. Chem. Soc.* **97**, 3496.
Guest, M. F., Hillier, I. H., and Saunders, V. R. (1972). *J. Chem. Soc., Faraday Trans.* **5**, 867.
Hankins, D., Moskowitz, J. W., and Stillinger, E. H. (1970). *J. Chem. Phys.* **53**, 4544.
Hehre, W. G., Stewart, R. F., and Pople, J. A. (1969). *J. Chem. Phys.* **51**, 2657.
Hehre, W. G., Ditchfield, R., Stewart, R. F., and Pople, J. A. (1970). *J. Chem. Phys.* **52**, 2769.
Hillier, I. H., and Saunders, V. R. (1970). *J. Chem. Soc. A* p. 3475.
Hopfinger, A. (1974). *In* "Molecular and Quantum Pharmacology" (E. D. Bergmann and B. Pullman, eds.), p. 131. Reidel Publ., Dordrecht, Netherlands.
Hylton, J., Christoffersen, R. E., and Hall, G. (1974). *Chem. Phys. Lett.* **24**, 501.
Johansen, H. (1974). *Theor. Chim. Acta* **32**, 273.
Jug, K. (1973). *Theor. Chim. Acta* **31**, 63.
McAloon, B. J., and Perkins, P. G. (1972). *Theor. Chim. Acta* **24**, 102.
Mulliken, R. S. (1955). *J. Chem. Phys.* **23**, 1833.
Perahia, D., Pullman, A., and Berthod, H. (1975). *Theor. Chim. Acta* **40**, 47.
Politzer, P., and Mulliken, R. S. (1971). *J. Chem. Phys.* **55**, 5135.
Port, G. N. J., and Pullman, A. (1973a). *FEBS Lett.* **31**, 70.
Port, G. N. J., and Pullman, A. (1973b). *Theor. Chim. Acta* **31**, 231.
Port, G. N. J., and Pullman, A. (1974). *Int. J. Quantum Chem., Quantum Biol. Symp.* **1**, 21.
Pullman, A. (1970). *Jerusalem Symp. Quantum Chem. Biochem.* **2,** 9.
Pullman, A., and Armbruster, A. M. (1974). *Int. J. Quantum Chem., Symp.* **8**, 169.
Pullman, A., and Armbruster, A. M. (1975). *Chem. Phys. Lett.* **36,** 558.
Pullman, A., and Pullman, B. (1975). *Q. Rev. Biophys.* **7**, 505.
Pullman, A., Alagona, G., and Tomasi, J. (1974). *Theor. Chim. Acta* **33**, 87.
Pullman, A., Berthod, H., and Gresh, N. (1975). *Chem. Phys. Lett.* **33**, 11.
Pullman, B. (1977). *Adv. Quantum Chem.* (in press).
Pullman, B., Pullman, A., Berthod, H., and Gresh, N. (1975). *Theor. Chim. Acta* **40,** 93.
Sinanoglu, O. (1974). *In* "The World of Quantum Chemistry" (R. Daudel and B. Pullman, eds.), p. 265. Reidel Publ., Dordrecht, Netherlands.
Thomas, J. O., Tellgren, R., and Almlof, J. (1975). *Acta Crystallogr., Sect. B* **31**, 1946.

18

An Aspect of Submolecular Biology

Quantum-Mechanical Exploration of Biomolecular Conformations—The Case of Phospholipids

BERNARD PULLMAN

It seems to me that on the path of scientific adventure Albert Szent-Györgyi was always one step ahead of the others. Thus in the late 1950's, when "molecular biology" was considered by many, if not by all, as the most important frontier of the biological sciences, Albert Szent-Györgyi was already dreaming about–in fact more then dreaming, fighting for–"submolecular biology." This was, of course, the natural attitude of an arduous thinker who knew–what everybody should have known–that molecular properties are determined at the submolecular level and that they result from the combined effects of the constituent elementary particles. For a deep comprehension of the molecular world, one must unravel the interplay of electrons and nuclei, provided, of course, that one knows how to find the way toward these particles and the way from them to the properties of the organized compounds which are common subjects of scientific exploration. At that time, the late 1950's, a few years after the discovery of the molecular structure of nucleic acids, Szent-Györgyi was perhaps the only biologist who was concerned about electrons elsewhere than in the electron transport system and who believed that the ultimate elucidation and comprehension of the laws of biology must be looked for at this level. Moreoever, for Albert Szent-Györgyi, *submolecular biology* meant essentially *quantum biology*.

Madame Pullman and myself had the good fortune of being invited by Albert Szent-Györgyi to spend the summer months of 1957–1961 in the exceptional intellectual atmosphere of the Marine Biology Laboratory at Woods Hole. Albert succeeded in gathering there during those summers a number of distinguished scientists interested in basic problems of biochemistry, biophysics, and biology. Seminars and discussions held in his home at Penzance Point resulted in concepts, proposals, and theories which since have become essential parts of submolecular biology. They were developed in a number of places all over the world, but their seeds may frequently be traced, in one way or another, to these pioneering endeavours. Few organized laboratories can compete with the Woods Hole beach for the birth of daring hypotheses.

More than 12 years have elapsed since those early meetings, and with the exception of Albert Szent-Györgyi, we have all grown older, as has submolecular biology. It is not possible for me to summarize in this limited space the important evolution of this fascinating branch of science. Rather than trying to accomplish such an impossible task, I shall simply present a particular example which will illustrate, particularly to those who knew the status of these researchers 15 years ago, the development and prospectives of this field of study.

THEORETICAL APPROACH TO CONFORMATIONS OF BIOMOLECULES

There are two fundamental aspects of the quantum-mechanical exploration of biological molecules, arising out of the electronic or conformational characteristics of the systems studied. The great majority of work carried out in the early phases of quantum biochemistry had been devoted primarily to the study of the electronic distributions and properties of biomolecules. It is only in the past 3–4 years that studies have tended to develop toward the theoretical determination of the *conformational basis* of biological structures and mechanisms. This extension is due largely to the concomitant development of the all-valence electrons and all-electrons methods for studying molecular structures and of the big electronic computers, both factors being indispensable for the success of theoretical conformational analysis. Once started, the development of work in this direction has been rapid so that today important literature is available on the quantum-mechanical explorations of the conformational properties of a large variety of fundamental biological compounds, such as proteins and their constituents (Pullman and Pullman, 1974), nucleic acids and their constituents (Pullman and Saran, 1976) and also a large number of important pharmacological substances (B. Pullman, 1974, 1975; A. Pullman and Pullman, 1975).

What is the nature and the scope of these studies? We shall try to answer this question with an example chosen from the more recent and, to some extent, the most developed investigations in this field.

The example chosen is the *phospholipids,* which as everybody knows are among the most important constituents of biological membranes. Their general structure is represented by the formula in Fig. 1 from which it is immediately obvious that these substances contain a very large number of degrees of freedom which make the theoretical explorations of their conformational properties particularly difficult. It is also, however, particularly interesting, if only because of the fact that up till a short time ago no x-ray crystallographic study was available on these compounds. Even today only one such study seems to have been published (Hitchcock *et al.*, 1974), although a number of results exist for different model compounds of the different fragments of the phospholipids. Solution studies are available both for some phospholipids and model compounds but their conclusions frequently lack quantitative precision.

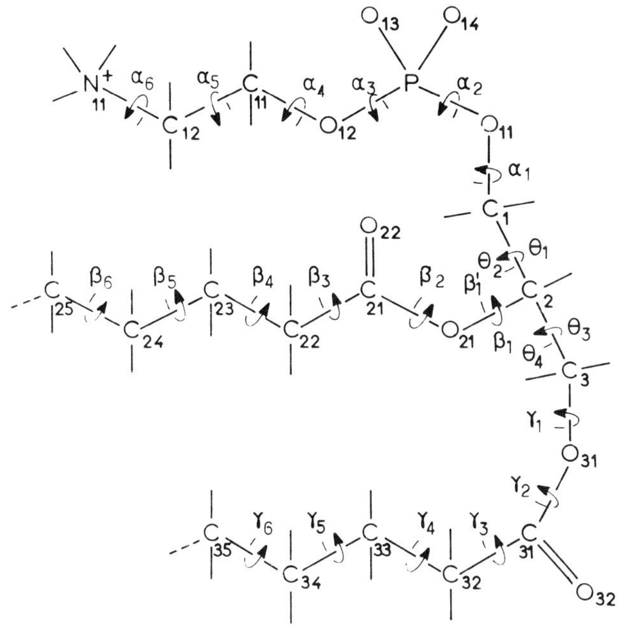

Fig. 1. The structure of phospholipids and Sundaralingam's notations (Sundaralingam, 1972) for the torsion angles. Note that the torsion angle about bond j is considered positive for a right-handed rotation; when looking along the bond j, the far bond j+1 rotates clockwise relative to the near bond j–1. The 0° value corresponds to the *cis*-planar arrangement of bonds j+1 and j–1.

THE CONFORMATION OF PHOSPHOLIPIDS

We have studied extensively, by refined quantum-mechanical methods, the principal conformational possibilities and preferences of this type of molecule and have thus investigated practically all the torsion angles which are indicated in Fig. 1 (Pullman and Berthod, 1974; A. Pullman *et al.*, 1975; Pullman and Saran, 1976). In this report we will, however, concentrate mainly on the conformational problem related to the so-called polar head of phospholipids represented essentially by the α-chain of Fig. 1. This chain is frequently considered as being conformationally relatively independent of the two other chains, β and γ, and has the particular feature of being immersed in a water environment.

The conformational properties of this chain are defined by the series of the six torsion angles, α_1–α_6, for which we shall adopt here the definitions proposed by Sundaralingam (1972). A usual practice in the study of a conformational problem involving such a multitude of torsion angles is to construct conformational energy maps as a function of the different couples of adjacent angles, with preselected values fixed for the torsion angles not involved in the particular map under construction. The procedure is developed toward optimal self-consistency.

In the present case the problem may be simplified by two observations:

1. It is relatively easy to attribute reasonable preselected values to the torsion angles α_1 and α_6 (Pullman and Saran 1975). The terminal cationic group may be expected to be preferentially staggered with respect to the C_{-12} methylene group (α_6 = 60° or 180° or 300°) and α_1 may also be shown to prefer a value around 180° (Pullman and Saran, 1975).

2. The remaining four torsion angles α_2–α_5 obviously fall into two groups: (1) the two torsion angles about the two P–O bonds α_2 and α_3, which are directly relevant to the conformation of the phosphate group and whose significance thus extends to other systems involving this group, in particular to polynucleotides and nucleic acids; and (2) the couple α_4, α_5 whose values define the overall conformation of the polar head in the sense that they determine the mutual arrangement of the positive and negative poles of this zwitterionic fragment. We shall deal with these two pairs of torsion angles.

The Conformational Properties of the Phosphate Group: The Conformation about α_2–α_3

As stated above, the problem of the conformational properties of the phosphodiester linkage (torsion about α_2 and α_3) transcends the field of phospholipids and is of equal importance, particularly in regard to polynucleotides and nucleic acids, where the usual designation of the two P–O torsion angles is ω and ω', following again a notation proposed by Sundaralingam (1969). It is therefore not surprising that these properties have been amply investigated in many theoretical studies, using both classical potential energy functions and various quantum-mechanical methods and also utilizing different model compounds: anions of dimethyl phosphate, disugar phosphate, dinucleoside phosphate, etc. For a review of the computations carried out in relation to the nucleic acids and their components one may consult Pullman and Saran (1975). Computations were also performed more explicitly in relation to the phosphate group in the polar head of the phospholipids using the partitioned potential functions of Yathindra and Sundaralingam (1974) and Vanderkooi (1973a,b), the quantum-mechanical EHT procedure of Gupta and Govil (1972), and the quantum-mechanical PCILO method of Pullman and Saran (1975b).

In fact the essential results, relevant to the intrinsic conformational properties of the phosphodiester linkage, seem to be quite evident already in the simple model of dimethyl phosphate (DMP^-), and are relatively only slightly modified in more complex models. [For a discussion, see Perahia *et al.* (1974).] On the other hand, they depend strongly on the method of study used, and from that point of view the results may be classified into two groups:

1. Those obtained by the classical potential energy functions (Sasisekharan and Lakshminarayanan, 1969; Yathindra and Sundaralingam, 1974) and by the simple quantum-mechanical EHT method (Saran and Govil, 1971). They are schematically exemplified by the conformational energy map of Fig. 2. Such computations generally indicate the existence of seven *equivalent* regions of energy minima located around (α_2, α_3) or (ω, ω') = (60°, 60°), (300°, 300°), (60°, 180°), (180°, 60°), (180°, 300°), (300°, 180°) and (180°, 180°), which means that they essentially predict energy equivalence between the so-called *gauche-gauche* (*gg*) forms (the first two in the above enumeration), the *gauche-trans* or *trans-gauche* (*gt* or *tg*) forms (the next four in the above enumeration), and the *trans-trans* (*tt*) form (the last in the above enumeration). Attempts to refine some of the empirical computations carried out recently by different authors on more complex models, generally relevant to the structure of the nucleic acids, give results which disfavor the *gg* forms with respect to others. Thus, computations by Olson and Flory (1972a,b) on a model of dinucleoside monophosphate show a global minimum for the *tt* conformation. Yathindra and Sundaralingam (1974) using a similar model favor one of the *gt* conformations. On the other hand, in the computations by Gupta and Govil (1972) (EHT method) explicitly devoted to the α-chain of phospholipids, some of the minima merge into larger zones but the essential image remains as in Fig. 2.

2. Results obtained by the more refined quantum-mechanical methods include: *ab initio* (Newton, 1973; Perahia *et al.*, 1974), PCILO (Pullman and Saran, 1975; Perahia *et al.*, 1974), and CNDO (Tewari *et al.*, 1974). Their main general features are illustrated on the PCILO map of dimethyl phosphate presented in Fig. 3. This map clearly indicates a net preference for the *gg* form

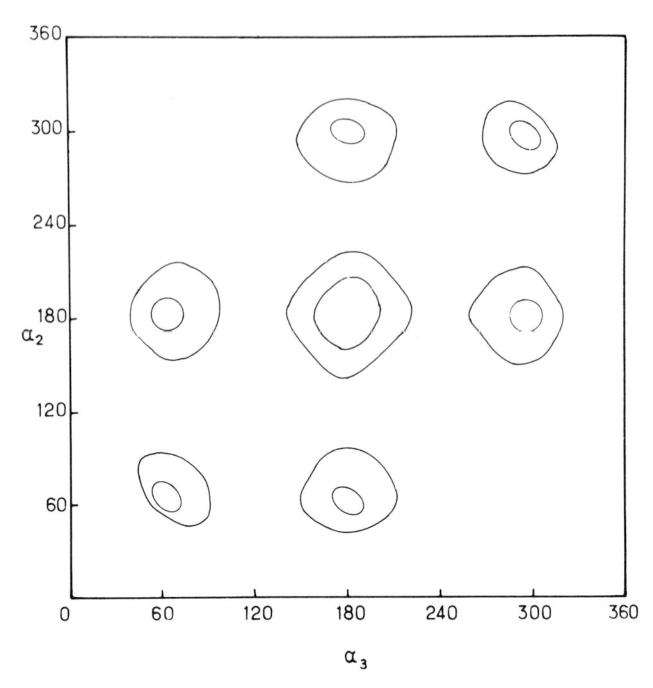

Fig. 2. Schematic indication of the energy minima on a α_2, α_3 conformational energy map in typical empirical or EHT computations for the diester linkage.

and predicts a relative instability for the *tt* form. *Ab initio* results are essentially identical to those of Fig. 3 and show an even greater energy difference between the *gg* and *tt* forms. More complex models give basically similar indications, and when asymmetrical lead to a distinction between the two *gg* forms, g^+g^+ ($\alpha_2 = \alpha_3 = 60°$) and g^-g^- ($\alpha_2 = \alpha_3 = 300°$). Rules have been proposed which permit prediction of the preponderance of one of these *gg* forms as a function of other structural characteristics of the group (Perahia *et al.*, 1974) or of its environment (Pullman and Saran, 1975). In particular, as we shall discuss later, a relation exists in the phospholipids between the preference for the g^+g^+ or g^-g^- conformations about the α_2, α_3 torsion angles and the conformations about the α_4, α_5 angles. Explicit α_2, α_3 maps indicating these dependencies are given by Pullman and Saran (1975).

In view of such disagreement between two groups of theoretical procedures, it seems natural to turn to experimental data for the estimation of the validity of these procedures in this field. The most abundant and precise data come from x-ray crystal structure studies, which show a very strong preference for the phosphate group for a *gg* conformation and a total absence of a compound with a *tt* conformation about the P—O bonds. These studies were performed on

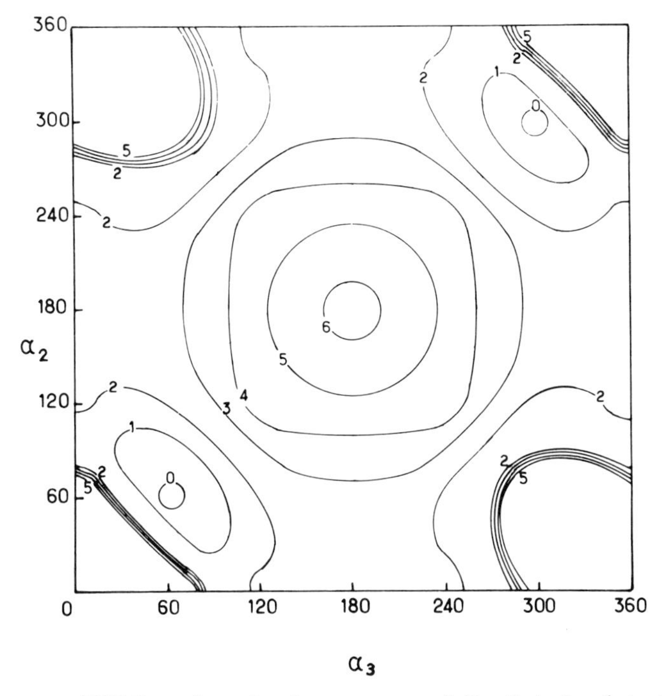

Fig. 3. α_2–α_3 PCILO conformational energy map of dimethyl phosphate. Isoenergy curves in kcal/mole with respect to the global energy minimum taken as energy zero.

simple phosphate diesters, e.g., diethyl phosphates, or more complex oligo- and polynucleotides (for a review, see Pullman and Saran, 1976), on model compounds of the polar head of phospholipids and on the one available crystal of a phospholipid (for a review, see Pullman and Saran, 1975). The results for the model compounds of the polar head of phospholipids are reproduced in Table I. The chemical formulas of the model compounds CP, EP, GPC, GPE and of the 1,2-dilauroyl-DL-phosphatidylethanolamine (DPE) are indicated in Fig. 4. Although the computations as described above have been carried out for the free molecules while the experimental results quoted refer to the solid state, the remarkable agreement between the predictions of the refined quantum-mechanical methods and the findings in the crystals seem to us to indicate that the phosphate group has an intrinsic preference for a *gg* conformation about the P–O bonds and that this preference is maintained in the crystals. This appears to be a more appropriate viewpoint than that expressed by some of the proponents of empirical computations who, by finding in their calculations a preference of the phosphate group for other conformations, call upon subsidiary factors to account for the experimental predominance of the *gg* form. It would be simpler

TABLE I
Conformation of the Polar Head Group in X-Ray Crystal Studies of Constituents of Phospholipids

Compound	α_1	α_2	α_3	α_4	α_5	α_6	θ_1	θ_2	θ_3	θ_4
GPC-1	164.6	288.6	301.1	222.0	72.6	56.8 174.1 292.2	172.3	288.9	297.1	180.5
GPC-2	188.4	63.9	65.5	139.9	284.6	68.2 184.4 306.2	297.0	60.6	290.5	169.4
GPC $CdCl_2$	169.0	291.0	287.0	178.0	73.0	54.0 171.0 287.0	171.0	288.0	59.0	300.0
GPE	186.3	287.9	278.8	163.8	55.5	45.3 164.6 276.3	166.8	286.7	298.9	179.9
1,2-Dilauroyl DL-phosphatidylethanolamine	211.0	51.0	64.0	101.0	77.0		310.0	64.0	182.0	73.0

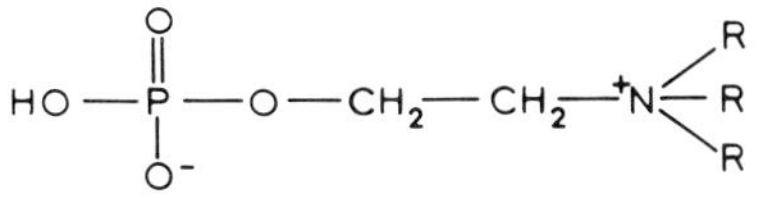

$HO-P(=O)(O^-)-O-CH_2-CH_2-N^+R_3$

I

(a) $R = CH_3$: choline phosphate (CP)

(b) $R = H$: ethanolamine phosphate (EP)

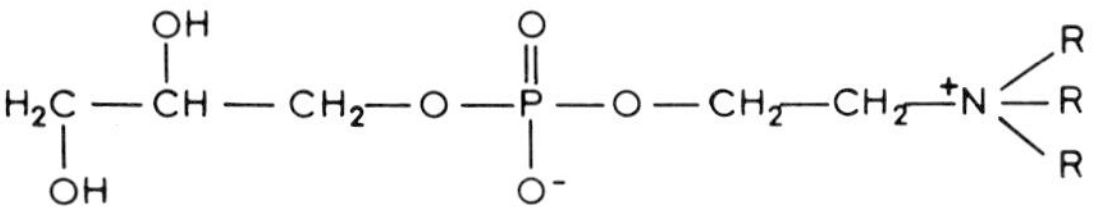

$H_2C(OH)-CH(OH)-CH_2-O-P(=O)(O^-)-O-CH_2-CH_2-N^+R_3$

II

(a) $R = CH_3$: L-α-glycerylphosphorylcholine (GPC)

(b) $R = H$: L-α-glycerylphosphorylethanolamine (GPE)

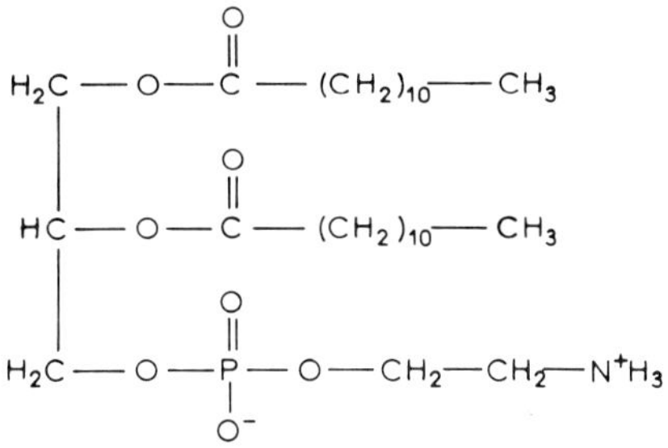

$H_2C-O-C(=O)-(CH_2)_{10}-CH_3$

$HC-O-C(=O)-(CH_2)_{10}-CH_3$

$H_2C-O-P(=O)(O^-)-O-CH_2-CH_2-N^+H_3$

III

1,2-dilauroyl-DL-phosphatidylethanolamine (DPE)

Fig. 4. Chemical formulas of some compounds mentioned in Table I.

to recognize the insufficiency of the empirical (and EHT) computations for these types of compounds.

This viewpoint is further strengthened by a series of recent computations carried out by the *ab initio* and PCILO methods on the model compound of the dimethyl phosphate anion (DMP^-) and which show that:

1. The preference for the *gg* form persists on hydration of this model compound although it is less pronounced than in the free molecule (A. Pullman *et al.*, 1975; B. Pullman *et al.*, 1975c). The computations were carried out in the "supermolecule" approximation for the study of solvation effects as described in this volume by Madame Pullman (see also Pullman and Pullman, 1975).

2. The preference for the *gg* form also persists on cation binding (Na^+, Mg^{2+}) to these model compounds and, in fact, seems even strengthened by it (B. Pullman *et al.*, 1975b). Such is at least the result of *ab initio* computations.

CNDO calculations (Nanda and Govil, 1975) predict, surprisingly, that cation binding should shift the conformational preference of DMP^- to the *tt* form. This most probably is due to an artifact in the method. Artifacts in the application of the CNDO method to cation binding by organic or biological ligands have been observed before (Perricaudet and Pullman, 1973).

It thus seems highly plausible that the *gg* conformation about the α_2, α_3 torsion angles represents an intrinsic preference of the phosphate group, relatively independent of the groups linked to the alkyl carbons attached to the esteric oxygens and also of the hydration of this group and cation binding to it.

The Conformation about $\alpha_4 - \alpha_5$

We shall now consider the conformational properties with respect to the α_4 and α_5 torsion angles whose values have a decisive significance for the mutual orientation of the cationic $-N^+H_3$ or $N^+(CH_3)_3$ head and the anionic phosphate group and thus for the overall conformation of the polar head.

Crystallographic x-ray studies on EP (Kraut, 1961), GPE-monohydrate (DeTitta and Craven, 1973), GPC (Abrahamsson and Pascher, 1966), GPC cadmium chloride trihydrate (Sundaralingam and Jensen, 1965), summarized in Table I, as well as those on related systems, e.g., serine phosphate (see Sundaralingam, 1972), indicate that in all these compounds α_4 assumes a value around 180° (with an appreciable spreading from 140° to 220°), while α_5 is localized in the vicinity of 60° and 300°. The conformations with respect to these two torsion angles can thus be described as *trans-gauche.*

The *gauche* conformation about α_5 is also found in solution by NMR studies on CP (Andrieux *et al.*, 1972; Culvenor and Ham, 1974) and on dipalmitoyllecithin (Bridsall *et al.*, 1972). Parallel, theoretical computations carried out by the empirical potential functions procedure (McAlister *et al.*, 1973) and by the molecular orbital Extended Hückel Theory (Gupta and Govil, 1972) indicated that the most stable conformations of these molecular systems should correspond to α_4 = 180° and α_5 = 60°, 180° and 300°.

The satisfactory agreement between the theoretical results and the above quoted experimental ones has led to the impression that the conformations observed either in the solid state or in water for the polar heads of phospholipids in their constituents (presumed to occur also in membranes) correspond to their *intrinsically most stable* conformation. However, more refined quantum-mechanical computations (Pullman and Berthod, 1974; B. Pullman *et al.*, 1975a) present evidence that this is not so. Figure 5 presents the results of the PCILO computations of the conformational energy map of EP as a function of α_4 and α_5, the remaining torsion angles being fixed in their crystallographic values. The global energy minimum representing the intrinsically most stable conformation corresponds to α_4 = 270°, α_5 = 30°. It represents a seven-membered hydrogen-

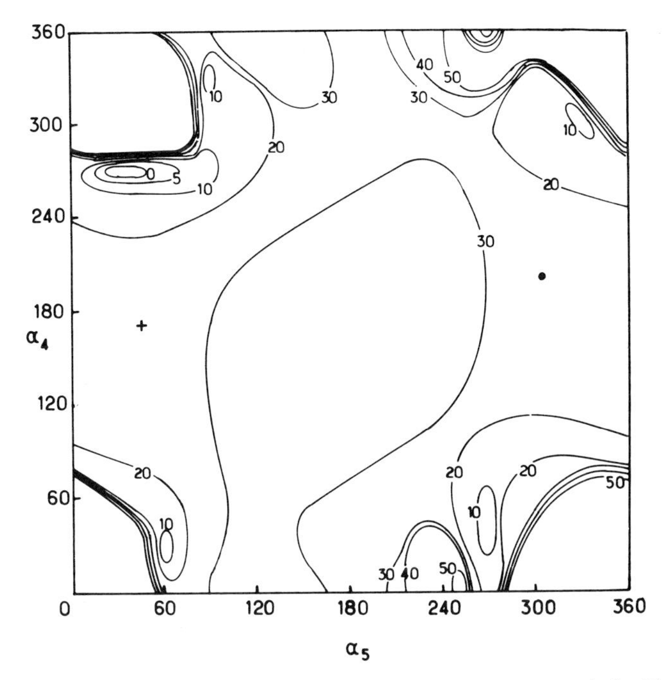

Fig. 5. PCILO conformational energy map for EP. Isoenergy curves in kcal/mole with respect to the global energy minimum taken as energy zero. Crystallographic conformations; circle = EP (Kraut, 1961), plus sign = GPE (DeTitta and Craven, 1973).

bonded ring structure, as depicted in Fig. 6, with the H-bond between an N^+–H of the cationic head and O^- of the phosphate group. The results of *ab initio* computations confirm the PCILO results (Pullman and Berthod, 1974; B. Pullman *et al.*, 1975a).

The PCILO results for CP are indicated in Fig. 3, and show a general analogy to those of EP. The conformationally allowed space within the same isoenergy limit is somewhat more restricted, owing to the increase of steric hindrance. Nevertheless the global energy minimum corresponds to a folded conformation similar to that found for EP: *gauche* (synclinal) with respect to α_5 (=300°), somewhat more extended (anticlinal) with respect to α_4 (=120°). The "inversion" of the global minimum between Figs. 5 and 7 from the upper-left corner to the lower-right corner of the map, respectively, is not due to the fact that one of the molecules is an ethanolamine and the other a choline but to the differences in the values of the torsion angles α_2 and α_3 in the crystallographic data of GPC (g^+g^+) and GPE (g^-g^-) which are used as input data. This situation shows the announced relation between the α_2, α_3 and the α_4, α_5 couples of torsion angles.

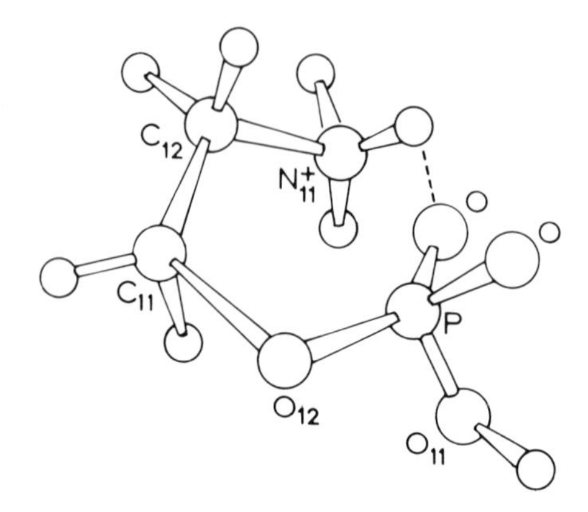

Fig. 6. The preferred conformation of the polar head.

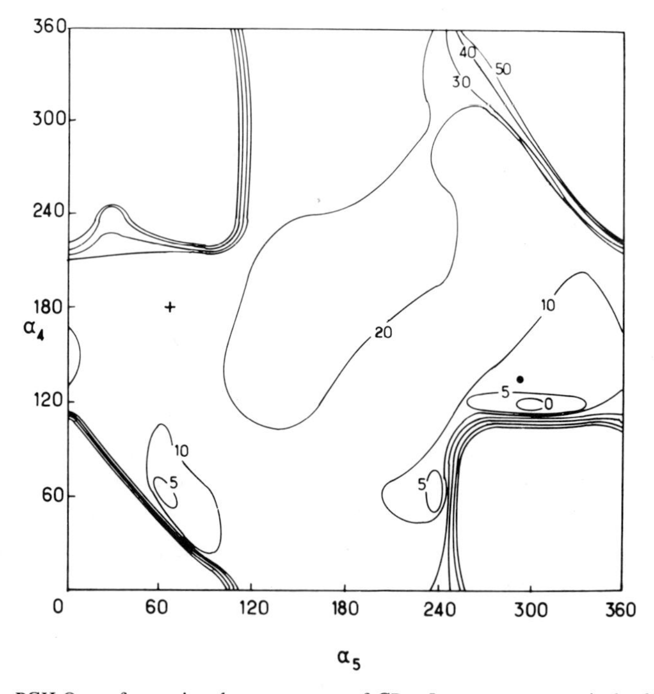

Fig. 7. PCILO conformational energy map of CP. Isoenergy curves in kcal/mole with respect to the global energy minimum taken as energy zero. Experimental conformations; circle = GPC (Abrahamsson and Pascher, 1966), plus sign = GPC cadmium chloride trihydrate (Sasisekharan and Lakshminarayanan, 1969).

The completely extended form corresponding to $\alpha_4 = \alpha_5 = 180°$ is 34 kcal/mole for EP and 20 kcal/mole for CP above the preferred conformations. The crystallographic conformation of GPE (DeTitta and Craven, 1973) at $\alpha_4 = 164°$, $\alpha_5 = 55°$, lies 26 kcal/mole above the intrinsically most stable one. In GPC the crystallographic conformation (Abrahamsson and Pascher, 1966) at $\alpha_4 = 140°$, $\alpha_5 = 285°$, is close to the theoretical global energy minimum and energetically only 6 kcal/mole above it. In the related GPC-cadmium chloride trihydrate (Sundaralingam and Jensen, 1956), the crystallographic conformation at $\alpha_4 = 178°$, $\alpha_5 = 73°$ is 13 kcal/mole above the computed global energy minimum. The findings in the last two crystals, when considered in conjunction with the theoretical results, suggest that the value of $\alpha_4 = 140°$ in GPC should be considered rather as distorted from the global energy value at 120° than from the usually considered 180°. This confirms to some extent *a strong intrinsic tendency* of free EP and CP for an anticlinal rather than extended conformation with respect to α_4. Under these circumstances, the existence of an extended (antiperiplanar) conformation in the solid state must be attributed to the action of crystal packing forces.

A striking confirmation of our viewpoint is being provided by the results of recent ^{31}P nuclear magnetic resonance studies on phospholipids in organic solvents (Henderson *et al.*, 1974). These studies lead to the conclusion that the hydrogen-bonded seven-membered ring, found by our computations to be the most stable arrangement, exists in fact in phosphatidylethanolamine. An electrostatic interaction between the positively charged terminal trimethylamino group and the negatively charged phosphate oxygens is suggested also by these studies for the related phosphatidylcholine. It may be considered that the molecule in organic solvents is the closest to its free state and confirms the intrinsic preference of the polar groups of phospholipids for a highly folded arrangement.

The polar head of the phospholipids is however immersed in water in biological systems, and it is, therefore, particularly important to consider the possible influence of hydration upon the preceding results.

In the polar heads of phospholipids there are obviously two principal hydration regions: the cationic $-N^+R_3$ end and the phosphate group. Information about the preferred hydration sites of these two groupings has been obtained from studies on the hydration of ammonium and alkylammonia (Port and Pullman, 1973; Pullman and Brochen, 1975) and of the anion of dimethyl phosphate (A. Pullman *et al.*, 1975; B. Pullman *et al.*, 1975c), respectively. It indicates the possibility of fixing three water molecules at the $-N^+H_3$ terminal group through strong hydrogen bonds along the N^+-H axes. The energy of binding of water molecules about the $-N^+(CH_3)_3$ terminal group is appreciably reduced with respect to that around the $-N^+H_3$ group, although the attraction still exceeds slightly that between water molecules. Up to five or even six water molecules may be fixed, strongly bound around the phosphate group, in its preferred *gg* conformation about the α_2 and α_3 torsion angles.

The experimental data on the hydration of the polar head of phospholipids is relatively abundant (Klose and Stelzner, 1974; Jendrasiak and Hasty, 1974a,b; Walter and Hayes, 1971; Chapman, 1972; Phillips *et al.*, 1972; Veksli *et al.*, 1969; Misiorowski and Wells, 1973; Henrikson, 1970). Although it does not lead to a unique scheme and does not fix precisely the preferred sites of hydration, it indicates *a number* for the "bound" water molecules which is comparable with that suggested by the theoretical studies. Depending on the experimental conditions and techniques utilized, the number of water molecules in the primary hydration shell (which is most strongly bound) varies at the polar head of phosphatidylcholine from two to six. Phosphatidylethanolamine has not been intensively studied because of the difficulty in hydrating it possibly because of intermolecular interactions.

Let us look at the results of the computations on hydrated PE and PC.

Conformation of Hydrated PE

Two conformational energy maps have been constructed for hydrated PE. The first (Fig. 8) corresponds to the overall fixation of six water molecules upon the skeleton of PE; three at the cationic head along the N^+–H bonds and three at the phosphate group in the planes $O_{13}PO_{14}$, $O_{11}PO_{13}$ and $O_{12}PO_{14}$) in bridged positions with respect to the oxygens. The results indicate that the conformationally allowed space has somewhat decreased with respect to that of free PE; the global energy minimum still corresponds to a *gauche* conformation with respect to α_5 (=60°) but it became somewhat more extended with respect to α_4 (=240°).

As shown by A. Pullman *et al.* (1975) and B. Pullman *et al.* (1975c), up to five or six water molecules may be strongly bound to the phosphate, and hence a second map was constructed for hydrated PE, involving five water molecules at the phosphate (the three previous ones plus two in the planes $O_{11}PO_{14}$ and $O_{12}PO_{13}$ with one of their hydrogens turned outside) and three on the cationic head, giving a total of eight water molecules. The results are presented in Fig. 9 which shows the conservation of the *gauche* conformation with respect to α_5 (=60°) and a further elongation of the structure with respect to α_4 now equal to 150°.

Conformation of Hydrated PC

Because the energy of binding of water molecules to a $-N^+(CH_3)_3$ cationic head is considerably reduced with respect to its value for N^+H_3, we did not consider the presence of "bound" water at the cationic end of PC and have limited our study to the successive fixation of three, four or five water molecules at the phosphate group alone.

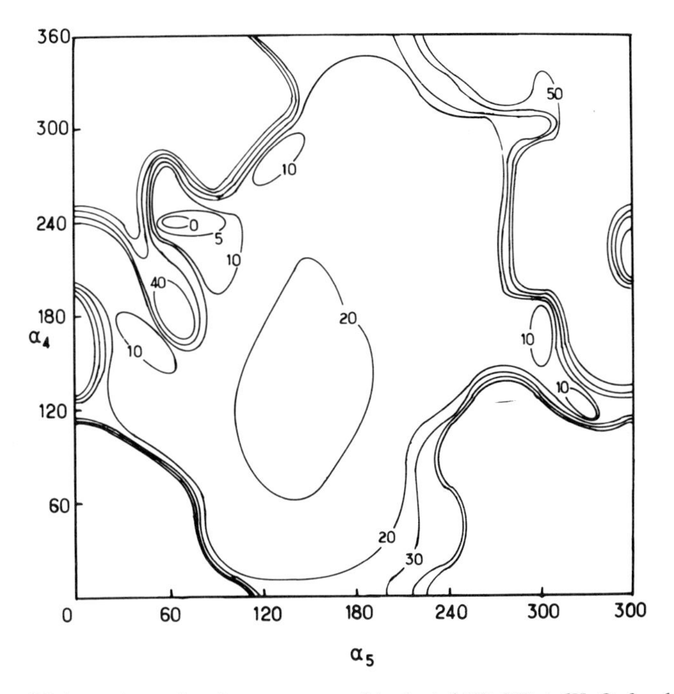

Fig. 8. PCILO conformational energy map of hydrated EP (EP + $6H_2O$; for details see text). Isoenergy curves in kcal/mole with respect to the global energy minimum taken as zero energy.

Figures 10 and 11 represent the results obtained for the tri- and penta-hydrate of PC, respectively (the results for the tetrahydrate are very similar to those for the trihydrate). The water molecules are fixed in the same positions as in the corresponding study on PE.

The results show the presence of a global energy minimum at $\alpha_5 = 300°$, $\alpha_4 = 120°$ in Fig. 10 and at $\alpha_5 = 300°$, $\alpha_4 = 180°$ in Fig. 11. Thus the preferred conformation of hydrated PC corresponds consistently to a *gauche* arrangement with respect to α_5 (as in the free molecule), and five water molecules are needed in the first hydration shell around the phosphate to produce an extension about α_4 to 180°.

All these results seem to indicate that the effect of water upon the conformation of the polar head of phospholipids consists essentially of extending the structure through the evolution of the torsion angle α_4 toward the value 180° (*trans*), leaving, nevertheless, the torsion angle α_5 at the *gauche* arrangement (60° or 300°). The *trans* conformation about α_4 occurs essentially for strong hydration of CP. In EP the conformation remains in between *gauche* and *trans* with respect to this torsion angle. It may be remarked that our results indicate

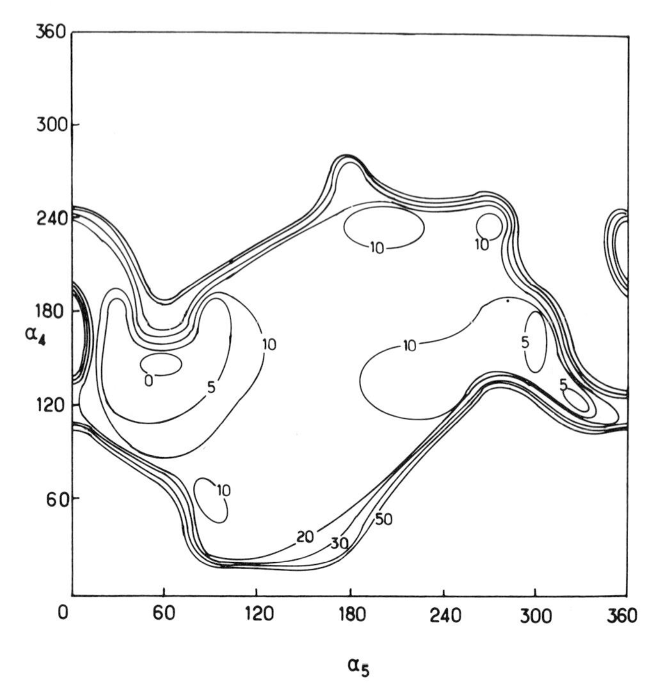

Fig. 9. PCILO conformational energy map of hydrated EP (EP + $8H_2O$; for details see text). Isoenergy curves in kcal/mole with respect to the global energy minimum taken as zero energy.

that no difference should be expected between EP and EC with respect to the conformation about the α_5 torsion angle, neither in the free nor in the solvated molecules. In this respect it is gratifying to record that contrary to previous statements (Andrieux *et al.*, 1972; Dufourcq and Lussan, 1972) more recent NMR studies (Richard *et al.*, 1974) confirm this analogy.

This evolution of α_4 and α_5 brings the results closer to the situation expected to be found in membranes ($\alpha_4 = 180°$, $\alpha_5 = 60$ or $300°$). It must, however, be clearly understood that these conformations are the result of environmental forces and not of an intrinsic tendency of the polar head.

In relation to the theoretical prediction that the value of α_4 of the solvated polar head of PE should correspond to an anticlinal rather than to an antiperiplanar arrangement, it may be interesting to draw attention to the recent result of the x-ray crystallographic study of 1,2-dilauroyl-DL-phosphatidyletholamine, which is, in fact, the first crystal study of a phospholipid (Hitchcock *et al.*, 1974). This compound exists in the crystal in a conformation corresponding to $\alpha_4 = 101°$, $\alpha_5 = 77°$. At first sight this result seems in striking agreement with the PCILO theoretical prediction. In fact the situation is more complex; α_2 and

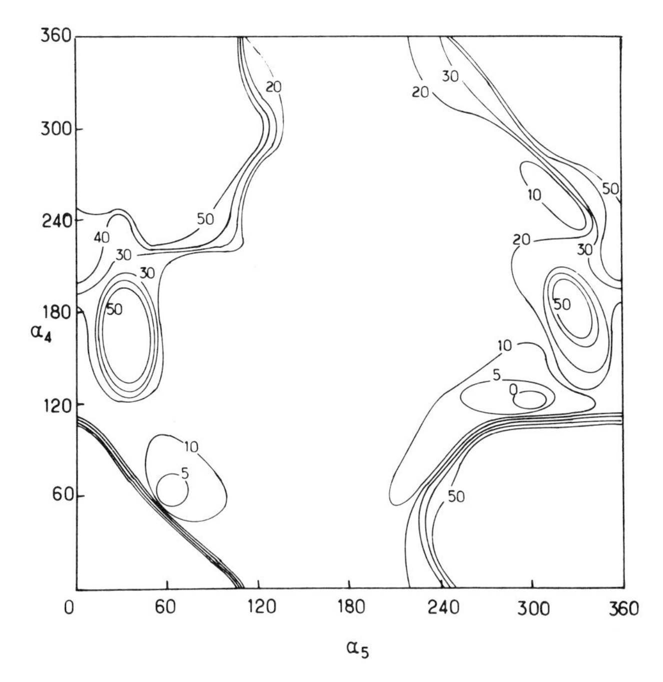

Fig. 10. PCILO conformational energy map of hydrated CP (CP + $3H_2O$; for details see text). Isoenergy curves in kcal/mole with respect to the global energy minimum taken as zero energy.

α_3 are close to 60° in this crystal and are therefore similar to that represented in Fig. 7. Under these conditions the value of α_4 is close to the expected one, but that of α_5 is "reversed" with respect to the expected one, which is ≅ 300°. This situation is summed up in Table II. The reason for this state of affairs is the existence in this crystal of an *intermolecular* hydrogen bond between the cationic head of one molecule and the negative oxygen of the phosphate group of the neighboring one, replacing the *intramolecular* hydrogen bond expected for the free molecule. Studies on the intermolecular interactions between phospholipid molecules leading to their overall organization in membranes are among the next steps in our program of research in this field.

CONCLUSIONS

This short, fragmentary account of the theoretical studies on the conformational properties of phospholipids, illustrates the great possibilities of the quantum-mechanical approach to this field of research, provided use is made of

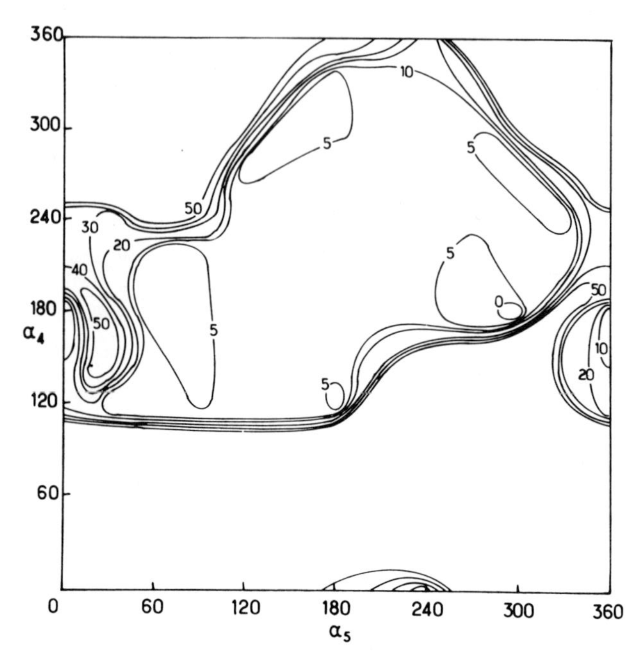

Fig. 11. PCILO conformational energy map of hydrated CP (CP + $5H_2O$; for details see text). Isoenergy curves in kcal/mole with respect to the global energy minimum taken as zero energy.

TABLE II
Preferred Conformations of the Polar Head of Phospholipids[a]

	α_1	α_2	α_3	α_4	α_5	α_6	Comments
Alister *et al.* (PPF)	180	60,180,300	60,180,300	180	60,180,300	180	*b,c*
Gupta and Govil (EHT)	180	60,180,300	60,180,300	180	60,180,300		
Pullman *et al.* (PCILO)	180	60,300	60,300	90,270	60,300	180	*d*
DEP *exp.* (Hitchock)	211	51	64	101	77		

a ⟷ Indicates preferred combinations.

*b*The computations of these authors indicate the value of 180° among the preferred ones for α_2, α_3 and α_5. The authors eliminate them from their final considerations only on empirical grounds.

*c*The preferential association of $(\alpha_2, \alpha_3) = (60°, 60°)$ and $(300°, 300°)$ does not appear either in the computations of these authors and is only adopted by them on empirical grounds again.

*d*The correlations indicated are all put into evidence by the computations. The preferred association of $(\alpha_2, \alpha_3) = (60°, 60°)$ correlates with $(\alpha_4, \alpha_5) = (90°, 300°)$, while the preferred association of $(\alpha_2, \alpha_3) = (300°, 300°)$ correlates with $(\alpha_4, \alpha_5) = (270°, 60°)$.

sufficiently refined methods. (PCILO, *ab initio*). The inappropriateness of empirical procedures for phosphate-containing compounds is evident. The explicit inclusion of the solvent effect represents an important step forward in theoretical conformational analysis, and the usefulness of the "supermolecule" model is evident. In compounds for which experimental results in the solid state or in solution are difficult to obtain, theoretical procedures represent competitive or complementary tools of investigation. The construction of conformational energy maps furnishes a much wider vision of the conformational possibilities of compounds than do other procedures. The distinction between the intrinsic conformational preferences of the free molecules and their behavior in given surroundings provides evidence for the role played by environmental factors. Theoretical investigations of molecular conformations have thus reached a stage of refinement in which they contribute efficiently to the exploration of submolecular biology.

REFERENCES

Abrahamsson, S., and Pascher, I. (1966). *Acta Crystallogr.* **11**, 79.

Andrieux, J., Dufourcq, J., and Lussan, C. (1972). *C. R. Hebd. Seances Acad. Sci.* **274**, 2358.

Bridsall, N. J. M., Feeney, J., Lee, A. G., Levine, Y. K., and Metcalfe, J. C. (1972). *J. Chem. Soc. Perkin Trans. 2* p. 1441.

Chapman, D. (1972). *Ann. N.Y. Acad. Sci.* **195**, 179.

Culvenor, C. C. J., and Ham, N. S. (1974). *Aust. J. Chem.* **27**, 2191.

DeTitta, G. T., and Craven, B. M. (1973). *Acta Crystallogr., Sect. B* **29**, 1354.

Dufourcq, J., and Lussan, C. (1972). *FEBS Lett.* **27**, 35.

Gupta, S. P., and Govil, G. (1972). *FEBS Lett.* **27**, 68.

Henderson, T. O., Glonek, T., and Myers, T. C. (1974). *Biochemistry* **13**, 623.

Henrikson, K. (1970). *Biochim. Biophys. Acta* **203**, 228.

Hitchcock, P. B., Mason, R., Thomas, K. M., and Shipley, G. G. (1974). *Proc. Natl. Acad. Sci. U.S.A.* **71**, 3036.

Jendrasiak, G. L., and Hasty, J. H. (1974a). *Biochim. Biophys. Acta* **337**, 79.

Jendrasiak, G. L., and Hasty, J. H. (1974b). *Biochim. Biophys. Acta* **348**, 45.

Klose, G., and Stelzner, F. (1974). *Biochim. Biophys. Acta* **363**, 1.

Kraut, J. (1961). *Acta Crystallogr.* **14**, 1146.

McAlister, J. M., Yathindra, N., and Sundaralingam, M. (1973). *Biochemistry* **12**, 1189.

Misiorowski, R. L., and Wells, M. A. (1973). *Biochemistry* **12**, 967.

Nanda, R. K., and Govil, G. (1975). *Theor. Chim. Acta* **38**, 71.

Newton, M. D. (1973). *J. Am. Chem. Soc.* **95**, 256.

Olson, W. K., and Flory, P. J. (1972a). *Biopolymers* **11**, 25.

Olson, W. K., and Flory, P. J. (1972b). *Biopolymers* **11**, 57.

Perahia, D., Pullman, B., and Saran, A. (1974). *Biochim. Biophys. Acta* **340**, 299.

Perahia, D., Pullman, A., and Berthod, H. (1975). *Theor. Chim. Acta* **40**, 47.

Perricaudet, M., and Pullman, A. (1973). *FEBS Lett.* **34**, 222.

Phillips, M. C., Finer, E. G., and Hauser, H. (1972). *Biochim. Biophys. Acta* **290**, 397.

Port, G. N. J., and Pullman, A. (1973). *Theor. Chim. Act* **31**, 231.

Pullman, A., and Brochen, P. (1975). *Chem. Phys. Lett.* **34**, 7.
Pullman, A., and Pullman, B. (1975). *Q. Rev. Biophys.* **7**, 505.
Pullman, A., Berthod, H., and Gresh, N. (1975). *Chem. Phys. Lett.* **33**, 11.
Pullman, B. (1974). *In* "Molecular and Quantum Pharmacology" (E. D. Bergmann and B. Pullman, eds.), p. 9. Reidel Publ., Dordrecht, Netherlands.
Pullman, B. (1977). *Adv. Quantum Chem.* (in press).
Pullman, B., and Berthod, H. (1974). *FEBS Lett.* **44**, 266.
Pullman, B., and Pullman, A. (1974). *Adv. Protein Chem.* **28**, 347.
Pullman, B., and Saran, A. (1975). *Int. J. Quantum Chem., Quant. Biol. Symp.* **2**, 71.
Pullman, B., and Saran, A. (1976). *Prog. Nucleic Acid Res. Mol. Biol.* **78**, 215.
Pullman, B., Berthod, H., and Gresh, N. (1975a). *FEBS Lett.* **53**, 199.
Pullman, B., Gresh, N., and Berthod, H. (1975b). *Theor. Chim. Acta* **40**, 71.
Pullman, B., Pullman, A., Berthod, H., and Gresh, N. (1975c). *Theor. Chim. Acta* **40**, 93.
Richard, H., Dufourcq, J., and Lussan, C. (1974). *FEBS Lett.* **45**, 136.
Saran, A., and Govil, G. (1971). *J. Theor. Biol.* **33**, 407.
Sasisekharan, V., and Lakhsminarayanan, A. V. (1969). *Biopolymers* **8**, 505.
Sundaralingam, M. (1969). *Biopolymers* **7**, 821.
Sundaralingam, M. (1972). *Ann. N.Y. Acad. Sci.* **195**, 324.
Sundaralingam, M., and Jensen, L. H. (1965). *Science* **150**, 1035.
Tewari, R., Nanda, R. K., and Govil, G. (1974). *J. Theor. Biol.* **46**, 229.
Vanderkooi, G. (1973a). *Jerusalem Symp. Quantum Chem. Biochem.* **5**, 469.
Vanderkooi, G. (1973b). *Chem. Phys. Lipids* **11**, 148.
Veksli, Z., Salsburg, N. J., and Chapman, D. (1969). *Biochim. Biophys. Acta* **183**, 434.
Walter, W. V., and Hayes, R. G. (1971). *Biochim. Biophys. Acta* **249**, 528.
Yathindra, N., and Sundaralingam, M. (1974). *Proc. Natl. Acad. Sci. U.S.A.* **71**, 3325.

CELL GROWTH AND CANCER

19

Reticuloendothelial Cells in Matrix-Induced Fibroblast–Transformation Ossicles*

CHARLES B. HUGGINS and A. H. REDDI

The matrix-induced fibroblast–transformation ossicle is a microcosm of differentiation which is reminiscent of the succession of developmental events occurring in the embryo. During the first 3 weeks of its formation, the matrix-induced ossicle is the site of rapid and dramatic cellular transition states which change from day to day on a precisely regulated schedule.

Transplantation of demineralized collagenous bone matrix from rat to allogeneic recipients rapidly causes the *de novo* formation of cartilage and, later, of bone (Reddi and Huggins, 1972). The experimental ossicle is the locus of formation of newly created hemopoietic bone marrow (Reddi and Huggins, 1975) first observed on transplantation day 12. Topography has a profound influence on the distribution of erythropoietic marrow in adult mammalia and

*This work was supported by grants from the American Cancer Society, The Jane Coffin Childs Memorial Fund for Medical Research, Grant USPHS CA11703-07 awarded by the National Cancer Institute, DHEW, and National Foundation March of Dimes, Grant #1-426.

avia. The primary influence on the location of erythropoiesis is thermal (Huggins and Blocksom, 1936; Huggins and Norman, 1936); the warm central bones contain red marrow whereas the cool outlying bones are yellow with marrow filled with lipocytes and devoid of blood formation.

The present paper describes the appearance of reticuloendothelial cells in matrix-induced ossicles. It was found that there is a good correlation between phagocytosis of ^{12}C- and ^{14}C-colloidal carbon with hemopoiesis in the experimentally created ossicles.

The surgical introduction of a small quantity of acid-insoluble bone matrix in powder form into the subcutaneous space of a rat rapidly causes coagulation of plasma; the granules of matrix are engulfed in the clot to make a discrete transformation plaque. In the newly formed pearly mass there ensues a spectacular cascade of interconnected biological effects occurring by timetable. At an early stage the transformation plaque is invaded by cells. The first wave of invaders consists of polymorphonuclear leukocytes; these soon disappear completely. A second cell-invasion comprises fibroblasts; those which penetrate the depths of the plaque are transformed by their proximity to the bone matrix under relatively anaerobic conditions (Reddi and Huggins, 1973). Transformation results from apposition of living cell with nonliving matrix. Critical physical factors in the transformation process include geometry of the bone matrix (Reddi, 1974), its mineral content, the electric charge (Reddi and Huggins, 1974) on its surface and the availability of oxygen (Reddi and Huggins, 1973) to the fibroblasts.

MATERIALS AND METHODS

Diaphyseal bone matrix was prepared from adult rats (Reddi and Huggins, 1972, 1975) and assayed as described earlier. In brief, rats of both sexes of Long-Evans strain, age 28–35 days, were used. Under sterile conditions a 1 cm incision was made in the skin and a subcutaneous pocket prepared by blunt dissection. A weighed knife-point-full of the demineralized matrix (10–20 mg) was transferred as a compact deposit on the floor of the surgically prepared space with the aid of a stainless steel spatula. A metallic skin clip was employed to close the incision. There were four sites in each rat; bilaterally symmetric transplants were made in upper thoracic and lower abdominal regions (Reddi and Huggins, 1975). The day of transplantation was designated as day 0.

On the designated days of harvest the rats were anesthetized and exsanguinated by cardiac puncture with a 19-gauge hypodermic needle followed by decapitation. Tissues for histological study were fixed in Bouin's fluid; paraffin sections were stained with hematoxylin–eosin.

Colloidal carbon was used for the identification and quantitation of reticuloendothelial cells in the matrix-induced bone marrow. The use of colloidal carbon in localization of phagocytic cells is well known (Vernon-Roberts, 1972). Colloidal carbon (Pelikan C11/1431A) was diluted to 25% (v/v) by physiological saline, 0.15 *M* NaCl. The carbon suspension* was administered intravenously via a caudal vein and the rats sacrificed 24 hr later. The entire transformation ossicle was fixed in neutral formalin and the distribution of carbon particles visualized by clearing the specimen by Spalteholz technique (Spalteholz, 1911; Huggins and Noonan, 1936). For histological identification of carbon the ossicles were fixed in Bouin's fixative, and paraffin sections were stained with hematoxylin-eosin.

^{14}C-Colloidal Carbon Assay

In order to quantitate the distribution of colloidal carbon in tissues a radiochemical assay was developed. ^{14}C-amorphous carbon (1 mCi = 1 mg) was purchased from New England Nuclear Company and was suspended in 5 ml of 25% (v/v) Pelikan colloidal carbon and equilibrated by constant stirring with a magnetic stir bar. The ^{14}C-colloidal carbon was centrifuged in a clinical centrifuge at 2000 rpm for 5 min. Depending on the particular batch of ^{14}C-carbon, 30–50% of radioactivity sedimented. The supernatant radioactive colloidal carbon was injected at a dose of 1 μCi/gm body weight; the rats were autopsied 24 hr later. Tissues for ^{14}C-colloidal carbon analysis were homogenized in 2 ml of 10% (w/v) trichloroacetic acid in a Polytron homogenizer for three 10-sec bursts at full setting. The volume was made up to 5.0 ml and kept ice-cold for 30 min. The tubes were centrifuged and the resulting precipitate was washed twice with 5 ml lots of 10% (w/v) trichloroacetic acid. The precipitates were solubilized in 88% (v/v) formic acid at 90° for 10 min. The formic acid suspension was counted in a liquid scintillation counter using a *p*-dioxane-based scintillation fluid (Reddi and Huggins, 1972). The results are expressed as cpm/mg tissue.

Alkaline Phosphatase Assay

Portions of harvested tissues were homogenized in ice-cold 0.15 *M* NaCl containing 3 m*M* $NaHCO_3$ in a Polytron homogenizer for three 10-sec bursts. Homogenates were centrifuged at 12,000 *g* for 15 min at 2°C. The enzyme assays were performed on supernatant solutions. Alkaline phosphatase [EC

*The carbon suspension (Pelikan C11/1431A, Günther Wagner, Hanover, Germany) contained 10% carbon black, 0.9% phenol and 4.3% fish glue. The particle diameter in histological sections was in the range 220–380 Å; mean diameter 280 Å (Bankston and De Bruyn, 1974).

3.1.3.1.; orthophosphoric-monoester phosphohydrolase (alkaline optimum)] was determined at pH 9.3; one unit of enzyme activity was defined as the activity that liberated 1 μmol of *p*-nitrophenol in 0.5 hr at 37°C (Huggins and Morii, 1961).

RESULTS AND DISCUSSION

^{14}C-Colloidal Carbon Assay

The method described herein for quantitation of phagocytosis by ^{14}C-colloidal carbon is useful. The reproducibility of the method was checked by determining the ^{14}C-carbon in eight samples of liver from an 8-week-old rat. Expressed as cpm/mg wet weight the mean and standard errors were: 4304±193. In other experiments the standard error ranged 4–8% of mean values.

Tissue Distribution

The pattern of ^{14}C-colloidal carbon in various tissues was determined in 56-day-old male rats. As depicted in Table I high values were found in liver, spleen and bone marrow. The general distribution correlated well with previous findings employing other methods (Vernon-Roberts, 1972). It is noteworthy that the celiac lymph nodes were more active in phagocytosis than iliac lymph nodes in agreement with observations made earlier using titanium dioxide

TABLE I
Phagocytosis of ^{14}C-Colloidal Carbon in Tissues of Rat

Tissue	^{14}C-Colloidal carbon (cpm/mg tissue)
Liver	4304
Spleen	3808
Lymph node, celiac	4409
Lymph node, iliac	405
Bone marrow, femur	1532
Adrenal	1065
Kidney	1083
Lung	897
Muscle, quadriceps	35
Testis	33
Thymus	26
Thyroid	1
Pituitary	30
Brain	3

(Huggins and Froelich, 1966). It is clear that ^{14}C-colloidal carbon is a valid method of assessing phagocytosis by reticuloendothelial cells.

Transformation Ossicles

The distribution of colloidal carbon in transformation ossicles is shown in Fig. 1. The carbon particles were exclusively localized in the core of the ossicles; the rim was clear. The clear-cut demarcation is noteworthy. In histological sections carbon was localized in cells lining the sinusoids in the bone marrow of the transformation ossicles from day 12 onwards.

The site of transplantation had a profound effect on activity of reticuloendothelial cells of the transformation ossicle. The demineralized matrix was transplanted in 8 male rats. The resultant transformation ossicles were harvested on days 23–28. The ^{14}C-colloidal carbon was administered a day prior to harvest. Thoracic ossicles were black in the gross; the abdominal ones were light gray in color. The ^{14}C-carbon values were significantly higher in the thoracic ossicles as seen in Table II. In other experiments similar results were obtained in the female

Fig. 1. Phagocytosis of colloidal carbon in reticuloendothelial cells in the core of a fibroblast–transformation ossicle; the rim of the ossicle is devoid of phagocytes. The sharp line of demarcation between core and rim is noteworthy. Spalteholz method. ×10.

TABLE II
Alkaline Phosphatase and Phagocytosis of ^{14}C-Colloidal Carbon in Transformation Ossicles

Site	Alkaline phosphatase (units/gm)	^{14}C-Colloidal carbon (cpm/mg)
Thoracic ossicle	26.7 ± 2.6[a]	869 ± 95
Abdominal ossicle	22.1 ± 3.0	309 ± 61[b]

[a]±, standard error of mean, n = 8–16 for all determinations.
[b]Significant difference from thoracic ossicle; $P < 0.01$.

rats. These observations are reminiscent of the differences in erythropoiesis between thoracic and abdominal sites (2). It would appear that there is a good correlation between phagocytic activity and hemopoiesis in general.

SUMMARY

A novel assay, based on the distribution of ^{14}C-colloidal carbon, has been described for quantitation of reticuloendothelial cell phagocytic activity. The appearance of reticuloendothelial cells in matrix-induced transformation ossicles was studied: these were evident on transplantation day 12. The ossicles in the thoracic region exhibited higher values for phagocytosis of colloidal carbon than did ossicles in lower abdominal region.

REFERENCES

Bankston, P. W., and De Bruyn, P. P. H. (1974). *Am. J. Anat.* **141**, 281–289.
Huggins, C. B., and Blocksom, B. H., Jr. (1936). *J. Exp. Med.* **64**, 253–274.
Huggins, C. B., and Froelich, J. P. (1966). *J. Exp. Med.* **124**, 1099–1106.
Huggins, C. B., and Morii, S. (1961). *J. Exp. Med.* **114**, 741–760.
Huggins, C. B., and Noonan, W. J. (1936). *J. Exp. Med.* **64**, 275–280.
Reddi, A. H. (1974). *Adv. Biol. Med. Phys.* **15**, 1–18.
Reddi, A. H., and Huggins, C. B. (1972). *Proc. Natl. Acad. Sci. U.S.A.* **69**, 1601–1605.
Reddi, A. H., and Huggins, C. B. (1973). *Proc. Soc. Exp. Biol. Med.* **143**, 634–637.
Reddi, A. H., and Huggins, C. B. (1974). *Proc. Natl. Acad. Sci. U.S.A.* **71**, 1648–1652.
Reddi, A. H., and Huggins, C. B. (1975). *Proc. Natal. Acad. Sci. U.S.A.* **72**, 2212–2216.
Spalteholz, K. W. (1911). "Ueber das Durchsichtmachen von menschlichen und tierischen Präparaten." Hirzel, Leipzig.
Vernon-Roberts, B. (1972). "The Macrophage." Cambridge Univ. Press, London and New York.

20

Biochemical Strategy of Gene Expression*

GEORGE WEBER

The mammalian genome is a self-instructive, replicating entity that contains information to direct and integrate macromolecular synthesis. My proposal that the strategy of the genome can be elucidated by exploring the pattern of gene expression as it is revealed in the activity, concentration and isozyme spectrum of certain enzymes is the thesis of this paper.

The enzymes through which the pattern of gene strategy is expressed and controlled I called the *key enzymes* (Weber, 1975a; Weber *et al.,* 1965). I propose, more specifically, that the strategy of gene expression is carried out, in a large part at least, through a genic program that integrates the pattern of reciprocal behavior of opposing key enzymes in antagonistic synthetic and catabolic pathways (Weber, 1975a, 1978).

Strategy is a military term that is defined as "maneuvering forces into the most advantageous position" (Webster's New World Dictionary of the American Language, 1962). The biochemical strategy of the genome must have such an ability to have survived the selective pressures of evolution and to have conferred selective advantages to the organism.

*This work was supported by United States Public Health, NCI Grants CA-05034 and CA-13526.

To reveal the quantitative and qualitative pattern in the strategy of gene expression there are two prerequisites which have to be met, at both the conceptual and the experimental level. These prerequisites entail (a) the identification of the key enzymes in various metabolic pathways and (b) the availability of biological models in which the alterations in gene expression can be conveniently and precisely analyzed.

CONCEPTUAL BACKGROUND

The Concept of Key Enzymes as Targets in Expression of the Biochemical Strategy of the Genome

On the basis of experimental evidence gained in systematic studies of endocrine regulation of carbohydrate metabolism it was recognized in this laboratory that in the modulation of gene expression the rate and direction of opposing metabolic pathways are determined through control of a few enzymes (Weber *et al.,* 1965). Thus, the regulatory mechanisms employ great economy in achieving regulation of opposing metabolic sequences. To test the concept that certain enzymes may be key to the regulation of opposing pathways of synthesis and catabolism, systematic studies were carried out in which enzymes predicted to play key roles were examined for their behavior under various conditions involving modulation of gene expression. The pattern of behavior of key enzymes was contrasted with that of other enzymes which were present in great excess and the regulation of which appeared to be less vital for the integration of the genetic programs. Subsequently, our investigations were also extended to testing the applicability of the key enzyme concept to enzymes of ornithine, cAMP, purine, pyrimidine and DNA pathways. As a result of such conceptual considerations and the experimental results, the identifying features and examples of certain key enzymes are summarized in Table I.

Biological Models for Analyzing the Biochemical Strategy of Gene Expression

The testing of the proposal that the strategy of gene expression is manifested in the reciprocal behavior pattern of opposing key enzymes in antagonistic synthetic and degradative metabolic pathways requires biological models in which the quantitative and qualitative aspects of gene expression can be modulated in a reversible and, in certain cases, in an irreversible fashion (Weber, 1977; Weber *et al.,* 1972a, 1977). Such conditions are provided by investigations of development (Weber *et al.,* 1971), endocrine regulation (Weber, 1972; Weber *et al.,* 1965), regeneration (Weber, 1975b), and neoplasia (Weber, 1974a; 1977).

TABLE I
Identification of Key Enzymes

Features of key enzymes	Examples in various metabolic areas
I. Place in pathway	
1. First in a reaction sequence	Pyruvate carboxylase; thymidine kinase
2. Last in a reaction sequence	Glucose-6-phosphatase; DNA polymerase
3. Pathways in themselves	Glucose-6-phosphatase; adenylate cyclase
4. Operate on both sides of reversible reaction pools	Thymidine kinase; dihydrothymine dehydrogenase
5. One-way enzyme opposed by another one-way enzyme	Phosphofructokinase; fructose diphosphatase
II. Regulatory properties	
6. Possesses relatively low activity in the pathway	Pyruvate carboxylase; DNA polymerase
7. Rate-limiting in pathway	PEP carboxykinase; ribonucleotide reductase
8. Target of feedback regulation	Thymidine kinase; glutamine PRPP amidotransferase
9. Target of multiple regulation	Pyruvate kinase; ribonucleotide reductase
10. Exhibits allosteric properties	Aspartate carbamoyltransferase
11. An interconvertible enzyme	Glycogen phosphorylase
12. Has isozyme pattern	Glucokinase–hexokinase; pyruvate kinases
III. Biological role	
13. Involved in overcoming thermodynamic barriers	Four key gluconeogenic enzymes
14. Final common path of two or more metabolic pathways	DNA polymerase; glucose-6-phosphatase
IV. Intracellular compartmentation	
15. Localized in particulate fraction when other enzymes of same pathway are in cytoplasm	Glucose-6-phosphatase

1. In *development* the program of gene expression is unfolded in a sequential manner, displaying growth and differentiation.

2. In *endocrine regulation* gene expression is modulated through acute (rapid) and chronic (slow) adaptive adjustments.

3. In *regenerating liver* the cells are triggered into expressing their potential replicative program in a self-limited fashion.

4. In *neoplasia,* using a spectrum of liver tumors of vastly different malignancy and growth rates, the expression of transformation-linked (all or none) and progression-linked (gradually emerging) enzymatic and metabolic imbalance is fixed in each separate line of transplantable neoplasms.

The Signal Systems: Tissue-Specific and Shared Signals

The strategy of gene expression depends on (a) cell- and tissue-specific, and (b) program-specific signals. The cell- and tissue-specificity is readily detectable by examining the pattern of behavior of key enzymes in differentiating organs, in hormonal targets and also in tissues that can display regenerative replication.

In the expression of neoplastic transformation and progression both cell specificity and program sharing have been documented. In cancer cells the cell- and tissue-specificity is indicated by the differentiated functions that may be retained, lost, increased or derepressed (Weber, 1974b). A more important and apparently more closely linked property of neoplastic transformation is characterized by the ubiquitous presence of alterations in the control of commitment to replication and in the regulation of replicative properties. This is a shared program that all tumor cells express in different degrees.

To carry out meaningful studies in analyzing the pattern of genetic logic in neoplasia, it is particularly important to meet the stringent prerequisites for the elucidation of such a pattern in cancer cells (Weber, 1974a, 1977, 1978). For this reason and because of the special interest of Professor Szent-Györgyi (1968) in cancer, I provide more details of our approach.

Special Features and Prerequisites for Studies in Neoplasia

There are many different types of neoplastic cells available in a wide variety of animals, primary or transplantable tumors and tissue culture systems with an enormous range of biological properties. However, from the beginning of my investigations, I considered it crucial to focus on biologically meaningful model systems where valid comparisons can be made with homologous normal tissues. It was also essential to me that the model system chosen should provide gradations in the expression of the degrees of biological malignancy that would allow the detection of the quantitative and qualitative biochemical aspects of neoplastic transformation and progression.

Thus, if one postulates that the biochemical strategy of the cancer cell may be identified by examination of the behavior of antagonistic key enzymes, one could make specific predictions that can be readily tested in a suitable biological model system which permits pattern recognition. Our introduction into this field of research of the rigorous approaches of testing predictions and falsifiability is also consonant with ideas expressed in other realms of investigation by Medawar (1970) and Popper (1968).

The Molecular Correlation Concept of Neoplasia: The Biological Model Systems

To gain insight into the strategy of gene expression I needed a model system that would provide certain well-defined biological properties which allowed the identification of clear-cut, ordered biochemical alterations. To focus the initial scope on one cell type and to assure the availability of homologous control tissues, from the many available tumors I selected a number of chemically produced, transplantable rat hepatocellular carcinomas, thus insuring the histological homogeneity of the model system (Weber, 1961, 1974a, 1977). I chose to examine those liver tumor lines that I could arrange into a spectrum of hepatomas that displayed vastly different degrees of malignancy as measured by the growth rate of the different transplantable tumor cell lines (Weber, 1961, 1974a, 1977).

This spectrum of hepatomas of different growth rates provided the model system in which the biochemical pattern of gene expression could be analyzed. Then with the guidance of a genetic-biochemical theory, predictions could be made and tested. This theory is the molecular correlation concept of malignancy (Weber, 1974a, 1977).

The Molecular Correlation Concept of Malignancy: The Conceptual and Experimental Approach

The objective of this concept is to discover the biochemical strategy of gene expression in neoplasia. This is accomplished through (1) the identification and elucidation of the behavior of key enzymes and isozymes in opposing synthetic and catabolic pathways, and (2) the elucidation of the linking of the biochemical alterations to neoplasia by establishing the relationship of the metabolic alterations to neoplastic transformation and progression as revealed in the degrees of expression of malignancy in the tumor spectrum.

This conceptual and experimental approach made the following predictions. Assuming that (a) the enzymes measured were key and (b) the enzyme assays were carefully adapted to the kinetic requirements of the liver and hepatoma systems, then the enzyme activity changes should signal alterations in the enzyme amount. This also can be verified by immunotitration that determines the enzyme concentrations independently from the assay of enzymatic activity. Thus, from determining the end product of the expression of the gene program, i.e., the concentration of catalytically active specific enzyme proteins, the strategy of gene expression can be pinpointed and the alterations measured. On the basis of what we knew of the clinical and biological behavior of neoplasia, three main classes for biochemical-biological relationships were expected.

Biological Strategy of the Cancer Cell

From what we understand of the *biological* strategy of the cancer cell, neoplasia is expressed by a transformation that is hereditable and characterizes all tumor cells. Thus, there is a shared program in all cancer cells, the *malignant transformation.*

Furthermore, the biological strategy of the cancer cell can be expressed in different degrees, revealing the manifestations of the malignant properties that range from mild through advanced to the full-blown status of neoplastic gene expression; this is termed *progression.*

Biochemical Strategy of the Cancer Cell. Pattern of Biological-Biochemical Linking: Relationship with Transformation and Progression

If the biological strategy of the cancer cell expresses itself in a biochemical strategy, the following main relationships are expected (Table II).

Progression-Linked Alterations

The theory predicts that there should be a pattern of key enzymes, the amounts of which should be altered in a way that parallels the degrees in the expression of tumor malignancy. Thus, the concentration of certain key enzymes of biosynthesis should increase and the opposing catabolic ones should decrease in parallel with the malignancy and growth rate of neoplasms. In consequence, the metabolic imbalance should become more and more pronounced with the progressive degrees in the expression of malignancy.

Transformation-Linked Alterations

Since there is a shared program in all transformed cells, and since all tumor cells are transformed, there should be a pattern of key enzymes that exhibits the same type of alterations in all tumors within the spectrum of neoplasms of different growth rates. Thus, the activities of certain key enzymes would be

TABLE II
Biochemical Strategy of the Cancer Cell: Pleiotropic Alterations Manifested in the Reprogramming of Gene Expression.

Indicators of reprogramming of gene expression: activities of enzymes, isozymes, pathways	Linking of biochemical alterations with malignancy
Activities that *correlate* with tumor malignancy: indicate reprogramming of gene expression that is linked with the different degrees of expression of the neoplastic transformation	Class 1: Progression-linked discriminants[a]
Activities that are increased or decreased *in all hepatomas*: indicate reprogramming of gene expression that is linked with the malignant transformation per se	Class 2: Transformation-linked discriminants[a]
Activities that do not relate to growth rate: random, coincidental alterations not connected with the neoplastic transformation	Class 3: Coincidental, nonessential alterations

[a]Biochemical parameters that discriminate the pattern of neoplasia from that of normal, fetal, differentiating or regenerating liver.

increased in all tumors, and those of other enzymes, that are functionally opposed to these, would be decreased in all neoplasms.

Coincidental Changes

The theory anticipated and accounted for any apparent *diversity* in the biochemical pattern of tumors. The theory predicts that an *ordered* pattern would be identified in the relationship of key enzymes that are stringently linked with the biological behavior of malignancy and growth rate. The randomness and diversity that may emerge for some biochemical parameters would be due to enzymes that are present in an excess, as the regulation of their concentration may not be stringently linked with the core of neoplastic transformation.

Predictions and Testing

Therefore, the theory predicted that the key enzymes in opposing metabolic pathways in cancer cells would be altered in concentration and activity and that this pattern of alteration would be stringently linked either with transformation or with progression or with both. The key enzymes in their behavior should reveal the ordered and neoplasia-linked reprogramming of gene expression. In contrast, the enzymes that are less stringently regulated because they are not key in the regulation of rate and direction of opposing pathways should not be closely linked with the essence of neoplasia and would be the ones that, in a random fashion, might alter or remain in normal range.

Thus what is important for the biochemical strategy of the cancer cell should be ordered and what is not might appear as the randomness and the diversity.

In what follows I describe briefly the main experimental observations that support the formulation of our present concepts of the biochemical strategy of gene expression in normal, differentiating, regenerating and neoplastic cells.

MATERIALS AND METHODS

The animals and tumor systems (Weber, 1974a), the techniques for analyzing endocrine regulation (Weber, 1972, 1975a; Weber *et al.,* 1965), differentiation (Weber, 1972) and regeneration (Weber, 1975b) were described in detail elsewhere. The techniques for measuring the activities of the various opposing metabolic pathways and enzymes were reported (Ferdinandus *et al.,* 1971; Weber, 1974a; Weber *et al.,* 1972a). The methods for expressing enzymatic activities and the evaluation and statistical analysis of the correlations and the biochemical alterations were cited elsewhere (Weber, 1974a, 1976, 1977, 1978). The conceptual and experimental methods employed in my laboratory were also discussed in a series of original papers and reviews (Weber, 1961, 1972, 1973, 1974a,b, 1975a,b, 1976; Weber *et al.,* 1965, 1971, 1972b, 1974, 1976a,b, 1977).

RESULTS AND DISCUSSION

Carbohydrate Metabolism: Biochemical Strategy of Gene Expression

For the discoveries made in this laboratory in carbohydrate metabolism, it would be fitting to cite one of the many famous incisive statements of Professor Szent-Györgyi: "Discovery is to see what everyone has seen, but think what no one has thought." The reason this seems applicable is because the metabolites, the reaction steps and the enzymes involved in glycolysis and what appeared to be its ready reversal, gluconeogenesis, were described in much detail in the 1940's and 1950's. Yet there was no insight into the behavior of the metabolic pathways and the enzymes under the conditions of metabolic regulation which we now know involve reprogramming of gene expression.

A breakthrough in the conceptual approach to the mechanisms of gluconeogenesis occurred when Krebs (1954) recognized in his famous Johns Hopkins lectures that gluconeogenesis, because of thermodynamic barriers, cannot be simply a reversal of glycolysis. He pointed out that these barriers are overcome by the insertion of enzymes, such as two specific phosphatases, glucose-6-phosphatase and fructose-1,6-diphosphatase, and at the early stage of this process, pyruvate carboxylase and phosphoenolpyruvate carboxykinase. Our work at that time concerned the behavior of carbohydrate metabolism in normal and neoplastic conditions. We recognized that the behavior of the various enzymes in carbohydrate metabolism may be better understood by assuming that the logic and economy of gene expression would entail a coordinated regulation of two sets of antagonistic enzymes, namely the four enzymes particularly relevant for gluconeogenesis and the triad of enzymes of glycolysis that opposed them, glucokinase, phosphofructokinase and pyruvate kinase. Since from our studies it appeared that the regulation of the concentration and activity of these opposing enzymes may be key to metabolic regulation and gene expression, I termed such enzymes *key enzymes* (Weber, 1975a; Weber *et al.*, 1965). Thus it was predicted that the regulation of gene expression relevant to control of rate and direction of glycolysis and gluconeogenesis in liver would be achieved through modulation of the concentration and activity of the four key gluconeogenic enzymes and the three key glycolytic enzymes. The properties of these seven enzymes were characteristic in that they governed one-way reactions. In turn, the remaining enzymes in these pathways catalyzed reversible reactions that were shared by the processes of both glycolysis and gluconeogenesis. Since these enzymes could function in both directions, I termed them bifunctional enzymes, (e.g., phosphohexoseisomerase, aldolase, lactate dehydrogenase). In contrast with the key gluconeogenic and glycolytic enzymes, the bifunctional enzymes were present in great excess, and it appeared that they were not likely to become rate-limiting under the various regulatory conditions (Weber, 1975a; Weber *et al.*, 1965).

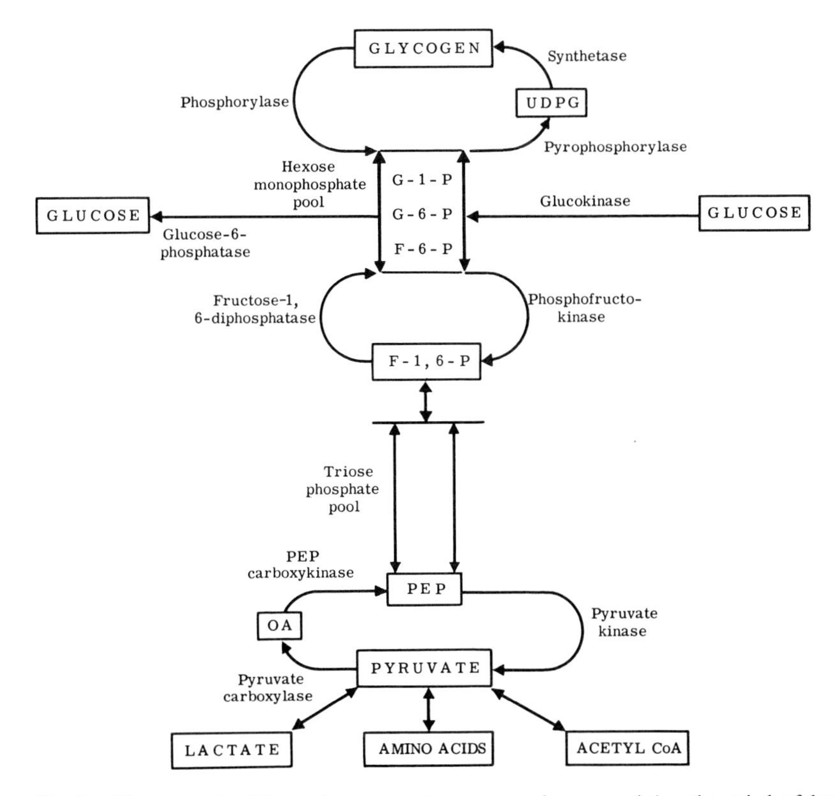

Fig. 1. The quartet of key gluconeogenic enzymes is opposed by the triad of key glycolytic enzymes. These opposing key enzymes govern one-way reactions on both sides of metabolic pools of reversible reactions. From Weber *et al.* (1965).

The advantage of this concept was that it provided predictions for the reciprocal behavior of the opposing key enzymes that can be readily tested, verified or disproved.

Now a brief outline will be given that indicates that the strategy of gene expression in carbohydrate metabolism in differentiation, hormonal control and neoplasia indeed is accomplished through control of the concentration and activity of opposing key enzymes.

For the development of our concepts on carbohydrate metabolism in the liver and the role of key enzymes, the reader is referred to the original publications (Weber, 1975a,b; Weber *et al.*, 1965). The current view is pictured in Fig. 1.

Differentiation

The behavior of the four key enzymes of gluconeogenesis and the three key enzymes of glycolysis is compared in Fig. 2. The activities of the key gluconeogenic enzymes are low in fetal rat liver and they rise sharply in the immediate

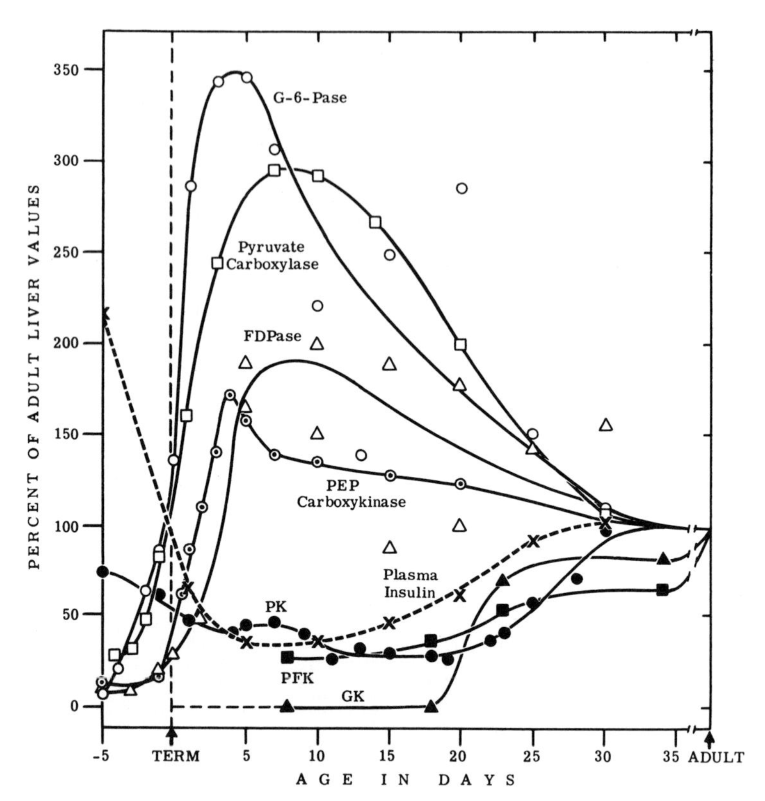

Fig. 2. Developmental pattern of the opposing key glycolytic and gluconeogenic enzymes in rat liver. From Weber *et al.* (1971).

perinatal period and subsequently slowly decline during development to the activity observed in the liver of the adult animal. This pattern contrasts with that of the key glycolytic enzymes which rise slowly in postnatal differentiation to the activities of the adult liver. The behavior of the opposing key enzyme activities gives roughly a mirror picture, but that of the bifunctional enzymes such as phosphohexoseisomerase or lactate dehydrogenase shows no such pattern. Thus, the biochemical strategy of differentiation is expressed by the reciprocal behavior of opposing key enzymes in antagonistic pathways of glycolysis and gluconeogenesis (Weber *et al.*, 1971; Weber, 1972).

Endocrine Regulation

I pointed out earlier that questions of hormonal regulation of intermediary metabolism may be translated at the molecular level to inquiries that formulate the problem in a way that makes it solvable. Thus, the question as to how insulin increases hepatic glycolysis and suppresses gluconeogenesis can be expressed:

how does insulin achieve the predominance of the activities of the key glycolytic enzymes over those of the opposing gluconeogenic ones? Our studies indicated that insulin achieves its integrative action on intermediary metabolism through determining the amount, isozyme pattern and activity of opposing key enzymes of gluconeogenesis and glycolysis (Dunaway and Weber, 1974; Weber, 1972, 1975a; Weber *et al.,* 1965). Two examples of the experimental evidence in support of this conclusion will now be given.

As Fig. 2 demonstrates, the behavior of opposing key enzymes in differentiation follows a reciprocal pattern and then the forces of gene expression appear to settle in a homeostatic balance. However, the potential of the genome can be unleashed. In Fig. 3 is plotted the behavior of two key enzymes characteristic of the gluconeogenic and glycolytic groups of enzymes. Thus, the activity of glucose-6-phosphatase decreases and that of pyruvate kinase increases to the adult level. When adult rats are made diabetic by injection of alloxan, the activity of the gluconeogenic enzymes, as exemplified here by glucose-6-phosphatase, increases and the glycolytic enzymes, as represented here by pyruvate kinase, decreases. When the diabetic rats are treated with insulin, the activity of the phosphatase decreases to normal and that of pyruvate kinase increases to normal. The degradation of phosphatase is not affected by actinomycin, but the rise of pyruvate kinase is blocked by this inhibitor of DNA-di-

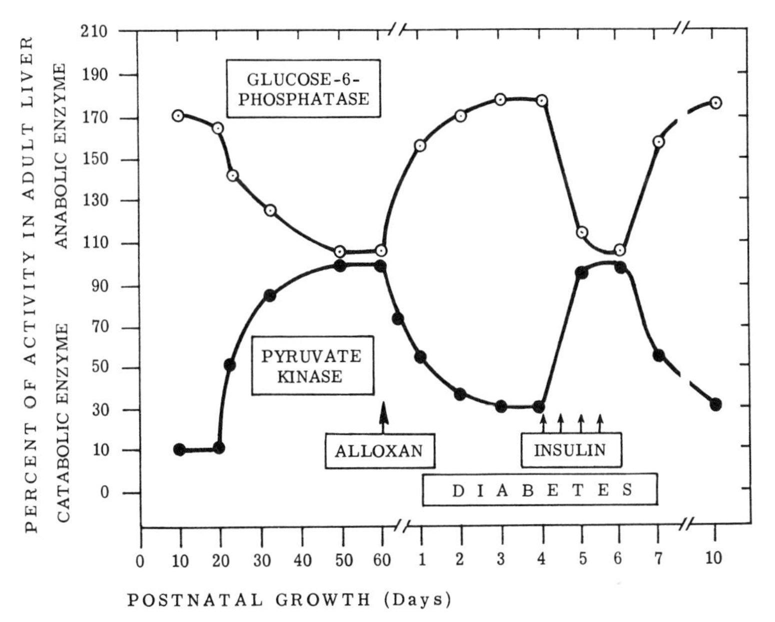

Fig. 3. Pattern of programming and reprogramming of gene expression in differentiation, diabetes and insulin treatment. From Weber (1972).

rected RNA synthesis. If insulin administration is stopped, the gluconeogenic enzyme increases and the glycolytic one decreases roughly to the postnatal levels. Thus, in hormonal modulation the biochemical strategy of the genome is accomplished through reciprocal adjustment of the concentrations of opposing key enzymes of antagonistic pathways (Weber, 1972, 1975a; Weber *et al.,* 1965, 1972a).

Immunological Evidence for Reprogramming of Gene Expression in Insulin Action. To gain insight into this mechanism we obtained extensive evidence by measuring the activity, and independently, the immunoprecipitable isozyme concentration of phosphofructokinase in diabetes and after insulin administration. The results shown in Fig. 4 demonstrate that the decrease of activity and concentration of phosphofructokinase in diabetes and the rise after insulin administration are due to reprogramming of gene expression as manifested in the altered concentrations of this key glycolytic enzyme (Dunaway and Weber, 1974).

Pleiotropic Action of Insulin Effects. Extensive evidence obtained in my laboratory and in other centers indicated to me that the reprogramming of gene expression, achieved by the integrative action of insulin, operates through determining the amount and activity of an array of enzymes involved in glycolysis, gluconeogenesis, pentose phosphate pathway, lipogenesis, thymidine metabolism and the urea cycle (Weber, 1972; Weber *et al.,* 1974). The evidence has been reviewed (Weber, 1972). The pleiotropic action of insulin is shown in Fig. 5 and the opposing action of this hormone on gene expression is summarized in Table III.

The term pleiotropy as it refers to reprogramming of gene expression through hormonal influences was discussed elsewhere (Weber *et al.,* 1974). The emergence of coordinated multienzyme alterations is relevant for the present discussion because it characterizes the biochemical strategy of the normal cell as it is expressed through regulation of key enzymes under endocrine conditions (Weber, 1972; Weber *et al.,* 1974).

These studies in differentiation and hormonal regulation were carried out in model systems where the pattern of alterations in the activity of opposing key enzymes has undergone reversible processes of modulation. In examining the biochemical strategy of the genome, now I will turn to a model system where the different degrees of alterations in gene expression are fixed and appear to be largely or completely irreversible.

Neoplasia

Carbohydrate and Pentose Phosphate Metabolism. The first testing ground of the molecular correlation concept of neoplasia was in studies on carbohydrate metabolism in hepatomas of different growth rates (Weber, 1961). The results

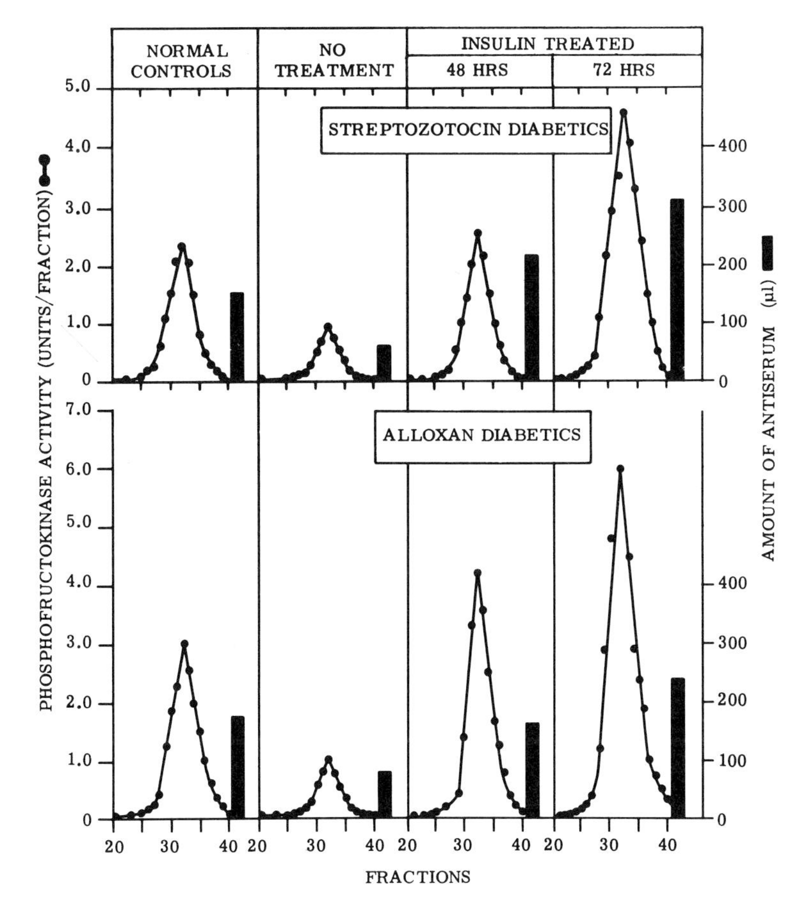

Fig. 4. The effect of diabetes and insulin administration on the major liver phosphofructokinase isozyme. A comparison of the behavior of enzyme concentration is shown as determined by measurement of enzyme activity and by titration of the immunoprecipitable enzyme concentration. This isozyme is 85% of the total phosphofructokinase activity. From Dunaway and Weber (1974).

provided support for the operation of a pattern anticipated by the molecular correlation concept (Weber, 1974a). The key enzymes of glycolysis increased and concurrently the key enzymes of gluconeogenesis decreased in parallel with the increase in tumor growth rate in the various hepatoma lines. Thus, these enzymes belong to Class 1 of the molecular correlation concept and their behavior is linked with the degrees in the expression of malignant properties.

In contrast, the two key enzymes leading to pentose phosphate biosynthesis, glucose-6-phosphate dehydrogenase and transaldolase, were increased in all the

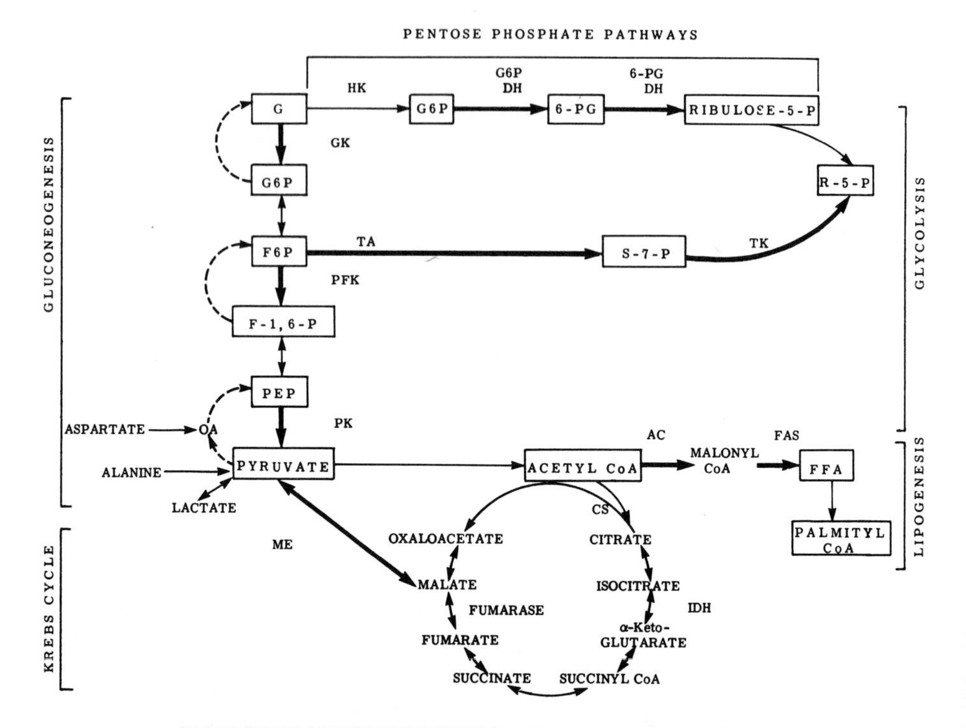

GK=GLUCOKINASE, PFK=PHOSPHOFRUCTOKINASE, PK=PYRUVATE KINASE, HK=HEXOKINASE, G6P DH=GLUCOSE-6-PHOSPHATE DEHYDROGENASE, 6-PG DH=6-PHOSPHOGLUCONATE DEHYDROGENASE, IDH=ISOCITRATE DEHYDROGENASE, ME=MALIC ENZYME, AC=ACETYL CoA CARBOXYLASE, FAS=FATTY ACID SYNTHETASE, CS=CITRATE SYNTHASE, TA=TRANSALDOLASE, TK=TRANSKETOLASE.

Fig. 5. The integrative action of insulin at the hepatic enzyme level. Insulin action reprograms gene expression, leading to a suppression of the biosynthesis of the four key gluconeogenic enzymes and the induction of biosynthesis of the key glycolytic enzymes, free fatty acid synthesizing enzymes and the enzymes of the oxidative and nonoxidative pathways of pentose phosphate biosynthesis. From Weber (1978).

hepatomas (Weber *et al.*, 1974, Heinrich *et al.*, 1976). The behavior of these two enzymes appears to be linked with the neoplastic transformation itself and their ubiquitous increase in all hepatomas groups them in Class 2.

The enzymes that were present in great excess, such as phosphohexoseisomerase, aldolase and lactate dehydrogenase, showed no relationship with growth rate; they belong to Class 3, the coincidental alterations.

Since the activities of the carbohydrate enzymes were proportionate to their concentration, the changes in enzyme activities in the tumors reflected alterations in enzyme concentration, indicating the emergence of an altered gene expression. In the case of phosphofructokinase (Dunaway *et al.*, 1974), the change in enzyme concentration was also confirmed by measuring the enzyme

TABLE III
Insulin: Integration of Hepatic Metabolism at the Enzyme Level

Functions increased	Enzymes induced
Glycogenesis	Glycogen synthetase
Glycolysis	Glucokinase
	Phosphofructokinase
	Pyruvate kinase
Lipogenesis	Citrate cleavage enzyme
	Acetyl-CoA carboxylase
	Fatty acid synthetase
NADPH production	Glucose-6-phosphate dehydrogenase
	Malate enzyme
Thymidine incorporation into DNA	DNA polymerase
Thymidine degradation to CO_2	Thymidine phosphorylase
Functions decreased	**Enzymes suppressed**
Gluconeogenesis	Glucose-6-phosphatase
	Fructose-1,6-diphosphatase
	Phosphoenolpyruvate carboxykinase
	Pyruvate carboxylase
Urea cycle	Ornithine carbamoyltransferase
	Arginine synthetase
	Argininosuccinase
Orotate incorporation into RNA	

protein by immunotitration, thus providing an independent corroboration of the conclusions.

It was also shown that there was an isozyme shift from glucokinase to hexokinase and from liver-type to muscle-type pyruvate kinase in the hepatomas (Fig. 6), resulting in the emergence of low K_m enzymes that largely or completely lost sensitivity to regulation of their amount and activity by nutritional, hormonal or feedback regulatory signals (Weber, 1974a). Thus, the biochemical strategy of the cancer cell entails both *quantitative and qualitative* reprogramming of gene expression.

The increase in the activities of the two key enzymes of pentose phosphate biosynthesis provides a heightened capacity for channeling precursors into pentose phosphate biosynthesis. Recent studies in my laboratory provided evidence that the increased glucose-6-phosphate dehydrogenase activity indeed represented an increased concentration of the enzyme protein as demonstrated by immunotitration (Selmeci and Weber, 1976).

Isotope studies confirmed the conclusions drawn from enzyme assays indi-

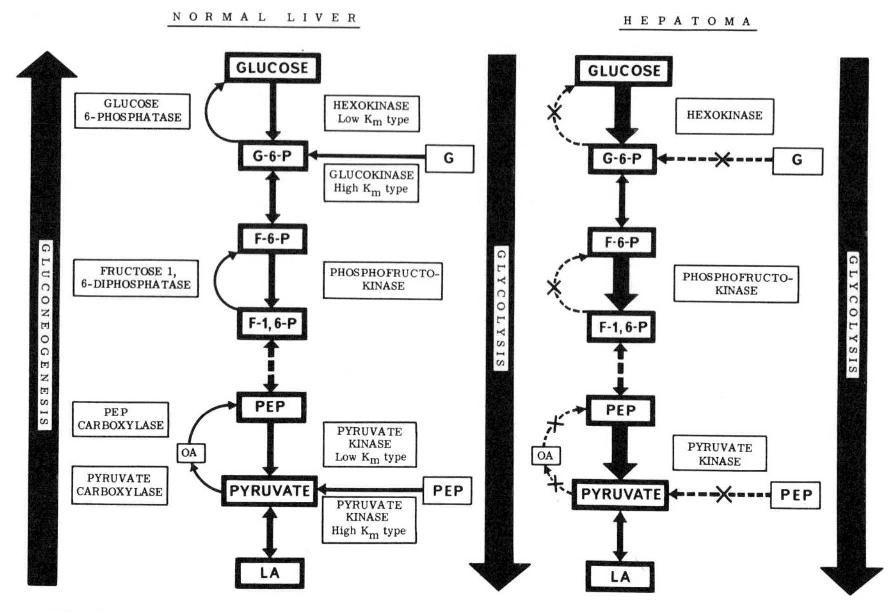

Fig. 6. The three main changes in carbohydrate metabolism in the cancer cell are illustrated, revealing the reprogramming of gene expression manifested in the cancer cell. These alterations involve decrease in activities of the key gluconeogenic enzymes, increase in activities of the key glycolytic enzymes and a shift in isozyme pattern from production of the high K_m regulatory isozymes to the biosynthesis of the nonhepatic low K_m isozymes. From Weber (1974a).

cating an increased potential for glycolysis and pentose phosphate pathway activity and a decrease for gluconeogenesis (Sweeney *et al.*, 1963).

Selective Advantages Conferred. The discovery of this pattern in reprogramming of gene expression in the cancer cell throws light on the mechanisms employed by the strategy of gene expression. Through the reciprocal alterations in the concentrations of the opposing key enzymes, the glycolytic potential markedly increased and the opposing gluconeogenic one declined. With the isozyme shift the responsiveness to physiological regulatory signals of the prevailing glycolytic enzymes was decreased. These alterations in gene expression should confer selective advantages to the cancer cells.

Purine Metabolism: Biochemical Strategy of the Cancer Cell

The increased potential for pentose phosphate biosynthesis in the hepatomas (Heinrich *et al.*, 1976; Selmeci and Weber, 1976; Sweeney *et al.*, 1963; Weber *et al.*, 1974) has drawn our attention to the utilization of ribose 5-phosphate in purine biosynthesis. This metabolic area received much attention in chemotherapy of cancer in man, but it seemed relatively poorly understood in terms of the behavior of key enzymes and opposing metabolic pathways.

Recent studies in my laboratories demonstrated that the enzyme that channels ribose 5-phosphate to PRPP biosynthesis, PRPP synthetase, was increased in the rapidly growing hepatomas (Hemrich *et al.*, 1974). The subsequent enzyme, glutamine-PRPP amidotransferase, the first committed enzyme of the *de novo* biosynthesis of purine, was increased in all the hepatomas (Prajda *et al.*, 1975). In turn, the key enzyme of purine degradation, xanthine oxidase, the rate-limiting catabolic enzyme, was decreased in all the liver tumors (Prajda and Weber, 1975; Prajda *et al.*, 1976). The activity of the final enzyme of purine degradation, uricase, was also decreased in all the hepatomas (Weber *et al.*, 1976b). Thus, the biochemical strategy of gene expression of the cancer cell in purine metabolism is also manifested in reciprocal changes of the activities of antagonistic enzymes in opposing pathways of IMP synthesis and degradation. The alterations in the activities of the purine enzymes reflect changes in enzyme concentration, since the velocities were proportionate to the enzyme added in the assay procedure. In the case of amidotransferase, current experiments provide independent confirmation of the increase of the concentration of this enzyme by immunotitration (Tsuda *et al.*, 1977).

These results indicate that PRPP synthetase is a Class 1 enzyme, as it relates to progression, whereas glutamine-PRPP amidotransferase, xanthine oxidase and uricase are Class 2 enzymes and their behavior appears to be linked with the neoplastic transformation.

Selective Advantages Conferred. Since the imbalance in purine metabolism was reflected in the increased concentration of synthetic and the decreased concentration of catabolic enzymes, the biochemical strategy of the cancer cell as manifested in the reprogramming of gene expression in purine metabolism should confer biological advantages to the cancer cell (Fig. 7).

IMP Utilization: Biochemical Strategy of the Cancer Cell

Recent and current studies indicate that IMP dehydrogenase which channels IMP into GMP biosynthesis was increased 2- to 3-fold in the slow-growing hepatomas and its activity was further elevated in the more rapidly growing tumors in parallel with tumor growth rate (Jackson *et al.*, 1975b). The activities of adenylosuccinate synthetase (Jackson *et al.*, 1975a) and adenylosuccinase (Jackson *et al.*, 1976) that channel IMP into AMP biosynthesis were increased in all the hepatomas.

Thus, all three enzymes of IMP utilization fit into Class 2, since their concentration is increased in all hepatomas; they are transformation-linked markers of neoplasia. IMP dehydrogenase which has the lowest activity of all the

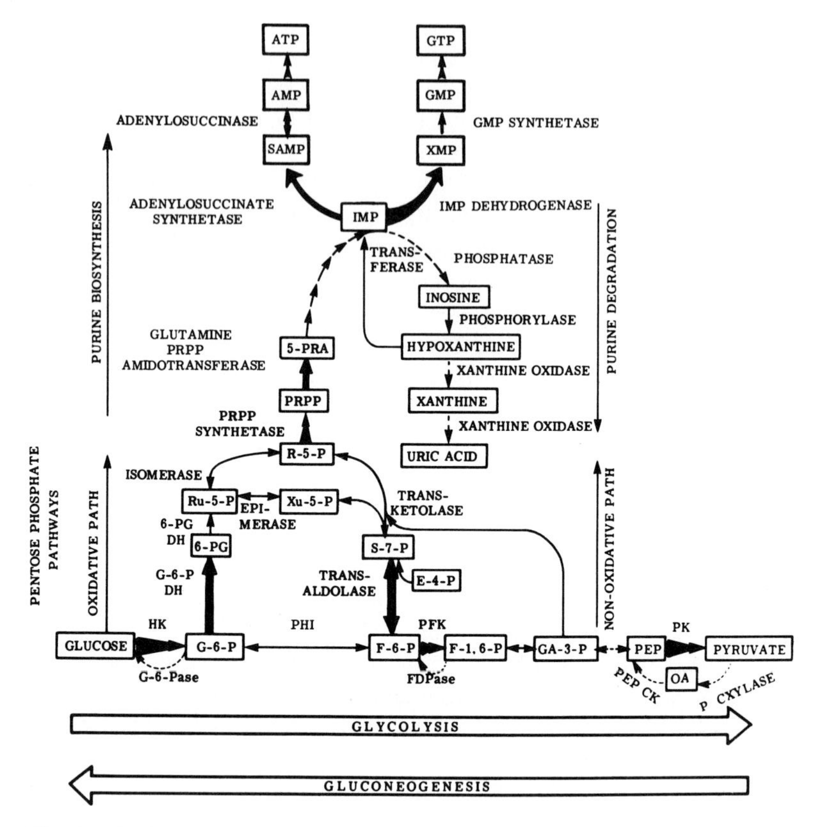

Fig. 7. Biochemical strategy of the cancer cell as revealed in the integrated pattern of imbalance in the activities of the key enzymes of glycolysis, gluconeogenesis, pentose phosphate pathways, purine synthesis, utilization and degradation. From Weber (1978).

purine enzymes we studied so far is a marker of both transformation and progression. The increase of this enzyme is the highest among all the purine enzymes, reaching 10- to 12-fold elevations in the rapidly growing tumors.

Selective Advantages Conferred. The increased potential of IMP utilization, as reflected in the ubiquitous increase of enzyme activities in all the hepatomas, should provide an increased capacity for the precursors for the biosynthesis of adenine and guanine nucleotides in the tumors. This reprogramming of gene expression should confer selective advantages to the cancer cells (Prajda *et al.*, 1975; Weber, 1977; Weber *et al.*, 1976b). The integrated features of the imbalance in glycolysis, gluconeogenesis, pentose phosphate metabolism, and the biosynthesis, catabolism and synthetic utilization of IMP are summarized in Fig. 7.

Pyrimidine and DNA Metabolism: Biochemical Strategy of the Cancer Cell

A systematic investigation of the pathways of pyrimidine metabolism demonstrated that the activities of the key enzymes of *de novo* and salvage pathways of the synthetic utilization of pyrimidines to DNA were increased and those of the catabolic pathways were decreased in parallel with the growth rate of liver tumors (Weber, 1974a, 1977). Isotope examination of the synthetic utilization of thymidine and the catabolism of this precursor to CO_2 confirmed the enzymatic indications that the antagonistic pathways of thymidine and DNA metabolism behave in an opposing fashion (Ferdinandus *et al.*, 1971; Weber *et al.*, 1972a; Baliga and Borek, 1975).

Thus, the activities of a number of key enzymes of pyrimidine synthesis and catabolism correlated with tumor growth rate. The behavior of these enzymes falls into Class 1 and indicates a linking with tumor progression and growth rate.

Recent work demonstrated that there is an enzyme in pyrimidine metabolism, UDP kinase, the activity of which is increased in every one of the hepatomas (Williams *et al.*, 1975). Thus, the behavior of this enzyme is linked with the malignant transformation and may be used as an indicator of malignancy in rodent hepatomas and also in human primary hepatomas (Weber *et al.*, 1976a).

An integrated picture of the behavior of the synthetic and catabolic enzymes of pyrimidine synthesis and degradation is summarized in Fig. 8.

Selective Advantages Conferred. The pattern of reprogramming gene expression in DNA metabolism also indicates the operation of reciprocal alterations in the activities of opposing key enzymes of synthetic and catabolic pathways. In parallel with tumor growth rate the potential for DNA synthesis was increased, whereas the capacity for degrading pyrimidine precursors was decreased (Ferdinandus *et al.*, 1971; Weber *et al.*, 1971). These alterations in gene expression are in line with those observed in carbohydrate, purine and other metabolic pathways and they should confer selective advantages to the cancer cell.

Ornithine Metabolism: Biochemical Strategy of the Cancer Cell

There is also an imbalance in the pattern of ornithine metabolism described earlier, revealing an increased utilization of ornithine to polyamine biosynthesis (Williams-Ashman *et al.*, 1972), in contrast with a decreased utilization of this amino acid for metabolism in the urea cycle (Weber *et al.*, 1972b). This imbalance is reflected in the increased ratio of ornithine decarboxylase:ornithine carbamoyltransferase which correlated positively with tumor growth rate. These alterations in gene expression should confer selective advantages to the cancer cells.

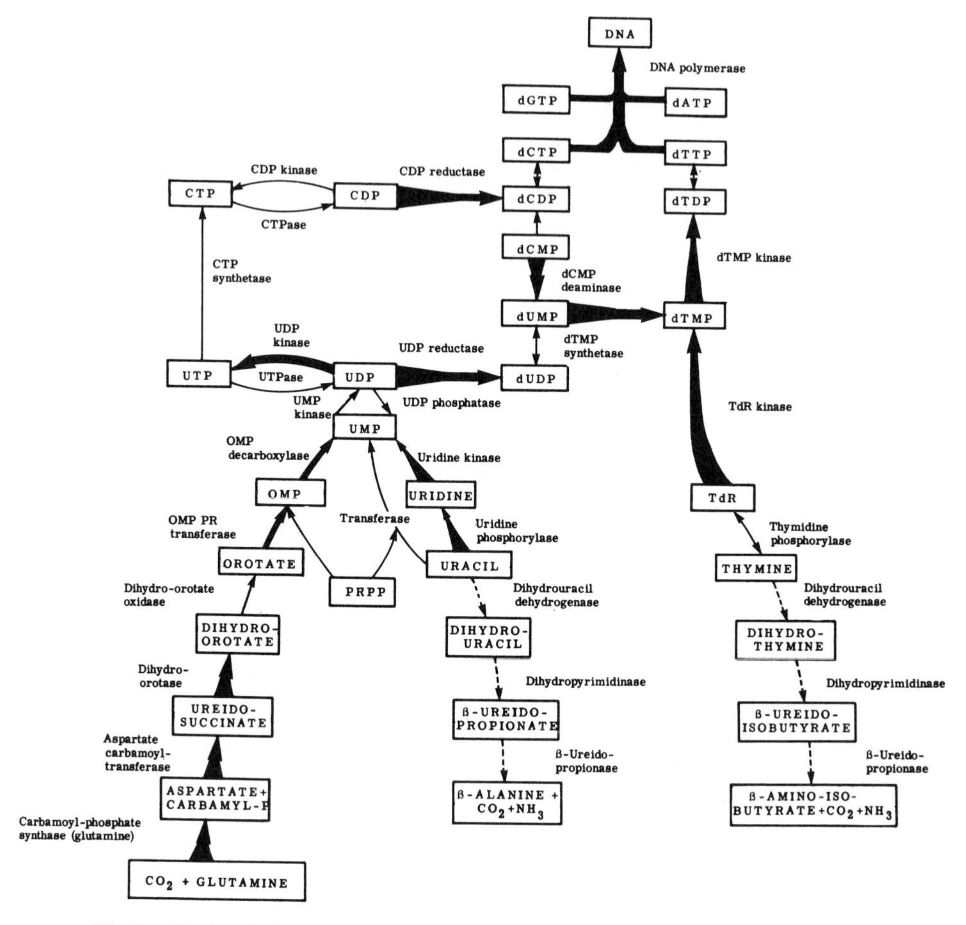

Fig. 8. Biochemical strategy of the cancer cell as revealed in the integrated reprogramming of gene expression manifested in the imbalance of the key enzymes of pyrimidine biosynthesis and degradation and the *de novo* and salvage pathways of DNA biosynthesis. From Weber (1978).

Membrane cAMP Metabolism: Biochemical Strategy of the Cancer Cell

In the spectrum of hepatomas of different growth rates the plasma membrane cAMP phosphodiesterase activity increased, and concurrently adenylate cyclase activity decreased in parallel with the rise in tumor growth rate. Thus, the opposing enzymes of cAMP degradation and synthesis behaved in a reciprocal fashion. In consequence, the ratio of the activities of cAMP phosphodiesterase: adenylate cyclase correlated positively with the growth rate of the liver tumors over 2-log scales. The opposing behavior of the antagonistic enzymes of membrane cAMP metabolism provides further evidence for the operation of the biochemical strategy of the cancer cell and these alterations should confer selective advantages to the cancer cell (Weber *et al.,* 1974).

Specificity to Neoplasia of the Biochemical Strategy Observed in the Cancer Cells

The strategy displayed by the gene expression of the cancer cell reveals that in neoplasia as in differentiation, hormonal regulation, and also in regeneration which is discussed in detail elsewhere (Weber, 1972, 1974a, 1975b) the pattern involves reciprocal behavior of antagonistic enzymes and metabolic pathways. Whereas the strategy of gene expression, as we identified it, is an overall pervasive method of genetic logic, nevertheless each pattern, whether it involves differentiation, hormonal regulation, regeneration or neoplastic transformation and progression, is characteristic of its own biological program and can be clearly differentiated from all other programs. Thus, the program of neoplastic transformation and progression is specific to the cancer cell and no similar pattern is observed in the other genic programs that characterize the other conditions of regulation or proliferation. The specificity of the program for neoplasia and the differences from the programs of differentiation and regeneration as characterized by quantitative and qualitative discriminants were outlined in detail elsewhere (Weber, 1973, 1974a, 1975b, 1977).

Biochemical Strategy of Gene Expression: General Characteristics and Possible Mechanisms

Evidence was presented and cited in this paper supporting my proposal that the strategy of the genome is revealed in the behavior of the integrated activity, amount and isozyme pattern of opposing and competing key enzymes. The evidence was examined in biological model systems appropriate for such investigations and a characteristic pattern was observed in development, in endocrine regulation, in the regenerating liver and in neoplasia.

In the case of *neoplasia,* the molecular mechanisms one would consider for testing must account for a number of manifestations of the ordered biochemical pattern that characterizes the strategy of the cancer cell (Weber, 1976, 1977; Weber *et al.,* 1974). Thus, the mechanism should account for the hereditability of the pattern and for the stability of the gene pool as expressed in the constancy of biological malignancy and the biochemical pattern in the different tumor lines. It should account for the fact that we observe reciprocal alterations in opposing key enzymes and metabolic pathways and some of these changes are expressed in increases in quantum jumps in relation to tumor growth rate. The explanation should account for the qualitative alterations which are characterized by the emergence of the isozyme shift. The molecular mechanism should account for the linking of a number of biochemical alterations with the neoplastic transformation, since these are all-or-none changes characteristic of the transformation itself. In turn, the molecular mechanism should account for the fact that a number of the biochemical parameters are linked positively or negatively with tumor growth rate and thus they characterize the degrees in the expression of malignancy and are markers of progression of neoplasia. The irreversibility of the enzymatic imbalance and isozyme shift must be included in the explanatory mechanism and the fact that there is an altered responsiveness of the enzyme activity and the enzyme amount in neoplasia. An important part of the explanation is the requirement for a mechanism which accounts for multienzyme changes involving many genes in different chromosomes. Finally, an accounting is needed for the specificity to neoplasia of the complex program of integrated metabolic imbalance that confers selective advantages to the cancer cells. One may add a further criterion of the explanatory mechanism, that it should be testable in a meaningful fashion.

In the case of the endocrine regulation some of the points considered would apply. However, the pattern of biochemical imbalance is not hereditable and it responds to hormonal replacement, and thus the alterations are reversible.

The mechanisms that may underlie the integrated pleiotropic behavior discussed in this paper were examined elsewhere in some detail (Weber, 1978; Weber *et al.*, 1974). Therefore, only the main conclusions are given here. Control mechanisms, that would account for the pattern we discussed, involved genetic or epigenetic mechanisms. Attention was drawn to the possibility that the control mechanisms may operate through master or integrative genes that would control large groups of functionally linked genes (Weber *et al.*, 1974). This concept would have to account for the fact that in endocrine regulation the alterations are not hereditable, whereas in neoplasia they are hereditable.

Regulation which could provide integrated pleiotropic effects may be achieved by the selective destruction or processing of mRNA. Evidence that mRNA is produced as a giant macromolecule in mammalian systems and that only part of it carries translatable information (Georgiev, 1972) is compatible with a proposal that the nontranslatable part may well carry recognition sites conveying information that dictates the fate of the mRNA under the various physiological or pathological conditions. As a result, the nontranslatable regulatory information would direct the translation or nontranslation of the different species of mRNA's that code for functionally integrated key enzymes. Thus, the key glycolytic enzymes, and the key enzymes of purine, pyrimidine and DNA biosynthesis in their respective mRNA molecules would be encoded for their specific protein structure. However, the respective mRNA molecules would all carry similar regulatory sites that would dictate translation in the transformed cells. The opposing enzyme systems would carry a negative instruction for translation in the transformed cells. Thus, neoplastic transformation and progression may be achieved through controlling the action of master genes or the pattern of translation of mRNA molecules. Because of the availability of the suitable biological model systems analyzed in this paper, these hypotheses for the mechanism of operation of gene expression should be testable.

PERSONAL COMMENTS

In discussing the biochemical strategy of the genome, my thoughts return to the time when as a first year medical student I had the privilege to be present at the first lecture Professor Szent-Györgyi gave at the University of Budapest as the new Chairman of the Institute of Biochemistry. That lecture, given 31 years ago in the post-war semester of 1945, is well remembered, and since that time I have followed with great interest and fascination the career of the Professor. It seems to be the rule that all those who meet Szent-Györgyi feel a deep admiration and a personal interest in the discoveries and scientific journey of this great man.

A Hungarian contemporary of Professor Szent-Györgyi, the famous historian and literary critic, Karl Kerényi (1970), said of Homer's Ulysses

> The fascination of Ulysses, which has held men spellbound for thousands of years, is the way in which he mastered his destiny, and the fact that he did so.

The fascination of Szent-Györgyi's life in its perilous journey toward the Ithaca of the understanding and conquering of nature has held us spellbound, because of the way he, a great scientist, has mastered his destiny and the fact that he did so.

REFERENCES

Baliga, B. S., and Borek, E. (1975), *Adv. Enzyme Reg.* **13,** 27–36.

Dunaway, G. A., Jr., and Weber, G. (1974). *Arch. Biochem. Biophys.* **162,** 629–637.

Dunaway, G. A., Jr., Morris, H. P., and Weber, G. (1974). *Cancer Res.* **34,** 2209–2216.

Ferdinandus, J. A., Morris, H. P., and Weber, G. (1971). *Cancer Res.* **31,** 550–556.

Georgiev, G. P. (1972). *Curr. Top. Dev. Biol.* **7,** 1–60.

Heinrich, P. C., Morris, H. P., and Weber, G. (1974). *FEBS Lett.* **42,** 145–148.

Heinrich, P. C., Morris, H. P., and Weber, G. (1976). *Cancer Res.* **36,** 3189–3197.

Jackson, R. C., Morris, H. P., and Weber, G. (1975a). *Biochem. Biophys. Res. Commun.* **66,** 526–532.

Jackson, R. C., Weber, G., and Morris, H. P. (1975b). *Nature (London)* **256,** 331–333.

Jackson, R. C., Morris, H. P., and Weber, G. (1976). *Life Sci.* **18,** 1043–1048.

Kerényi, K. (1970). *In* "The Adventures of Ulysses" (E. Lessing, ed.), p. 15. Dodd, Mead, New York.
Krebs, H. A. (1954). *Bull. Johns Hopkins Hosp.* **95**, 19–33.
Medawar, P. B. (1970). "Induction and Intuition in Scientific Thought." Methuen, London.
Popper, K. R. (1968). "The Logic of Scientific Discovery." Harper, New York.
Prajda, N., and Weber, G. (1975). *FEBS Lett.* **59**, 245–249.
Prajda, N., Katunuma, N., Morris, H. P., and Weber, G. (1975). *Cancer Res.* **35**, 3061–3068.
Prajda, N., Morris, H. P., and Weber, G. (1976). *Cancer Res.* **36**, 4639–4646.
Selmeci, L. E., and Weber, G. (1976). *FEBS Lett.* **61**, 63–67.
Sweeney, M. J., Ashmore, J. Morris, H. P., and Weber, G. (1963). *Cancer Res.* **23**, 995–1002.
Szent-Györgyi, A. (1968). *Adv. Enzyme Regul.* **7**, 5–11.
Tsuda, M., Katunuma, N., and Weber, G. (1977). To be published.
Weber, G. (1961). *Adv. Cancer Res.* **6**, 403–494.
Weber, G. (1972). *Isr. J. Med. Sci.* **8**, 325–343.
Weber, G. (1973). *Adv. Enzyme Regul.* **11**, 79–102.
Weber, G. (1974a). *In* "The Molecular Biology of Cancer" (H. Busch, ed.), pp. 487–521. Academic Press, New York.
Weber, G. (1974b). *In* "Differentiation and Control of Malignancy of Tumor Cells" (W. Nakahara *et al.*, eds.), pp. 151–180. Univ. of Tokyo Press, Tokyo.
Weber, G. (1975a). *In* "Mechanism of Action and Regulation of Enzymes" (T. Keleti, ed.), Vol. 32, pp. 237–251. Elsevier North Holland, New York.
Weber, G. (1975b). *In* "Liver Regeneration after Experimental Injury" (R. Lesch and W. Reutter, eds.), pp. 103–117. Stratton Intercontinental Medical Book Corp., New York.
Weber, G. (1977). *N. Engl. J. Med.* **296**, 486–493, 541–551.
Weber, G. (1978). "Biochemical Strategy of the Cancer Cell". Pergamon, New York. (to be published).
Weber, G., Singhal, R. L., and Srivastava, S. K. (1965). *Adv. Enzyme Regul.* **3**, 43–75.
Weber, G., Queener, S. F., and Ferdinandus, J. A. (1971). *Adv. Enzyme Regul.* **9**, 63–95.
Weber, G., Ferdinandus, J. A., Queener, S. F., Dunaway, G. A., Jr., and Trahan, J.-P. (1972a). *Adv. Enzyme Regul.* **10**, 39–62.
Weber, G., Queener, S. F., and Morris, H. P. (1972b). *Cancer Res.* **32**, 1933–1940.
Weber, G., Trevisani, A., and Heinrich, P. C. (1974). *Adv. Enzyme Regul.* **12**, 11–41.
Weber, G., Malt, R. A., Glover, J. L., Williams, J. C., Prajda, N., and Waggoner, C. D. (1976a). *In* "Biological Characterization of Human Tumors," pp. 60–72. Excerpta Med. Found, Amsterdam.
Weber, G., Prajda, N., and Jackson, R. C. (1976b). *Adv. Enzyme Regul.* **14**, 3–24.
Weber, G., Williams, J. C., and Jackson, R. C. (1977). *Adv. Enzyme Regul.* **15**, 53–77.
Webster's New World Dictionary of the American Language. (1962). College Edition. World Publ. Co., New York.
Williams, J. C., Morris, H. P., and Weber, G. (1975). *Nature (London)* **283**, 567–569.
Williams-Ashman, H. G., Coppoc, G. L., and Weber, G. (1972). *Cancer Res.* **32**, 1924–1932.

21

A Possible Role of Transglutaminase in Tumor Growth and Metastases

K. LAKI, G. CSAKO, S. T. YANCEY, and E. F. WILSON

Dr. Szent-Györgyi presented at the Symposium his concept concerning the malignant transformations of cells. In this paper we explore some of the factors that may influence the proliferation of malignant cells. Previous reports from our laboratories indicated the involvement of transglutaminase in the growth of YPC-1 tumor in CAF_1 mice (Laki *et al.,* 1966; Laki and Yancey, 1968; Yancey and Laki, 1972). The experiments presented here further explore the problem by comparing the transglutaminase activity of the various organs and tumors of different strains of mice to the malignancy of the tumors. Special attention is paid to the metastatic growth of the YPC-1 tumor grown in CAF_1 mice in relation to the transglutaminase activities. The results indicate the involvement of transglutaminase in tumor growth and spread.

MATERIALS AND METHODS

Animals

Adult male mice weighing 20–25 gm (CAF_1, C57BL/6N and CDF_1) were supplied by the NIH Laboratory Animal Genetic Center.

Tumors

A number of solid tumors and a leukemia were transplanted sc into the mice: plasma-cell tumor YPC-1 and another spontaneously growing tumor $YCAF_1$ into CAF_1 mice; adenocarcinoma CA_{755} and Lewis lung carcinoma into C57BL/6N mice; acute lymphocytic leukemia YLL; stroma-cell tumor of the ovaries K50932; melanoma S_{91} and Lewis lung carcinoma into CDF_1 mice. All the tumors are syngenic with the single exception of the Lewis lung carcinoma which is allogenic to CDF_1 mice.

These tumors are being maintained in the Laboratory of Chemical Pharmacology, Division of Cancer Treatment, National Cancer Institute. The tumor fragments inoculated subcutaneously represented 5–10 mg tissue.

Determination of Tissue Transglutaminase

The transglutaminase activity of the tissue homogenates was measured by the enzyme-catalyzed incorporation of labeled methylamine into the synthetic dipeptide, benzyloxycarbonyl-L-glutaminylglycine (Chung, 1972; Folk, 1972). The reaction is depicted as follows:

$$-\text{Gln}- + H_2N{-}R \rightarrow \underset{}{-\overset{\overset{\text{NHR}}{|}}{\text{Glu}}-} + NH_3$$

In the absence of a primary amine, hydrolysis of the amide takes place. Plasma transglutaminase (activated Factor XIII) carries out a similar reaction between glutamine residues of one fibrin molecule and the amino group of the lysine residue of another molecule (Doolittle *et al.*, 1972), but is inactive on this synthetic substrate (Folk and Chung, 1973). This difference between the two enzymes makes it possible to study tissue transglutaminase without the interference of the plasma transglutaminase. The organs or tumors from the animals killed by cervical dislocation were excised, chilled until weighing, then were homogenized with Waring blender into 5 volumes of 0.25 *M* sucrose solution. After centrifuging the homogenates for 30 min at 17,000 rpm, the supernatants were tested for transglutaminase activity. From these activities, the transglutaminase content was calculated by comparing these to the activities of pure guinea pig liver transglutaminase. The pure guinea pig liver transglutaminase was kindly supplied by Dr. S. Chung (National Institute of Dental Research).

To 1 ml of the homgenate, 1 μl of 5.0% DFP was added to prevent proteolysis during incubation. The composition of the buffer mixture was as follows: 10 volumes of Tris acetate buffer (1.0 *M*) pH 7.5; 3 volumes of $CaCl_2$ (1.0 *M*), and 1 volume of EDTA (0.1 *M*). 15 μl of this buffer mixture were combined with 25 μl 0.2 *M X*-GlnGly (benzyloxycarbonyl) (*Z*)-L-glutaminylglycine, a gift from Dr. J. E. Folk, National Institute of Dental Research), and 50 μl tissue homogenate. The reaction started with the addition of 10 μl CH_3NH_2 (^{14}C labeled methylamine HCl, 0.25 mCi, 1.9 mg in 2.5 ml ethanol, New England Nuclear, Boston, Massachusetts). After 1 hr incubation at 37°C, the reaction was arrested with 10 μl 0.10 *N* HCl. The arrested reaction mixture was then applied to a 2.0 × 0.6 cm column of Dowex 50W-X8 in the hydrogen cycle. The reaction product of *Z*-GlnGly and $^{14}C\text{-}H_2NH_2$ was washed off the column five times with 1.0 ml H_2O into counting vials. Blanks containing no homogenate were treated the same way. Activities are expressed as counts per minute. In these experiments, we are measuring initial velocities, since only 6% of the substrate (CH_3NH_2) appears in the reaction product. (The total cpm of the substrate added is 1,500,000 of which about 100,000 cpm appears in the reaction product catalyzed by 50 μl of the liver homogenate.)

Estimation of Metastases of the YPC-1 Tumor

At the start and the end of the experimental period eight normal mice were sacrificed and the spleen, kidneys, liver, lung, heart and brain were removed to be weighed. Similarly, at various times the organs and the tumor of eight tumor-bearing animals were removed and weighed. The weight increment, the difference with respect to the normal controls, was considered as the apparent mass of tumor in the organs. The presence of the tumor in different organs at various times was verified also by histological examination using Zenker's fixation and hemotoxylin–eosin staining.

RESULTS

Some of our experiments are summarized in Table I. The upper half of the table shows the transglutaminase content of the organs in three different strains of mice. It is quite apparent that the level and distribution is different in the animals examined.

In the lower half of the table the tumors are listed in decreasing order

TABLE I
Transglutaminase Activity and Content of the Organs of Different Strains of Mice and of Various Tumors Including Average Survival Times of the Animals

	CAF_1 $n = 16$		C57BL/6N $n = 8$		CDF_1 $n = 16$	
Organ or tumor	cpm[a] ($\times 10^6$)	TGase content[a] (μg)	cpm ($\times 10^6$)	TGase content (μg)	cpm ($\times 10^6$)	TGase content (μg)
Liver	9.5	114.0	6.4	76.8	2.6	31.2
Spleen	1.2	14.4	1.8	21.6	0.8	9.6
Kidney	1.1	13.2	1.2	14.4	0.4	4.8
Lung	1.0	12.0	0.8	9.6	0.4	4.8
Heart	0.5	6.0	0.6	7.2	0.3	3.6
Brain	0.2	2.4	0.2	2.4	0.1	1.2
YPC-1	12	14.4				
(day 12)[b]	(15 days)[c]					
$YCAF_1$	0.4	4.8				
(day 28)	(35–40 days)					
CA 755			0.4	4.8		
(day 14)			(15 days)			
YLL					1.2	14.4
(day 14)					(15 days)	
K 50932					0.7	8.4
(day 14)					(14–20 days)	
S_{91} melanoma					0.5	6.0
(day 14)					(14–20 days)	
Lewis lung			0.2	2.4	0.2	2.4
(day 21)			(30 days)		(52–55 days)[d]	

[a]cpm and TGase content/gm wet tissue
[b]time of tumor excised
[c]average survival time
[d]allogenic

according to their transglutaminase activity and content together with the day of harvesting the tumor and the average survival times of the animals implanted with the corresponding tumor.

Inspection of the data in Table I shows an inverse correlation between transglutaminase content of the tumors grown in a given strain of animals and the average survival time. It is seen that the higher the tumor's transglutaminase content the shorter is the life span of the animals bearing the tumor. Correspondingly, the Lewis lung tumor growing in C57BL/6N and in CDF_1 mice kills the C57BL/6N mice having higher tissue transglutaminase level faster.

The blood-borne spread of the YPC-1 tumor was followed by the weight increase of the organs at various times after implantation. The weight increments of the liver, spleen, kidney and the lung are shown in Fig. 1. Up to the fifth day, the curves seem to be exponential.

The tumor weight increases in the case of the liver and spleen apparently fall on a Gompertzian curve [describing tumor growth retardation (Frindel *et al.*, 1967; Laird, 1964)]. In lung and kidney, the growth levels off or even decreases after the fifth day. On the basis of the histological examination, this is undoubtedly due to tissue damage, especially in kidneys. Even in the case of liver, tissue damage has to be considered since transglutaminase activity of this organ decreases with the increase of tumor mass (Fig. 2) as discussed below.

Inspection of the data (Fig. 1) shows that during the first 5 days after implantation, the apparent doubling time of the tumor at the implantation site

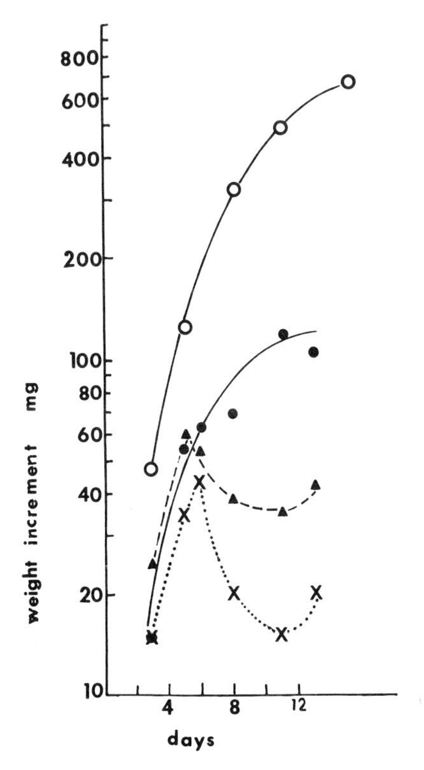

Fig. 1. The weight increment of different organs of CAF_1 mice bearing sc YPC-1 tumor. The average weight increment of the organs due to metastasis is plotted against the time elapsed after tumor implantation. Symbols: liver, open circle; spleen, closed circle; lung, triangle; kidney, cross.

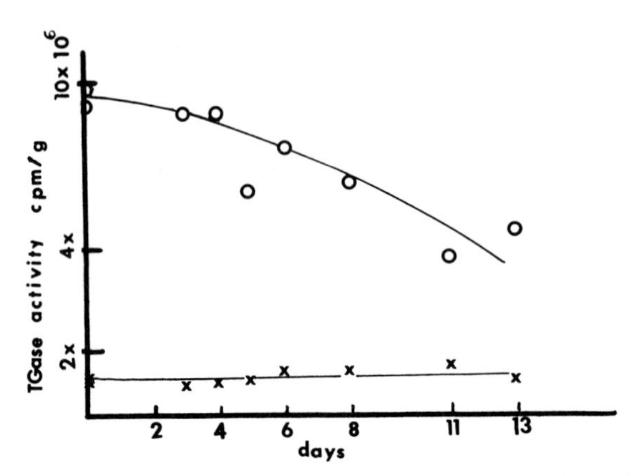

Fig. 2. The course of transglutaminase activity in the liver and spleen of the CAF_1 mice during metastases of the YPC-1 tumor. Symbols: liver, circle; spleen, cross.

in the liver, in the spleen, and in the lung is 1.2 days, in the kidney, 1.8 days. The somewhat longer doubling time in the kidney may be due to an early damaging effect of tumor penetration. Extrapolation of the curves indicates that within hours after implantation, the tumor quickly spreads in these organs and, in fact, the total mass of tumor in these organs is about the same at the fifth day as the weight of the tumor at the implantation site. However, at the eighth day, sc tumor is larger (596.6 mg) than the metastatic tumors together, reaching 2200 mg at the thirteenth day.

Histological examination (Fig. 3) shows the presence of tumor cells in the organs discussed before, thus supporting our argument of the weight increment to be a measure of tumor penetration. In the spleen, many small microscopic nodules of tumor cells are detectable. The liver exhibits diffuse infiltration of tumor cells. In the lung, tumor cells are seen lodged within alveolar vessels. In the kidney, the damage is apparent from the presence of hyaline material in renal tubules. On the other hand, no histological changes are seen in the heart and the brain.

The transglutaminase content of the organs of the CAF_1 mice and the YPC-1 tumor is given in Table II. It is seen that the transglutaminase activity of the tumor is about the same as that of the spleen and much smaller than the transglutaminase activity of the liver. This lower activity of the tumor explains the drop of activity of the liver as increasing amounts of tumor penetrate the liver (Fig. 2).

The mass of the tumor after 5 days of implantation in the organs examined is also given in the table. The data in Table II indicate a rough correlation between transglutaminasc activity and tumor penetration. It is seen that the YPC-1 tumor does not penetrate into the organs of low transglutaminase content.

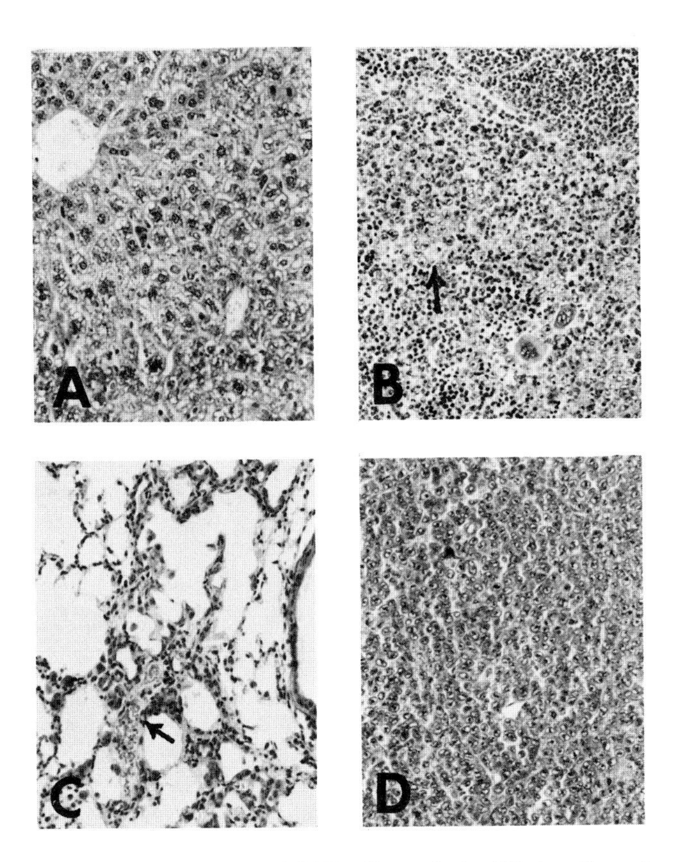

Fig. 3. Histological examinations of CAF_1 mice at the twelfth day after sc implantation of YPC-1 plasma cell tumor. ×140. (A) *Liver.* It becomes pale in appearance by advanced tumor growth. Microscopically, sinusoids are diffusely infiltrated by large cells with basophilic cytoplasm and pale lobulated or indented nuclei (neoplastic plasma cells). (B) *Spleen.* Red pulp is heavily infiltrated by cells resembling those of the sc tumor. Several foci of plasmatic tumor cells (arrow) can also be observed. (C) *Lung.* A number of plasmatic tumor cells are seen lodged along alveolar vessels (arrow). (D) *Sc plasma cell tumor.* Numerous closely packed tumor cells. Moderately pleomorphic nuclei are variable in staining, but usually have abundant chromatin, often concentrated in a central chromatin knot and at the nuclear margin. Many nuclei are eccentric in location. Moderate amount of deeply basophilic and finely granular cytoplasm is characteristic for the cells. Scant stroma; frequent small areas of necrosis are visible.

TABLE II
Transglutaminase Content of the Organs of Normal CAF_1 Mice and the Weight Increment Due to the Metastases of YPC-1 Tumor in the Organs at the Fifth Day Following Implantation

Organ or tumor	TGase (μg)/gm wet tissue	Weight increment (mg) of the organs at fifth day after sc implantation of YPC-1 tumor: Total increase	Weight increment (mg) of the organs at fifth day after sc implantation of YPC-1 tumor: Increase/gm wet tissue
Liver	114.0	125	120
Spleen	14.4	55	674
Kidney	13.2	35	110
Lung	12.0	55	318
Heart	6.0	0	0
Brain	2.4	0	0
YPC-1	14.4		
		270 mg	

DISCUSSION

It has been reported from this laboratory that the frequency of metastases of human breast cancer and melanoma into various organs show a correlation to the transglutaminase content of the organs (Laki, 1975). The data presented in this paper show that in mice there is also a correlation of the growth and the spread of tumors into the various organs to the transglutaminase content of the tumors and the organs.

Metastasis undoubtedly depends on many factors, such as the compatibility of the immune system (Kim *et al.,* 1975), the affinity of the tumor cell for a particular organ (Nicholson and Winkelhke, 1975), for example. Our findings indicate that transglutaminase may also be a factor that influences tumor proliferation.

It has been proposed (Laki, 1975) that not only normal tissues, but also tumors, exhibit great variations in their transglutaminase content. The organs and tumors investigated in this work provide examples of variations in the transglutaminase activities of different tissues.

The role of transglutaminase in the tissues is unknown. It is difficult to visualize that such an enzyme, active in connecting proteins with isopeptide bonds, could exist in the cells, unless it acts in the interior of the cell only on some specific protein yet to be identified, or its role is to act outside the cell. It is worth pointing out, however, that actin is a substrate for transglutaminase (Derrick and Laki, 1966), and actin is now recognized as an ubiquitous cell constituent involved in karyokinesis during cell division (Sanger, 1975). Thus, actin may turn out to be the specific protein in the cell on which transglutaminase acts, just as fibrin is the specific substrate for plasma transglutaminase (activated Factor XIII).

It is also conceivable that transglutaminase is manufactured to be released into the surroundings to interact with fibrinogen. Tissue transglutaminase also cross-bonds fibrinogen or fibrin with activities comparable to plasma transglutaminase (Farrell and Laki, 1970). It is possible that in promoting metastatic growth, transglutaminase acts in conjunction with fibrinogen (Laki, 1974). This would then be a situation analogous to wound healing where proliferation of granulation tissue requires a fibrin network on which transglutaminase has acted (Beck *et al.,* 1961, 1962; Bohn, 1972; Marktl and Rudas, 1974). The presence of fibrin in tumors (Cliffton and Agostino, 1965, 1974) supports an analogy of tumor proliferation with the proliferation of granulation tissue in wound healing. According to current concept, the presence of cross-bonded fibrin or fibrinogen is a necessary substrate for the capillarization of the granulation tissue (Laki, 1974).

A third possibility may be that transglutaminase masks antigenic sites on the membrane of the tumor cell by connecting them to fibrinogen or some other protein. As a result, the immune system would become inoperative and the tumor could grow unopposed.

The finding, that in different species (man and mouse) with different tumors, a correlation appears between the extent of metastasis and transglutaminase content of the host organs, suggests that the correlation is not coincidental.

REFERENCES

Beck, E., Duckert, F., and Ernst, M. (1961). *Thromb. Diath. Haemorrh.* **6,** 485–491.

Beck, E., Duckert, F., Vogel, A., and Ernst, M. (1962). *Z. Zellsforsch. Mikrosk. Anat.* **57,** 327–346.

Bohn, H. (1972). *Ann. N.Y. Acad. Sci.* **202,** 256–272.

Chung, S. I. (1972). *Ann. N.Y. Acad. Sci.* **202,** 240–255.

Cliffton, E., and Agostino, D. (1965). *Vasc. Dis.* **2,** 43–52.

Cliffton, E., and Agostino, D. (1974). *J. Med.* **5,** 7–148.

Derrick, N., and Laki, K. (1966). *Biochem. Biophys. Res. Commun.* **22,** 82–88.

Doolittle, R. F., Cassman, K. G., Chen, R., Sharp, J. J., and Wooding, G. L. (1972). *Ann. N.Y. Acad. Sci.* **202,** 114–126.

Farrell, J., and Laki, K. (1970). *Blood* **35,** 804–808.

Folk, J. E. (1972). *Ann. N.Y. Acad. Sci.* **202,** 59–76.

Folk, J. E., and Chung, S. E. (1973). *Adv. Enzymol.* **38,** 109–191.

Frindel, E., Malaise, E. P., Alpen, E., and Tubiana, M. (1967). *Cancer Res.* **27,** 1122–1131.

Kim, U., Baumler, A., Carruthers, C., and Bielat, K. (1975). *Proc. Natl. Acad. Sci. U.S.A.* **72,** 1012–1016.

Laird, A. K. (1964). *Br. J. Cancer* **18,** 490–502.

Laki, K. (1974). *J. Med.* **5**, 32–37.

Laki, K. (1975). Faktor XIII und Metastasierung. *Thromb. Diath. Haemorrh. Suppl.* **62**, 61–71. "Transglutaminese and Metastasis." Deutschen Arbeitsgemeinscheft für Blutgerinnungsforschung e. V., München.

Laki, K., and Yancey, S. T. (1968). *In* "Fibrinogen" (K. Laki, ed.), pp. 359–367. Dekker, New York.

Laki, K., Tyler, H. M., and Yancey, S. T. (1966). *Biochem. Biophys. Res. Commun.* **24**, 776–781.

Marktl, W., and Rudas, B. (1974). *Thromb. Diath. Haemorrh.* **32**, 575–581.

Nicholson, G. L., and Winkelhke, J. L. (1975). *Nature (London)* **255**, 230–232.

Sanger, J. W. (1975). *Proc. Natl. Acad. Sci. U.S.A.* **72**, 2451–2455.

Yancey, S. T., and Laki, K. (1972). *Ann. N.Y. Acad. Sci.* **202**, 344–348.

22

Some Problems and Approaches in Viral and Cancer Chemotherapy*

SEYMOUR S. COHEN

It is not surprising that in a book dedicated to Dr. Albert Szent-Györgyi, one or more of the contributions will be concerned with some aspect of the cancer problem. Many biochemists, including the Prof, have sought and are seeking specific aberrations in tumors and in cancer cells as clues to the origins of the pathology. However, I shall discuss some problems in the development of a chemotherapy for this pathology, and indicate some recent approaches to improving our capabilities in this area. Chemotherapists are, I think, in a cul-de-sac in our present direction of effort, and I shall indicate how our present course arose and how we might escape to a more hopeful path of work. In so doing it will be obvious that I think also that to effect a successful chemotherapy, we must know more of the origins of the development of uncontrolled cell multiplication.

*The author is an American Cancer Society Professor of Microbiology. The most recent work to which he will refer is supported by a grant AI-11636 from the National Institute of Allergy and Infectious Disease.

Since I spent many years on the biochemistry of phage multiplication, you may see no obvious link connecting that work to my present studies on the development and use of chemotherapeutic agents. My interest in the problems of both basic and clinical science began in 1953 with a study of a thymine-requiring strain of *Escherichia coli.* This study had been spurred by a search for the path to the biosynthesis of the unique phage pyrimidine, 5-hydroxymethylcytosine (Cohen, 1953). Unexpectedly, this bacterium demonstrated two phenomena: the first, clarified in the course of 10 years of subsequent effort, was that infection by T2 permitted the infected cells to synthesize thymine and viral DNA (Barner and Cohen, 1954). Reactions for several of the enzymes, including thymidylate synthetase and dCMP hydroxymethylase (Cohen, 1961), whose synthesis we detected after infection are presented in Fig. 1. These observations led to the demonstration and eventually the generalization that viruses introduced viral genes which permitted infected cells to synthesize new enzymes and to expand their metabolic equipment (Cohen, 1961, 1968).

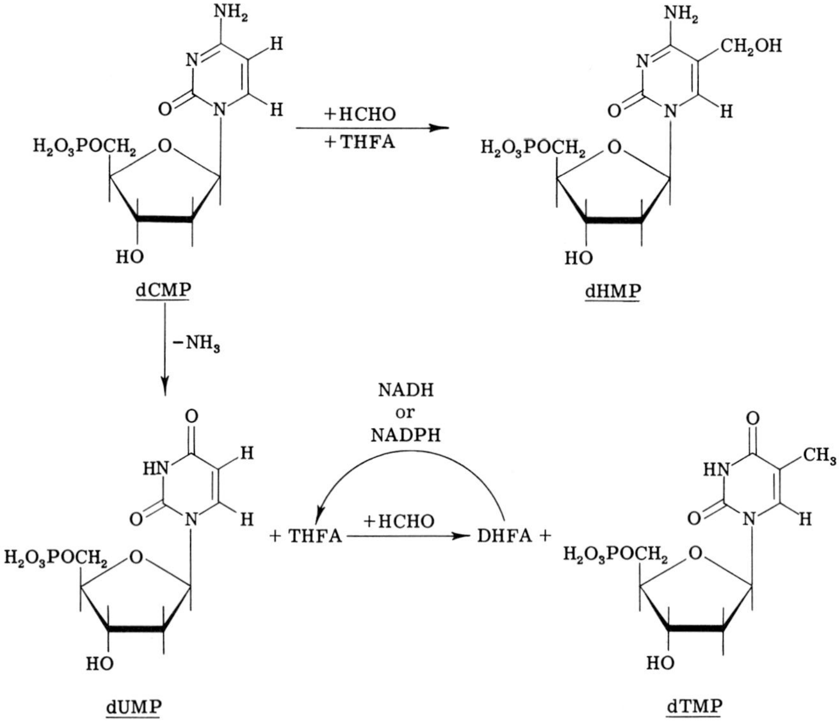

Fig. 1. Some reactions for which enzymes are induced in T-even virus infection of *E. coli.*

EXPLOITATION OF VIRUS-INDUCED ENZYMES

In developing briefly the chemotherapeutic consequences of this generalization, I shall refer to herpesvirus, which has been shown to be capable of transforming various animal cells to the oncogenic state. As we know, all large DNA viruses induce the production of new enzymes fairly early, enzymes which expand the capabilities of virus-infected cells to make virus components, including DNA. Herpesvirus infection increases at least six enzymatic functions related to DNA synthesis, and these may relate to the formation of new proteins determined by viral genes. These functions include those of thymidine kinase, deoxycytidine kinase, thymidylate kinase, deoxycytidylate deaminase, ribonucleotide reductase and DNA polymerase. Although the data suggest strongly that some of these proteins are indeed virus-determined, e.g., thymidine kinase, it will be important to be able to prove this as rigorously as it has been proven for some bacteriophage-induced enzymes. If a virus-infected cell does indeed contain an essential enzyme lacking in the uninfected cell, it is theoretically possible to block virus replication specifically and to minimize a generalized toxicity which may be produced by the inhibitor.

One chemical agent which is active in this way is 5-iododeoxyuridine (IUdR) (Fig. 2). The chemotherapeutic use (Kaufman and Maloney, 1963) of this compound in corneal infections by herpesvirus (herpetic keratitis) is an outstanding example of the exploitation of the herpesvirus-induced production of a

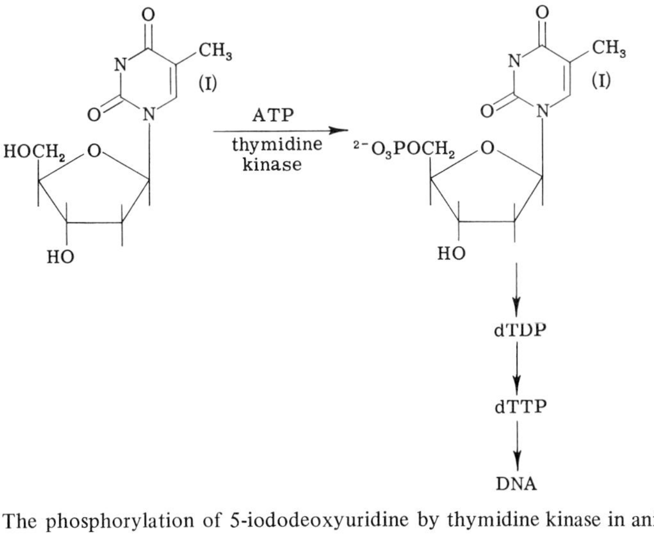

Fig. 2. The phosphorylation of 5-iododeoxyuridine by thymidine kinase in animal cells infected by herpesvirus or vaccinia virus.

new enzyme, thymidine kinase. Corneal cells lack this enzyme and uninfected corneal cells are not significantly affected by IUdR. In infected cells, the compound is phosphorylated to the monophosphate and then to the triphosphate which is incorporated into viral DNA. Such a DNA does not appear to be assembled and packaged well and infectious virus does not accumulate. Virus mutants, deficient in the ability to produce thymidine kinase, may be selected for by this treatment; such emergent viral strains are insensitive to continuing treatment with IUdR but are sensitive to other inhibitors.

Recently 5′-aminoiododeoxyuridine has been shown to selectively inhibit viral-specific DNA synthesis in both infected cells in tissue culture (Cheng *et al.*, 1975) and in herpesvirus keratitis in rabbits. This compound does not inhibit the *in vitro* activity of a purified HSV-1 DNA polymerase. However the inhibitor is phosphorylated by the virus-induced thymidine kinase specifically.

It has been reported that the unusual compound, phosphonacetic acid, active against herpes dermatitis in mice and herpetic keratitis in rabbits, specifically inhibits the virus-induced DNA-dependent DNA polymerase (Overby *et al.*, 1974). It may be asked if the very high GC content of viral DNA does not ensure marked differences from host proteins in the polypeptides determined by herpes DNA, thereby facilitating the discrimination effected by antiherpes agents. Whatever the reason for the surprising plethora of viral-specific inhibitors, it may be hoped that such effects will extend also to the large number of as yet unidentified virus-specific functions coded by viral DNA, which will surely affect the activities involved in membrane penetration, uncoating, membrane biosynthesis and virus exit.

It may be possible to exploit the phenomenon of virus-induced enzymes in tumor cells if it can be shown that such cells continue to carry these evidences of an earlier virus infection. As is well-known, it is far from clear that any human tumor does arise as a result of virus infection. For this reason my previous remarks may have no direct bearing on cancer chemotherapy, although they are clearly relevant to the problem of antiviral chemotherapy.

INHIBITION OF DNA SYNTHESIS AND CELL TOXICITY

The second phenomenon detected initially in the thymineless bacterium was the lethality of thymine deprivation (Barner and Cohen, 1954). This was termed "thymineless death" and was shown to require a relatively specific inhibition or prevention of DNA synthesis under conditions of a continuing synthesis of certain cell polymers, notably RNA. At first sight, and as was subsequently proved, this effect bore considerable resemblance to the action of certain antifolates, such as aminopterin and amethopterin, which produce an inhibition of DNA synthesis. Further, it could be demonstrated that sulfanilamide, a substance thought of as bacteriostatic, could provoke thymineless death in bacteria in the presence of other metabolites, such as purines and methionine, which require folic acid in their biosynthesis (Cohen and Barner, 1956).

It seems appropriate that we should be presenting this explanation of antifolate toxicity in Boston, the town in which Sidney Farber had first discovered the therapeutic effects of aminopterin in childhood leukemia in 1948. Here also Emil Frei is now pursuing the therapeutic effects of amethopterin to its permissible extreme, i.e., host toxicity short of death.

THE MODE OF ACTION OF 5-FLUOROURACIL

While we were gaining some experience with the control of the lethality provoked by thymine deficiency, and simultaneously were analyzing the extracts of virus-infected bacteria for enzymes, such as thymidylate synthetase, which were markedly induced by phage, the antitumor effects of 5-fluorouracil were noted by Heidelberger and Duschinsky (Duschinsky *et al.,* 1957). On their invitation, we studied the toxicity of this analogue in bacteria and demonstrated that the compound was lethal. A bacteriostatic effect attributable to formation of the ribonucleotide and possibly to incorporation of flourouracil into RNA was blocked by exogenous uracil. However, the lethal effect was blocked by exogenous thymine, and it appeared that the compound provoked thymineless death. We then converted fluourouracil to the deoxyribonucleotide (F-dUMP) and showed that this nucleotide was a powerful inhibitor of the phage-induced thymidylate synthetase (Cohen *et al.,* 1958). In animal cells the lethal effects of fluorouracil are prevented by thymidine and their thymidylate synthetase is also inhibited by F-dUMP. Thus, the lethal effect of this analogue does appear to simulate thymineless death in bacteria, and in animal cells, as well as in its use in clinical therapy.

For obvious reasons, I became greatly interested at this time in the clinical use of the compound. It was evident that most of the fluorouracil administered was rapidly metabolized, and it has slowly appeared that its clinical effects can be improved by inhibition of the degradation of the analogue (W. Waddell, personal communication). If the beneficial effects of 5-fluorouracil are to be attributed mainly to the creation of a specific thymine deficiency, it would appear that the analogue should be given in a form which would minimize incorporation into ribonucleotides and RNA. The obvious compound for this role would be 5-fluorodeoxyuridine, i.e., FUdR; however, there has not usually been a therapeutic benefit in the use of the deoxyribonucleoside over that of free fluorouracil. This occurs because of the rapid cleavage of the deoxyribonucleoside to free fluorouracil. No successful effort has been made to block the one or more nucleoside phosphorylases that effect this cleavage, but it is evident that appropriate inhibitors for this cleavage should be obtained.

We have approached the problem of maximizing the lethality of fluorouracil in another way. In bacteria we were able to show that nucleotides are dephosphorylated but that the resulting nucleoside is metabolized differently from exogenously supplied nucleoside (Lichtenstein *et al.*, 1960). The slowly generated nucleoside derived from exogenous nucleotide is rapidly rephosphorylated intracellularly and is not cleaved significantly to the free base. Taking advantage of this molecular drip mechanism into the cell, exogenous 5-fluorodeoxycytidylate served to maintain high lethal concentrations of 5-fluorodeoxyuridylate within the cells. As a result, cells exposed to the exogenous nucleotide die until the culture is virtually sterilized, whereas the lethal effect of the readily detoxified nucleoside is quite ephemeral. We shall return below to the problem of the use of nucleotides as chemotherapeutic agents.

D-ARABINOSYL NUCLEOSIDES AS THERAPEUTIC AGENTS

We began work with D-arabinose in 1947, in extending the work of Warburg and Dickens on the pentose synthesized by phosphogluconate dehydrogenase. Lipmann had pointed out in 1936 that D-arabinose 5-phosphate might be a product of the oxidative decarboxylation of phosphogluconate (Lipmann, 1936) and we developed methods to distinguish ribose and D-arabinose in the enzymatic reaction products. McNair Scott and I detected ribose phosphate among these products, and Horecker and his colleagues subsequently discovered that ribulose 5-phosphate was the primary pentose generated.

In the course of these studies we also found that prokaryotic cells carry out the isomerization:

$$\text{D-Ribulose 5-phosphate} \rightleftharpoons \text{D-arabinose 5-phosphate}$$

but that eukaryotic cells lack this enzyme (Lim and Cohen, 1966). In some microbial genera, D-arabinose 5-phosphate is taken to 2-keto-3-deoxyoctonate 8-phosphate in many bacteria, whereas in Mycobacteria, Corynebacteria and Nocardia, D-arabinose is a major component of the cell wall polysaccharide. It is clear that a specific inhibitor for this pentosephosphate isomerase might be a useful and specific chemotherapeutic agent for such diseases as tuberculosis, leprosy and diphtheria.

Shortly after the discovery of thymineless death, Bergmann (Bergmann and Burke, 1955) at Yale detected D-arabinosylthymine in a unique marine sponge. We then showed that the thymineless bacterium liberated thymine slowly from the sponge nucleoside. Fox synthesized D-arabinosylfluorouracil and we observed that the 5′-phosphate (F-araUMP) is a strong inhibitor of thymidylate synthetase (Cohen, 1961). Although less active than F-dUMP, it is far more inhibitory than F-UMP, and gave an early indication that ara derivatives more nearly simulated deoxyribosyl than ribosyl derivatives.

In any case, this ara nucleoside (D-arabinosylfluorouracil) is cleaved in bacteria and in humans at a significant rate. Although it has a beneficial effect on animal tumors, serious comparisons of the clinical effectiveness of araFU and free FU have not been made.

araC AND TETRAHYDROURIDINE

Thus sensitized to the properties of D-arabinosyl derivatives and of pyrimidine nucleosides, we began to look at an early sample of D-arabinosylcytosine and noted its rapid deamination by cytidine deaminase (Pizer and Cohen, 1960). As we know now, this potent antileukemic drug has a half-life in human serum of about 15 min as a result of deamination. As soon as the Upjohn group had detected the marked effect of D-arabinosylcytosine on a mouse leukemia and its antiviral activity with several DNA viruses (Evans *et al.,* 1961), I urged these workers to search for an inhibitor of human cytidine deaminase.

The ability of a group of skilled workers to design and test such an inhibitor deserves our notice. In that group, Hanze (1967) made tetrahydrouridine, which Camiener (1968) demonstrated to be a potent inhibitor of the human enzyme as in Fig. 3, and Neil and colleagues (1970) then showed that tetrahydrouridine markedly potentiated the activity of araC. Indeed in the presence of this deaminase inhibitor, both araC and 5-azacytidine even show activity when administered orally to leukemic or tumor-bearing animals. Tetrahydrouridine has essentially no toxicity when administered to animals, but does increase the level of phosphorylation of araC in animal and human tumor cells.

Although it would seem obvious that the efficacy of araC should be tested in combination with tetrahydrouridine, it has taken 10 years to convince the National Cancer Institute that it would be useful to study the toxicology of this mixture in humans. Such a study appears to have been completed recently with promising results, and clinical trials of the efficacy of the combination may be in preparation, if not actually under way. In my estimation there is an enormous gap between the most elementary principles of our biochemical knowledge and their application in clinical practice in the areas of chemotherapy. I think that the task of narrowing this gap warrants a major effort not only in overcoming problems of communication between basic and clinical scientists but also in eliminating administrative and bureaucratic roadblocks.

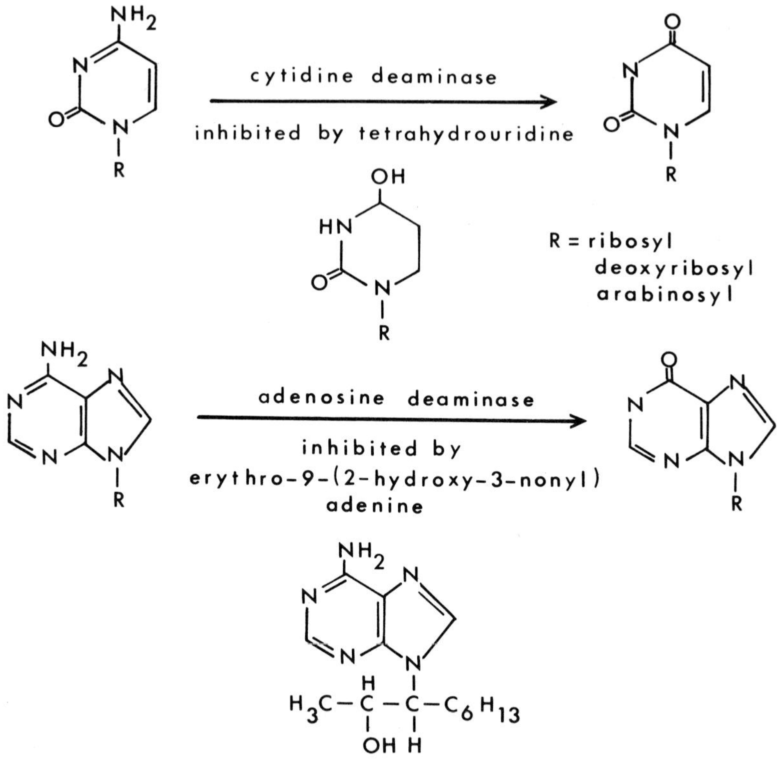

cytidine deaminase

inhibited by tetrahydrouridine

R = ribosyl
deoxyribosyl
arabinosyl

adenosine deaminase

inhibited by erythro-9-(2-hydroxy-3-nonyl) adenine

Fig. 3. The enzymatic deamination of cytosine and adenine nucleosides and an inhibitor of each deaminase.

THE MODES OF ACTION OF ara NUCLEOSIDES

In animal cell cultures, araC inhibits DNA synthesis specifically and is lethal to the cells, which show chromosome breaks. araC is phosphorylated to 5′-araCMP and this compound is converted in turn to the triphosphate. Furth and I (1968) showed that araCTP is an active inhibitor of mammalian and tumor DNA polymerase, competitive with dCTP. Interestingly araCTP is not inhibitory to *E. coli* DNA polymerase I, suggesting that *E. coli* is not always an appropriate guide to cancer problems.

In studies of the nucleotide pools of cells inhibited by araC, it has been found that deoxyribonucleotides accumulate. This result does not support the proposal that araCDP or araCTP inhibited ribonucleotide reductase, and indeed Moore and I (1967) could find no evidence of such an inhibition in earlier enzymatic studies. It does appear that araC is incorporated into DNA (Graham and Whitmore, 1970).

In paralleling these studies on araC, we have explored the properties of araA, a rather insoluble nucleoside first made in 1960 by Baker and his collaborators (Lee *et al.,* 1960). Hubert-Habart and I (1962) were able to detect a lethal activity in certain mutants of *E. coli.* Testing this compound on mouse fibroblasts, it was found initially that the toxicity of araA appeared markedly less than that of araC, i.e., killing cells at 2×10^{-4} *M* as compared to 10^{-6} *M.* Needless to say, araA is rapidly deaminated to the far less toxic D-arabinosylhypoxanthine. Nevertheless, in animal cells residual araA is phosphorylated to araAMP and then to araATP, which inhibits DNA synthesis. AraATP is inhibitory to mammalian DNA polymerase ($K_i = 10^{-6}$ *M*) (Furth and Cohen, 1968) and we have shown also that araATP is incorporated in cells mainly in internucleotide linkage in the small amount of DNA which continues to be synthesized (Plunkett and Cohen, 1975a). It is not clear then which of these effects is most directly related to the lethal activity of the nucleoside, whether a lack of DNA synthesis provokes a nuclease activity damaging to essential genes, as appears to be the case in thymineless death, or if the insertion of ara nucleotides into the DNA leads to irreversibly damaging events.

Other interesting aspects of the metabolism of araATP include the formation of cyclic araAMP by adenyl cyclase and the reversible addition of araAMP to tRNA, but it seems unlikely that these reactions contribute to lethality. There is no indication that araCTP participates in comparable side reactions; nevertheless, there are reports that high concentrations of araCTP do affect sialic acid metabolism and it will be of interest to know if such reactions are significant in the chemotherapeutic use of araC for tumor viruses, DNA viruses, and even some RNA viruses such as rabies virus.

THE POTENTIATION OF araA BY A DEAMINASE INHIBITOR

The K_i of araATP on DNA polymerase is comparable to that of araCTP, yet the lethal concentrations of araA and araC on mouse fibroblasts are in the range of 100:1. While part of this difference relates to the rapidity of transport and phosphorylation of the two nucleosides, it appears that the rapid deamination of araA by adenosine deaminase is a major factor in the relatively low toxicity of araA. Effective inhibitors of this enzyme ($K_i \leqslant 10^{-8}$ *M*) became available in the past year, and Plunkett and I (1975b) have exploited one such compound, as presented in Fig. 3. The inhibitory adenine derivative has been designated as EHNA. It is essentially nontoxic to cells or mice at 10^{-5} *M*.

When tested in our mouse fibroblast system, 10^{-5} *M* EHNA markedly potentiates the lethality of araA, as presented in Fig. 4. It can be seen that in the presence of the inhibitor, araA is at least 20 times as toxic as it is alone. Of great interest is the fact that whereas the survival of mice bearing an Ehrlich ascites

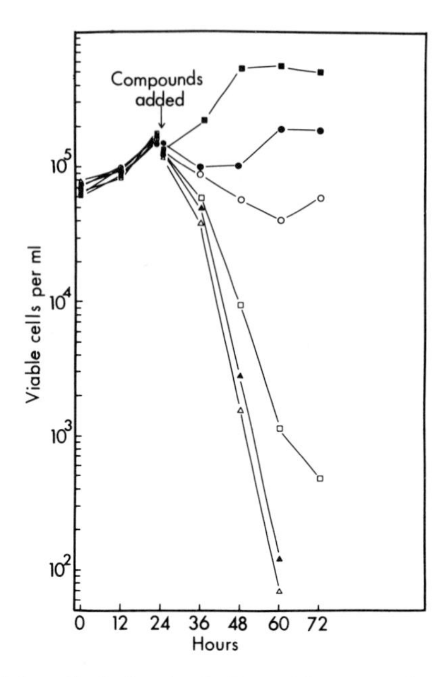

Fig. 4. Viability of L cells during incubation with araA and various concentrations of *erythro*-9-(2-hydroxy-3-nonyl)adenine (EHNA). Closed squares, control; closed circles, 1×10^{-4} *M* araA; open circles, 10^{-4} *M* araA plus 10^{-8} *M* EHNA; open squares, 10^{-4} *M* araA plus 10^{-7} *M* EHNA; closed triangles, 10^{-4} *M* araA plus 10^{-6} *M* EHNA; open triangles, 10^{-4} *M* araA plus 10^{-5} *M* EHNA (Plunkett and Cohen, 1975b).

carcinoma is increased from 14 to 20 days by araA (50 mg/kg) alone, the addition of 10^{-5} *M* EHNA to the araA increases survival further to about 30 days. It appears therefore that we are in a similar position with respect to the clinical development of this combination, as exists with respect to araC and tetrahydrouridine.

It may be mentioned that immunodeficient patients are frequently deficient in adenosine deaminase as well. It is possible that EHNA may provide an interesting approach to the causal relations of the two deficiencies.

THE POTENTIATION OF CORDYCEPIN

Cordycepin, or 3′-deoxyadenosine, was isolated initially as an antibiotic toxic to selected tumors. Its use as an antitumor agent was disappointing however, and like many antibiotics which never quite made it to widespread use in the clinic, cordycepin became a valuable addition to the armamentarium of the biochemically oriented research worker. Numerous papers describe its use in terminating poly A synthesis, as well as in partial inhibition of RNA synthesis. Very little effect could be detected on DNA synthesis, and this particularly intrigued us, in view of recent observations of the role of RNA primers in DNA synthesis. Adenosine deaminase effectively detoxifies cordycepin and we undertook to test the effect of EHNA on the potentiation of the toxicity of this nucleoside. It can be seen in Fig. 5 that the presence of EHNA in our mouse fibroblast system converts the effect of cordycepin from an apparently cytostatic activity to a profoundly lethal action (Plunkett and Cohen, 1975b), essentially sterilizing our cultures in a period of only about 40% of the S phase. Under these conditions the apparent concentration of intracellular cordycepin triphosphate is elevated severalfold over that in the absence of EHNA. RNA synthesis in this culture was completely inhibited after 30 min, whereas DNA synthesis was completely inhibited after an hour. It appears likely then that cordycepin does in fact terminate the RNA primers active in DNA synthesis.

Obviously such an experiment raises the questions of the possible effectiveness of numerous other adenosine analogues, whose deamination and possible detoxification have not been controlled.

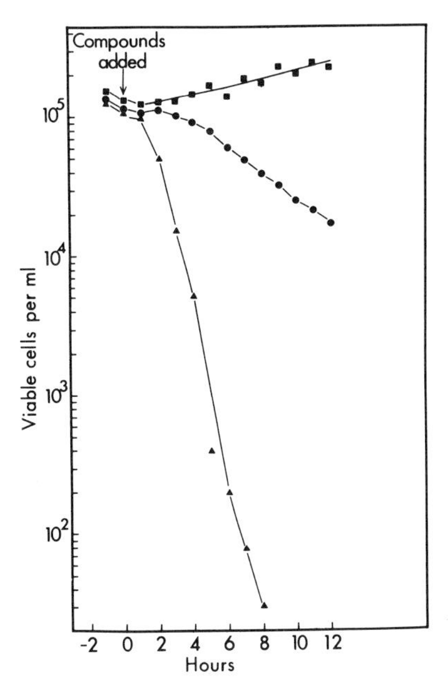

Fig. 5. Viability of L cells during incubation with cordycepin and EHNA. Closed squares, control; closed circles, 10^{-5} *M* cordycepin plus 10^{-5} *M* EHNA; closed triangles, 10^{-4} *M* cordycepin plus 10^{-5} *M* EHNA (Plunkett and Cohen, 1975b).

THE PENETRATION OF NUCLEOTIDES INTO CELLS

In the presence of araA alone at 2×10^{-4} *M*, the viability of cells falls to about 25% of the initial culture after 24 hr, and the surviving cells multiply thereafter at the rate of the control. This pattern parallels the course of deamination, since essentially all of the araA has been converted to the hypoxanthine derivative by 24 hr.

In the presence of the 5′-phosphate of araA (araAMP) at 2×10^{-4} *M*, the viability increases very slowly initially, but after 24 hr falls rapidly and continues to decline until almost all of the cells are killed (Fig. 6). This result is similar to that observed in treating bacteria with fluorodeoxycytidylate as compared to fluorodeoxycytidine; however, it seemed possible that with animal cells some intact araAMP slowly entered the cells. We have tested this possibility with $^{3}H,^{32}P$-labeled araAMP in the presence of a large excess of ^{31}P (Plunkett *et al.*, 1974).

The nucleotide survives in the medium for far longer periods than does the nucleoside. Furthermore, it was found that the tritium-labeled base and the phosphate entered the cell at similar slow rates for several hours and both

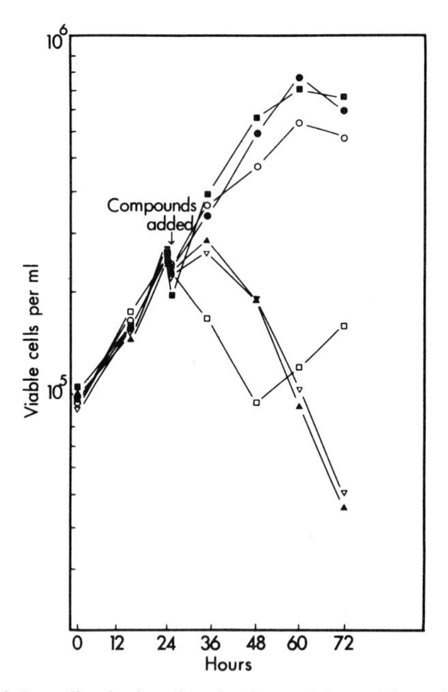

Fig. 6. Viability of L cells during incubation with arabinosyl nucleotides. All compounds were added to a final concentration of 2×10^{-4} *M*. Closed squares, control; closed circles, araAMP cyclic 2′,5′-phosphate; open circles, 3′,5′-cAMP; open squares, araA; open triangles, 3′,5′-cyclic araAMP; closed triangles, araAMP (Plunkett and Cohen, 1975b).

isotopes could be isolated in acid-soluble components as well as in the nucleic acids. AraAMP containing ^{3}H and ^{32}P in a ratio close to that of the exogenous araAMP was isolated from the ATP fraction, whereas the AMP isolated from this fraction had a very small amount of tritium and was relatively much lower in ^{32}P. AraAMP high in both ^{3}H and ^{32}P was also isolated from DNA. Thus, araAMP is taken up intact at a rate 3–5% that of araA. Because of its great toxicity even this slow uptake of araAMP could be detected in our viability studies. It has been found by LePage *et al.* (1972) that circulating araAMP is more stable in the human that in other experimental animals. Humans lack a very active kidney phosphatase and it is possible therefore that araAMP will prove to be a more durable and effective antitumor and antiviral agent than araA alone.

Having detected a penetration of nucleotides into the cells of our test system, we have compared the toxicity of 2′,3′-dideoxyadenosine and that of its 5′-phosphate, ddAMP, in this system. We had demonstrated the lethality of the nucleoside in *E. coli* and had shown that the triphosphate of this compound terminates newly synthesized DNA (Plunkett and Cohen, 1975b). However, the nucleoside is not phosphorylated and is therefore essentially inactive in mouse fibroblasts. As can be seen in Fig. 7, the dideoxynucleotide, however, is quite lethal in this system.

DISCUSSION

By these methods then we think we have learned how to place three sugar analogues of adenosine and 2′-deoxyadenosine into mammalian cell DNA: cordycepin as a terminator of RNA primers, D-arabinosyladenine into internucleotide linkage in DNA chains, and 2′,3′-dideoxyadenosine as a terminator of this DNA. It is likely that these methods and analogues can become useful in a further dissection of the relations of structure and function in DNA.

The work on problems of chemotherapy, in which we have been engaged during the past 20 years, has led us to many problems of basic science relating to the synthesis, metabolism and mode of action of nucleosides and nucleotides. Our findings have led us in turn to many additional problems of applied biomedical science.

Superimposed on the frequent but relatively unimportant dilemma of presenting and categorizing this work to my biochemist colleagues or to clinical friends seeking some leads, I have felt overwhelmed by the difficulties imposed by rigid administrative divisions in school, institute, and hospital. These divisions establish separate functions, departments and empires in biomedical activity which unconscionably impede the work and our search for answers. For these reasons, I have welcomed the establishment of comprehensive cancer centers

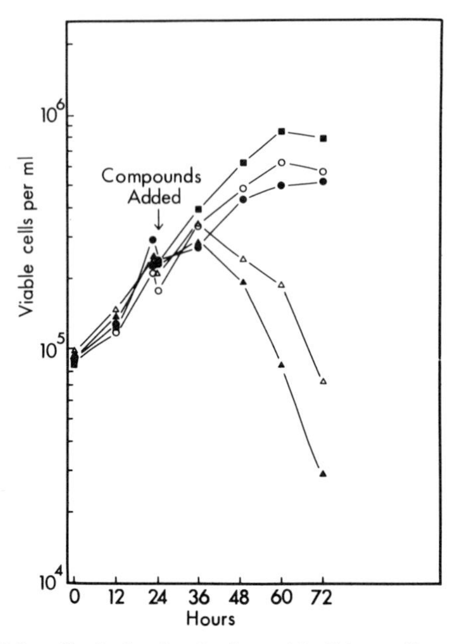

Fig. 7. Viability of L cells during incubation with dideoxyribosyl compounds. Closed squares, control; open circles, 4×10^{-4} *M* ddA; closed circles, 8×10^{-4} *M* ddA; open triangles, 4×10^{-4} *M* ddAMP; closed triangles, 8×10^{-4} *M* ddAMP (Plunkett and Cohen, 1975b).

which have accepted the charge of integrating their activities in education, research and treatment related to cancer. The expensive experiments of attempting to build such cancer centers remain experiments and will undoubtedly remain so for some time. Nevertheless, the institutions engaged in these efforts have at the least examined their own activities and have through group discussion arrived at proposals designed to overcome their previous institutional limitations. For a variety of reasons, both professional and human, the centers may be unable to solve the problems of pulling together their professional and human strengths to the end of integrating the many disciplines necessary to solve the complex cancer problems. We should watch these experiments with great interest and concern; I do believe further that these efforts deserve our most generous assistance.

Finally, I wish to propose a reasonable scientific solution to the problems of viral, if not cancer chemotherapy. I have pointed earlier to the inadequacy of our efforts to exploit virus-induced enzymes for purposes of chemotherapy. Anyone familiar with the efforts of the pharmaceutical firms to develop antiviral agents over the past three decades will agree that hundreds of millions of dollars have been spent in empirical screening efforts almost without success. It should be apparent that modern science is perfectly capable of solving these problems in a straightforward manner. The example I have given of the success of the Upjohn group in designing an effective inhibitor for human detoxification of araC may be relevant to this argument.

The route may be expensive but surely would be less expensive that that which has prevailed until now. Biochemical virologists can define the active sites and crucial sequences in specific virus-induced enzymes essential to the multiplication of herpesvirus, influenza virus, etc. Organic chemists can tailor compounds to inhibit these sites reversibly or irreversibly and in concert with pharmacologists can attempt to assure survival of the inhibitor and penetrability into tissues and cells. Capable clinical pharmacologists can solve the problems posed by the biological and biochemical defenses of humans. In short, I propose that chemotherapy become a far more integrated multidisciplinary activity than it is now. All of these problems may be solved by our present expensive fumbling and bumbling methods over the next few centuries, but I do not know anyone who can afford to wait for the next millenium.

REFERENCES

Barner, H. D., and Cohen, S. S. (1954). *J. Bacteriol.* **68,** 80–88.
Bergmann, W., and Burke, D. C. (1955). *J. Org. Chem.* **20,** 1501–1507.
Camiener, G. W. (1968). *Biochem. Pharmacol.* **17,** 1981–1991.
Cheng, Y. C., Goz, B., Neenan, J. P., Ward, D. C., and Prusoff, W. H. (1975). *J. Virol.* **15,** 1284–1285.
Cohen, S. S. (1953). *Cold Spring Harbor Symp. Quant. Biol.* **18,** 221–225.
Cohen, S. S. (1961). *Fed. Proc., Fed. Am. Soc. Exp. Biol.* **20,** 641–649.
Cohen, S. S. (1968). "Virus Induced Enzymes." Columbia Univ. Press, New York.
Cohen, S. S., and Barner, H. D. (1956). *J. Bacteriol.* **71,** 588–597.
Cohen, S. S., Flaks, J. G., Barner, H. D., Loeb, M. R., and Lichtenstein, J. (1958). *Proc. Natl. Acad. Sci. U.S.A.* **44,** 1004–1112.
Duschinsky, R., Pleven, E., and Heidelberger, C. (1957). *J. Am. Chem. Soc.* **79,** 4559–4560.
Evans, J. S., Musser, E. A., Mengel, G. D., Forsblad, K. R., and Hunter, J. H. (1961). *Proc. Soc. Exp. Biol. Med.* **106,** 350–353.
Furth, J. J., and Cohen, S. S. (1968). *Cancer Res.* **28,** 2061–2067.
Graham, F. L., and Whitmore, G. F. (1970). *Cancer Res.* **30,** 2636–2644.
Hanze, A. R. (1967). *J. Am. Chem. Soc.* **89,** 6720–6725.
Hubert-Habart, M., and Cohen, S. S. (1962). *Biochim. Biophys. Acta* **59,** 468–471.
Kaufman, H. E., and Maloney, E. D. (1963). *Am. J. Ophthalmol.* **69,** 626–629.
Lee, W. W., Benitez, A., Goodman, L., and Baker, B. R. (1960). *J. Am. Chem. Soc.* **82,** 2648–2649.
LePage, G. A., Linn, Y. T., Orth, R. E., and Gottlieb, J. A. (1972). *Cancer Res.* **32,** 2441–2444.
Lichtenstein, J., Barner, H. D., and Cohen, S. S. (1960). *J. Biol. Chem.* **235,** 457–465.
Lim, K., and Cohen, S. S. (1966). *J. Biol. Chem.* **241,** 4304–4315.
Lipmann, F. (1936). *Nature (London)* **138,** 588–589.
Moore, E. C., and Cohen, S. S. (1967). *J. Biol. Chem.* **242,** 2116–2118.

Neil, G. L., Maxley, T. E., and Manak, R. C. (1970). *Cancer Res.* **30,** 2166–2172.
Overby, L. R., Robishaw, E. E., Schleicher, J. B., Rueter, A., Shipkowitz, N. L., and Mao, J. C. (1974). *Antimicrob. Agents & Chemother.* **6,** 360–365.
Pizer, L., and Cohen, S. S. (1960). *J. Biol. Chem.* **235,** 2387–2392.
Plunkett, W., and Cohen, S. S. (1975a). *Cancer Res.* **35,** 415–422.
Plunkett, W., and Cohen, S. S. (1975b). *Cancer Res.* **35,** 1547–1554.
Plunkett, W., Lapi, L., Ortiz, P. J., and Cohen, S. S. (1974). *Proc. Natl. Acad. Sci. U.S.A.* **71,** 73–77.

23

Electronic Biology and Cancer

ALBERT SZENT-GYÖRGYI

Biology today is a molecular science. Its charm is in the wonderful subtlety of its reactions, the main actors of which are the protein molecules. It is difficult to believe that this subtlety could be brought about by clumsy macromolecules without the participation of much smaller and more mobile and reactive units which could hardly be anything but specially reactive electrons. One may wonder to what degree protein molecules are the actors, or the stage for the drama of life, and to what degree life itself is a molecular or electronic phenomenon. Such a basic problem can be approached only with the broadest philosophical outlook, beginning at the origin.

Life originated on a dark and airless globe, covered by dense water vapor. Having condensed from hydrogen, the atmosphere had to be strongly reducing, dominated by electron donors, reducing substances which tended to give off rather than take up electrons. At the resulting high electronic pressure, all allowed places on the energy bands of the protein molecule had to be occupied by electrons, leaving no room for mobility, making the protein dielectric, an insulator. The single orbitals had to be occupied by pairs of electrons which, spinning in opposite directions, compensated each others magnetic moments. The protein had to consist of well-balanced, stable, dielectric, closed-shell molecules, which had no unbalanced forces that could link them together to complex structures.

We can only philosophize that under such inhospitable conditions life could build only simple systems which performed the most basic vegetative functions that demanded no complex structures: fermentation and proliferation. Fermentation produced energy and proliferation made life perennial, proceeding as fast as conditions permitted.

This first, dark and anaerobic period I call the "α period," and the corresponding state of living systems the "α state." When, owing to the cooling of our globe, its water envelope condensed and light could reach its surface, life began to develop and differentiate. It used the energy of light to separate the elements of water, H and O. The hydrogen it fixed by linking it to carbon, creating foodstuffs. The oxygen it released into the atmosphere. Oxygen is a powerful oxidizing agent, an electron acceptor, and so, henceforth, life was no longer dominated solely by electron donors, but became dependent on the quotient D/A, the relation of donors to acceptors. Oxygen can oxidize, take electrons from other substances. It can also take electrons from protein. By taking electrons from energy bands it could make room for motion, transforming the protein from an insulator into a semiconductor. By taking single electrons it could uncouple electron pairs which upsets the balance, making reactive free radicals from the inert molecules, a "free radical" being a molecule containing an unpaired electron which can be detected by the signal it gives in the electron spin resonance (ESR) spectroscope. With their transformed proteins and their unbalanced forces, the living systems started to differentiate, to build increasingly complex structures with increasingly complex functions. This second period of life, which followed the appearance of light and oxygen I will call the "β period," and the corresponding physical state the "β state."

If it were actually the oxygen that initiated the transformation from the α to the β state by taking electrons from protein, then this transfer of electrons would have to be one of the most basic processes of biology. Its regulation will be my central theme.

OXYGEN AND DICARBONYLS

Electrons, in molecules, are paired and tend to go from one molecule to the other pairwise. Thus bivalent acceptors, which can take up two electrons, will always tend to take over two electrons, even if the two go over one by one. The transfer of an electron pair to oxygen is called "*oxidation*", "*burning.*" This transfer of an electron pair leaves a somewhat smaller, but still well-balanced molecule behind.

The situation will be different if the acceptor is monovalent, can take up but one electron. Taking one electron means that both molecules, the one which accepted the single electron and the one which donated it, become reactive free radicals. The transfer of a single electron from one molecule to another is "*charge transfer,*" which was hitherto looked upon by most biologists as a rare event, an oddity, an item of Nature's own curiosity shop.

The oxygen molecule, O=O, consists of two O atoms linked together by a double bond. If the double bond opens up, —O—O—, a bivalent electron acceptor is produced. If the —O—O— breaks up into single O atoms, these atoms again are bivalent, so the oxygen will tend to burn, take up electrons pairwise without creating free radicals. So oxygen, as such, could not transform the protein into free radicals. Nature solved this problem by linking oxygen atoms to carbon, C atoms, instead of linking them to one another. An oxygen linked to carbon by a double bond is C=O, a carbonyl. Carbonyls are monovalent acceptors. They are weak acceptors because the C=O group is too small to accommodate easily a whole additional electron. However, if two carbonyls are linked together to a dicarbonyl, then the π electronic systems of the neighboring (conjugated) double bonds fuse to a big π system, which easily takes up a whole electron, is a "strong" acceptor, but is, all the same, still a monovalent.

The simplest dicarbonyl is glyoxal, the first methyl derivative of which is methylglyoxal (Fig. 1). This makes our problem very exciting because more than 60 years ago two Englishmen and a German, H. D. Dakin and H. W. Dudley (1913) and C. Neuberg (1913), discovered a most active enzymatic system present in all living cells, which can transform the reactive methylglyoxal into an inactive D-lactic acid with amazing speed. This enzymatic system is called "glyoxalase." Nature does not indulge in luxuries, and if there is such a most active and widely spread enzymatic system, it must have something very important to do, but nobody knew what, because neither methylglyoxal, nor D-lactic acid were known as metabolites. Could, then, methylglyoxal have been responsible for transforming the proteins into conductors and free radicals by taking electrons out of them? Glyoxalase could act as its antagonist which prevents these changes by inactivating it. Methylglyoxal and the glyoxalase, together, could then be one of the most important tools of cellular regulation, one being the green, the other the red light.

Einstein said that Nature is simple but subtle. The methylglyoxal molecule is very simple, but has, all the same, very specific characteristics. The calculations of Alberte Pullman showed that it has a very low lying empty orbital (personal communication) which makes it a strong acceptor. This orbital has to be on the ketonic oxygen, while the aldehydic group could serve to establish a link with

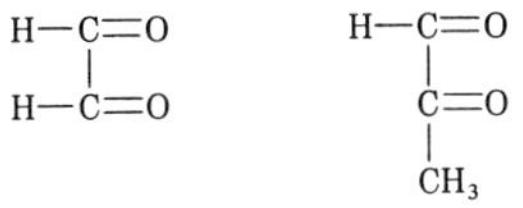

```
H—C=O        H—C=O
   |             |
H—C=O           C=O
                 |
                CH3
```

Fig. 1. Glyoxal and methylglyoxal.

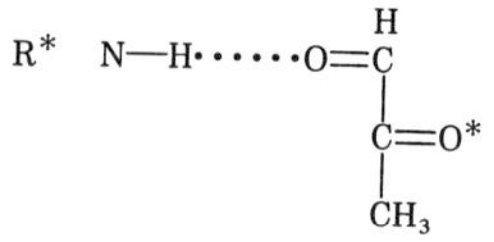

```
                    H
R*  N—H······O=C
                |
                C=O*
                |
                CH3
```

Fig. 2. H-bond between a peptide and methylglyoxal. The unpaired electron is symbolized by an asterisk. R stands for the peptide bond.

the protein. Linus Pauling showed, many years ago (1959) that the NH of proteins readily forms hydrogen bonds with carbonyls. In Fig. 2 the peptide bond of protein is symbolized by R. When its electron goes over to the ketonic group of methylglyoxal (marked by an asterisk), it has to pass six atoms, and an H-bond will thus be very far from its earlier pair, left behind on R. This is very important because it was thought that two radicals, formed by charge transfer. must dissociate to have their electrons uncoupled to give an ESR signal. In the case of methylglyoxal, however, the charge transfer complex can be expected to be uncoupled and give an ESR signal without dissociation. There is no need to suppose that protein and methylglyoxal dissociate after the electron transfer; the final product of the reaction can be a biradical, consisting of the positively charged protein and the negatively charged methylglyoxal radical. It is even difficult to see what could drive the two apart since the forces holding them together are only increased by the charge transfer energy and the attraction of the opposite charges.

AMINES AND DICARBONYLS

Proteins are very complicated substances, so before addressing myself to them I tried to collect some data by studying the interaction of simpler amines and methylglyoxal. The simplest amines are methyl and ethylamine, CH_3NH_2 and $CH_3CH_2NH_2$. I started with mixing a watery solution of these amines with the solution of methylglyoxal, expecting them to react and produce the vivid colors of free radicals. They slowly formed a so-called "Schiff base," and no radicals. Only the weak yellow color of the Schiff base developed which gave an absorption at the short wavelength end of the spectrum. Either no free radicals were formed, or else they were too unstable to accumulate and give visible colors. So I repeated the experiment using dimethylsulfoxide as solvent which, in my experience, stabilized free radicals. An intense red color developed, and in the spectroscope the solution gave two sharp peaks, one at 315 and another at 475 nm, and gave a spin resonance signal in Dr. H. Kon's ESR spectroscope (Kon and Szent-Györgyi, 1945), which indicated that a whole electron was passed from amine to carbonyls, leading to the formation of free radicals. By the addition of an amine, the dicarbonyl could thus be readily made into a reactive radical.

With this experience in hand I mixed protein with methylglyoxal, expecting colored radicals to be formed, but nothing happened, the dicarbonyl did not interact with protein at all, it could not bind to it. But if the free methylglyoxal did not bind to protein, perhaps the biradical it formed with amines did? When brought together with this biradical, the colorless protein changed into a dark brown substance which had the same color as the proteins which build the structure of the liver cells. Evidently, the biradical could bind to protein and take electrons from it, and so our next question was whether this protein–amine–dicarbonyl complex could also link with oxygen and pass on the electrons of protein to it. The complexes of the simple amines could not answer this question, but here an earlier experiment, performed with Jane McLaughlin, came to our aid (Szent-Györgyi and McLaughlin, 1975). We found that dicarbonyls readily form charge transfer complexes with biogenic amines, such as dopamine or serotonin, and these complexes turned black when forming charge transfer complexes with oxygen. We could thus see whether they interacted with oxygen. These dicarbonyl complexes of biogenic amines also readily complexed with protein, and could thus transfer electrons taken from it to O_2. These expectations could readily be tested in the experiment by incubating, for a short while (30 min), protein in the presence of biogenic amine, dicarbonyls, and O_2 (air): The result was a black protein complex which gave an ESR signal, indicative of a multiradical. The black color inseparable from the protein by reprecipitation was thus the color of the protein charge transfer complex in which the electrons of the protein were transferred to O_2. We can thus piece the story together and say that a bridge can be built of free radicals between protein and oxygen through which oxygen can desaturate the protein, make it into a free radical and semiconductor.

If the oxygen molecule O=O, enters into a complex with the dicarbonyl, it has to form a peroxide. That peroxides were actually formed could be shown by repeating the experiment with addition of peroxidase, which activates peroxides. In its presence the protein colored faster and stronger, indicating that the enzyme activated the peroxide, enabled it to accept a second electron. With two electrons, O=O forms (in the presence of water) hydrogen peroxide which then has to be decomposed in tissue by catalase. For the completion of the reaction, both catalase and peroxidase are needed. This gives meaning to the discovery of D. Keilin and E. F. Hartree, made 30 years ago (1945), that catalase has both these enzymatic activities. We can thus sum up the whole story by saying that protein can be transformed by passing its electrons to oxygen through a chain built of free radicals, all reactions involved being electronic and not molecular.

THE α-β TRANSFORMATION AND CANCER

To be able to answer the question whether all this has anything to do with cancer, I will have to return briefly to the α-β transformation, the profound change in the nature of living systems which followed the appearance of oxygen. As discussed before, when light and oxygen appeared, life began to develop, differentiate and build increasingly complex structures that performed increasingly complex functions. The growing complexity was incompatible with unbridled proliferation. Proliferation had to be arrested and subjected to regulation to maintain the harmony of the whole. It was arrested by two factors. The dicarbonyls inactivated the SH groups which are indispensible for cell division, forming hemimercaptals with them. It has been shown by L. Egyud and myself (1966a,b) that dicarbonyls arrest cell division reversibly in low concentration on the ribosomal level. The arrest can be eliminated by glyoxalase which decomposes the dicarbonyls and starts up cell division again.

The other factor which must have helped to arrest cell division was the semisolid structures built in the β period. Cell division involves a complete rearrangement of the cellular interior, which is impossible in a rigid, solid structure. To be able to divide, the cell has to dismount its structures to a great extent. The most striking example of this demolition is the dissolution of the cellular nucleus, the membrane of which is dissolved, and the chromatin of which is condensed to mobile chromosomes. The mitochondria are also partly dissolved, forcing the cell to derive its energy to a greater extent from fermentation, which demands no structure. So, in division, the cell dedifferentiates, dissolving its structure, going through the α-β transformation in reverse, returning to its proliferative fermentative α state.

What lends validity and major medical interest to these relations is that after it has completed its division, the cell has to find its way back to the resting β state, build up its free radical electron transport chain and structures again. Should it be unable to do so, it has to persist in the proliferative α state, continue to proliferate when no proliferation is needed, and a tumor has to result. The cancer cell is a cell stuck in the α state. The same will be the result if, for any reason, the β state becomes unstable and the cell is unable to maintain its electron transport chain. The free radicals building this chain, are colored, so their absence should declare itself by the lack of color. This suggested a comparison of the color of the structural proteins of normal tissues with those of cancer. Such a comparison has value only if we compare cancer with the homologous normal tissue. Owing to the kindness of Dr. G. Weber, I am in possession of a rapidly growing parenchymal liver tumor (Morris Hepatoma 3924A). The structural proteins of the normal liver were chocolate brown, those of cancer light yellow-green. That the difference was due to the lack of electron transport chain could be shown by adding an electron acceptor to the tumor proteins, whereupon they assumed the color of the corresponding preparation of the normal liver.

I want to conclude this paper by answering three questions.

Question I: How was all this overlooked before? The explanation is simple. The protein chemist needs crystals, and to produce them he needs protein solutions. So what he did was to extract the soluble proteins from the organism and called the extracted tissue "the residue" and sent it down the drain. As I have shown, the soluble proteins perform only the simplest vegetative functions that need no free radicals or conduction. These free radicals and conduction are needed only by the structures, but the structures cannot be crystallized and so have been disregarded by the chemist.

Question II: How does all this relate to viruses? Cell division and its regulation demands a complex chemical mechanism. Viruses cannot build such a mechanism. They can only disturb it and set proliferation going by interfering with regulation. The regulatory mechanism can be disturbed by an endless number of factors. Viruses are one of them. The cancer problem is much more complex than "virus or no virus."

Question III: Will my findings lead to a cure for cancer? They may or may not. What can be said with certainty is what Bernal told us: "That we can control only what we understand."

REFERENCES

Dakin, H. D., and Dudley, H. W. (1913). *J. Biol. Chem.* **14**, 155–157.

Egyud, L., and Szent-Györgyi, A. (1966a). *Proc. Natl. Acad. Sci. U.S.A.* **55**, 388–393.

Egyud, L., and Szent-Györgyi, A. (1966b). *Proc. Natl. Acad. Sci. U.S.A.* **56**, 203–207.

Keilin, D., and Hartree, E. F. (1945). *Biochem. J.* **39**, 293–301.

Kon, H., and Szent-Györgyi, A. (1973). *Proc. Natl. Acad. Sci. U.S.A.* **70**, 1030–1031 and 3139–3140.

Neuberg, C. (1913). *Biochem. Z.* **49**, 502–506.

Pauling, L. (1959). *In* "Symposium on Protein Structure" (A. Neuberger, ed.). Methuen, London.

Szent-Györgyi, A., and McLaughlin, J. A. (1975). *Proc. Natl. Acad. Sci. U.S.A.* **72**, 1610–1611.

EPILOGUE

24

Epilogue

MICHAEL KASHA

No one who knows Albert Szent-Györgyi would have expected the symposium celebrating his 82nd birthday to be a conventional experience. But with all the great expectations generated by the program plans, the symposium climaxed with an event which even the strongest imagination would have been taxed to forecast.

The build-up of warmth of personal response toward Albert Szent-Györgyi was steady, with each lecturer's recounting of his career experiences. The build-up of esteem for Szent-Györgyi grew hourly as every research specialist got an appreciative and learned glimpse of Szent-Györgyi's influence and discoveries in his other great areas of biochemical research. This impression was the more deeply engraved as various lecturers displayed Albert Szent-Györgyi's own words from his publications of the era of discovery. The lucidity, the prescience, the drive toward a new understanding, all came through with thrilling impact.

The invited lecturers had finished their 2 days of warm appraisal in the symposium, and now the stage was set for Albert Szent-Györgyi. There was a certain tension and excitement. What would our great friend do? Would he, too, reminisce? Would he falter? The only person not looking backward, not even casting a glance over his shoulder that day, was Albert Szent-Györgyi.

He started a little hesitantly, in a slightly hushed voice. He later confessed privately that he felt an enormous pressure to make (even for him!) an unusually dynamic presentation. He started to build up his case with physicochemical caution and solidity. It went well, and a physical chemist's ears heard music.

Then the experiments demonstrating his thesis began. Holding giant test tubes in his hands, first one above his head, then the other, gesturing with each to make his point, his face vividly animated, a brilliant red color sparkled as the liquids mixed. A dramatic sequence began, more experiments, color slides were projected, Szent-Györgyi danced about the platform, alternating his attentions on the audience and on his test tubes. It was a ballet, the "Ballet of Albert Szent-Györgyi," flushed with energy and scientific drama. The physical chemistry and biochemistry were too tightly knit, too convincingly demonstrated, to be taken lightly. It was an hour of triumph.

Theoretical physicists teach us that the magnetic monopole, if captured, would represent a field strength, an energy, unmatched in the physical world. That afternoon in Boston, October 17, 1975, at 5 PM the magnetic monopole seemed captured within the heart and soul of Albert Szent-Györgyi. The intense magnetism, the pulsating energy, the sheer radiance, were felt by all.

Index

A 7
B 8
C 9
D 0
E 1
F 2
G 3
H 4
I 5
J 6